R for Data Science Cookbook

Over 100 hands-on recipes to effectively solve real-world data problems using the most popular R packages and techniques

Yu-Wei, Chiu (David Chiu)

[PACKT] open source
community experience distilled
PUBLISHING

BIRMINGHAM - MUMBAI

R for Data Science Cookbook

First published: July 2016

Production reference: 1270716

Published by Packt Publishing Ltd.
Livery Place
35 Livery Street
Birmingham B3 2PB, UK.

ISBN 978-1-78439-081-5

www.packtpub.com

Credits

Author
Yu-Wei, Chiu (David Chiu)

Reviewer
Prabhanjan Tattar

Commissioning Editor
Veena Pagare

Acquisition Editor
Tushar Gupta

Content Development Editor
Pooja Mhapsekar

Technical Editor
Madhunikita Sunil Chindarkar

Copy Editor
Priyanka Ravi

Project Coordinator
Suzanne Coutinho

Proofreader
Safis Editing

Indexer
Tejal Daruwale Soni

Graphics
Jason Monteiro

Production Coordinator
Aparna Bhagat

Cover Work
Aparna Bhagat

About the Author

Yu-Wei, Chiu (David Chiu) is the founder of LargitData (`www.LargitData.com`), a startup company that mainly focuses on providing big data and machine learning products. He has previously worked for Trend Micro as a software engineer, where he was responsible for building big data platforms for business intelligence and customer relationship management systems. In addition to being a start-up entrepreneur and data scientist, he specializes in using Spark and Hadoop to process big data and apply data mining techniques for data analysis. Yu-Wei is also a professional lecturer and has delivered lectures on big data and machine learning in R and Python, and given tech talks at a variety of conferences.

In 2015, Yu-Wei wrote *Machine Learning with R Cookbook, Packt Publishing*. In 2013, Yu-Wei reviewed *Bioinformatics with R Cookbook, Packt Publishing*. For more information, visit his personal website at `www.ywchiu.com`.

I have immense gratitude for my family and friends for supporting and encouraging me to complete this book. I would like to sincerely thank my mother, Ming-Yang Huang (Miranda Huang); my mentor, Man-Kwan Shan; the proofreader of this book, Brendan Fisher; members of LargitData; Data Science Program (DSP); and other friends who have offered their support.

About the Reviewer

Prabhanjan Tattar is currently working as a senior data scientist at Fractal Analytics, Inc. He has 8 years of experience as a statistical analyst. Survival analysis and statistical inference are his main areas of research and interest, and he has published several research papers in peer-reviewed journals, as well as authoring two books on R: *R Statistical Application Development by Example, Packt Publishing*, and *A Course in Statistics with R, Wiley*. The R packages `gpk`, `RSADBE`, and `ACSWR` are also maintained by him.

Table of Contents

Preface vii

Chapter 1: Functions in R 1

- Introduction 1
- Creating R functions 2
- Matching arguments 5
- Understanding environments 7
- Working with lexical scoping 10
- Understanding closure 13
- Performing lazy evaluation 16
- Creating infix operators 18
- Using the replacement function 20
- Handling errors in a function 23
- The debugging function 28

Chapter 2: Data Extracting, Transforming, and Loading 37

- Introduction 38
- Downloading open data 38
- Reading and writing CSV files 42
- Scanning text files 44
- Working with Excel files 46
- Reading data from databases 49
- Scraping web data 52
- Accessing Facebook data 62
- Working with twitteR 68

Chapter 3: Data Preprocessing and Preparation 73
Introduction 73
Renaming the data variable 74
Converting data types 76
Working with the date format 78
Adding new records 81
Filtering data 83
Dropping data 87
Merging data 88
Sorting data 90
Reshaping data 92
Detecting missing data 95
Imputing missing data 98
Chapter 4: Data Manipulation 101
Introduction 101
Enhancing a data.frame with a data.table 102
Managing data with a data.table 106
Performing fast aggregation with a data.table 111
Merging large datasets with a data.table 115
Subsetting and slicing data with dplyr 120
Sampling data with dplyr 123
Selecting columns with dplyr 125
Chaining operations in dplyr 128
Arranging rows with dplyr 130
Eliminating duplicated rows with dplyr 131
Adding new columns with dplyr 133
Summarizing data with dplyr 134
Merging data with dplyr 138
Chapter 5: Visualizing Data with ggplot2 143
Introduction 143
Creating basic plots with ggplot2 146
Changing aesthetics mapping 150
Introducing geometric objects 153
Performing transformations 158
Adjusting scales 161
Faceting 164
Adjusting themes 167
Combining plots 169
Creating maps 171

Chapter 6: Making Interactive Reports 177
Introduction 177
Creating R Markdown reports 178
Learning the markdown syntax 182
Embedding R code chunks 186
Creating interactive graphics with ggvis 190
Understanding basic syntax and grammar 194
Controlling axes and legends 201
Using scales 206
Adding interactivity to a ggvis plot 208
Creating an R Shiny document 215
Publishing an R Shiny report 221
Chapter 7: Simulation from Probability Distributions 227
Introduction 227
Generating random samples 228
Understanding uniform distributions 231
Generating binomial random variates 233
Generating Poisson random variates 236
Sampling from a normal distribution 239
Sampling from a chi-squared distribution 245
Understanding Student's t-distribution 248
Sampling from a dataset 251
Simulating the stochastic process 253
Chapter 8: Statistical Inference in R 257
Introduction 257
Getting confidence intervals 258
Performing Z-tests 265
Performing student's T-tests 268
Conducting exact binomial tests 272
Performing Kolmogorov-Smirnov tests 274
Working with the Pearson's chi-squared tests 276
Understanding the Wilcoxon Rank Sum and Signed Rank tests 279
Conducting one-way ANOVA 282
Performing two-way ANOVA 287
Chapter 9: Rule and Pattern Mining with R 293
Introduction 293
Transforming data into transactions 294
Displaying transactions and associations 296
Mining associations with the Apriori rule 299

Pruning redundant rules 303
Visualizing association rules 305
Mining frequent itemsets with Eclat 307
Creating transactions with temporal information 310
Mining frequent sequential patterns with cSPADE 313
Chapter 10: Time Series Mining with R 319
Introduction 319
Creating time series data 320
Plotting a time series object 324
Decomposing time series 327
Smoothing time series 331
Forecasting time series 336
Selecting an ARIMA model 341
Creating an ARIMA model 346
Forecasting with an ARIMA model 348
Predicting stock prices with an ARIMA model 352
Chapter 11: Supervised Machine Learning 357
Introduction 358
Fitting a linear regression model with lm 358
Summarizing linear model fits 361
Using linear regression to predict unknown values 363
Measuring the performance of the regression model 366
Performing a multiple regression analysis 368
Selecting the best-fitted regression model with stepwise regression 371
Applying the Gaussian model for generalized linear regression 374
Performing a logistic regression analysis 375
Building a classification model with recursive partitioning trees 379
Visualizing a recursive partitioning tree 382
Measuring model performance with a confusion matrix 384
Measuring prediction performance using ROCR 387
Chapter 12: Unsupervised Machine Learning 391
Introduction 392
Clustering data with hierarchical clustering 392
Cutting tree into clusters 397
Clustering data with the k-means method 400
Clustering data with the density-based method 402
Extracting silhouette information from clustering 405
Comparing clustering methods 407

Recognizing digits using the density-based clustering method 410
Grouping similar text documents with k-means clustering methods 412
Performing dimension reduction with Principal Component Analysis (PCA) 414
Determining the number of principal components using a scree plot 418
Determining the number of principal components using the Kaiser method 420
Visualizing multivariate data using a biplot 422
Index 425

Preface

Big data, the Internet of Things, and artificial intelligence have become the hottest technology buzzwords in recent years. Although there are many different terms used to define these technologies, the common concept is that they're all driven by data. Simply having data is not enough; being able to unlock its value is essential. Therefore, data scientists have begun to focus on how to gain insights from raw data.

Data science has become one of the most popular subjects among academic and industry groups. However, as data science is a very broad discipline, learning how to master it can be challenging. A beginner must learn how to prepare, process, aggregate, and visualize data. More advanced techniques involve machine learning, mining various data formats (text, image, and video), and, most importantly, using data to generate business value. The role of a data scientist is challenging and requires a great deal of effort. A successful data scientist requires a useful tool to help solve day-to-day problems.

In this field, the most widely used tool by data scientists is the R language, which is open source and free. Being a machine language, it provides many data processes, learning packages, and visualization functions, allowing users to analyze data on the fly. R helps users quickly perform analysis and execute machine learning algorithms on their dataset without knowing every detail of the sophisticated mathematical models.

R for Data Science Cookbook takes a practical approach to teaching you how to put data science into practice with R. The book has 12 chapters, each of which is introduced by breaking down the topic into several simple recipes. Through the step-by-step instructions in each recipe, you can apply what you have learned from the book by using a variety of packages in R.

The first section of this book deals with how to create R functions to avoid unnecessary duplication of code. You will learn how to prepare, process, and perform sophisticated ETL operations for heterogeneous data sources with R packages. An example of data manipulation is provided that illustrates how to use the `dplyr` and `data.table` packages to process larger data structures efficiently, while there is a section focusing on `ggplot2` that covers how to create advanced figures for data exploration. Also, you will learn how to build an interactive report using the `ggvis` package.

This book also explains how to use data mining to discover items that are frequently purchased together. Later chapters offer insight into time series analysis on financial data, while there is detailed information on the hot topic of machine learning, including data classification, regression, clustering, and dimension reduction.

With *R for Data Science Cookbook* in hand, I can assure you that you will find data science has never been easier.

What this book covers

Chapter 1, Functions in R, describes how to create R functions. This chapter covers the basic composition, environment, and argument matching of an R function. Furthermore, we will look at advanced topics such as closure, functional programming, and how to properly handle errors.

Chapter 2, Data Extracting, Transforming, and Loading, teaches you how to read structured and unstructured data with R. The chapter begins by collecting data from text files. Subsequently, we will look at how to connect R to a database. Lastly, you will learn how to write a web scraper to crawl through unstructured data from a web page or social media site.

Chapter 3, Data Preprocessing and Preparation, introduces you to preparing data ready for analysis. In this chapter, we will cover the data preprocess steps, such as type conversion, adding, filtering, dropping, merging, reshaping, and missing-value imputation, with some basic R functions.

Chapter 4, Data Manipulation, demonstrates how to manipulate data in an efficient and effective manner with the advanced R packages `data.table` and `dplyr`. The `data.table` package exposes you to the possibility of quickly loading and aggregating large amounts of data. The `dplyr` package provides the ability to manipulate data in SQL-like syntax.

Chapter 5, Visualizing Data with ggplot2, explores using `ggplot2` to visualize data. This chapter begins by introducing the basic building blocks of `ggplot2`. Next, we will cover advanced topics on how to create a more sophisticated graph with `ggplot2` functions. Lastly, we will describe how to build a map with `ggmap`.

Chapter 6, Making Interactive Reports, reveals how to create a professional report with R. In the beginning, the chapter discusses how to write R markdown syntax and embed R code chunks. We will also explore how to add interactive charts to the report with `ggvis`. Finally, we will look at how to create and publish an R Shiny report.

Chapter 7, Simulation from Probability Distributions, begins with an emphasis on sampling data from different probability distributions. As a concrete example, we will look at how to simulate a stochastic trading process with a probability function.

Chapter 8, Statistical Inference in R, begins with a discussion on point estimation and confidence intervals. Subsequently, you will be introduced to parametric and non-parametric testing methods. Lastly, we will look at how one can use ANOVA to analyze whether the salary basis of an engineer differs based on his job title and location.

Chapter 9, Rule and Pattern Mining with R, exposes you to the common methods used to discover associated items and underlying frequency patterns from transaction data. In this chapter, we use a real-world blog as example data so that you can learn how to perform rule and pattern mining on real-world data.

Chapter 10, Time Series Mining with R, begins by introducing you to creating and manipulating time series from a finance dataset. Subsequently, we will learn how to forecast time series with HoltWinters and ARIMA. For a more concrete example, this chapter reveals how to predict stock prices with ARIMA.

Chapter 11, Supervised Machine Learning, teaches you how to build a model that makes predictions based on labeled training data. You will learn how to use regression models to make sense of numeric relationships and apply a fitted model to data for continuous value prediction. For classification, you will learn how to fit data into a tree-based classifier.

Chapter 12, Unsupervised Machine Learning, introduces you to revealing the hidden structure of unlabeled data. Firstly, we will look at how to group similarly located hotels together with the clustering method. Subsequently, we will learn how to select and extract features on the economy freedom dataset with PCA.

What you need for this book

To follow the book's examples, you will need a computer with access to the Internet and the ability to install the R environment. You can download R from `http://www.cran.r-project.org/`. Detailed installation instructions are available in the first chapter.

The examples provided in this book were coded and tested with R version 3.2.4 on Microsoft Windows. The examples are likely to work with any recent version of R installed on either Mac OS X or a Unix-like OS.

Who this book is for

R for Data Science Cookbook is intended for those who are already familiar with the basic operation of R, but want to learn how to efficiently and effectively analyze real-world data problems using practical R packages.

Sections

In this book, you will find several headings that appear frequently (Getting ready, How to do it, How it works, There's more, and See also).

To give clear instructions on how to complete a recipe, we use these sections as follows:

Getting ready

This section tells you what to expect in the recipe, and describes how to set up any software or any preliminary settings required for the recipe.

How to do it...

This section contains the steps required to follow the recipe.

How it works...

This section usually consists of a detailed explanation of what happened in the previous section.

There's more...

This section consists of additional information about the recipe in order to make the reader more knowledgeable about the recipe.

See also

This section provides helpful links to other useful information for the recipe.

Conventions

In this book, you will find a number of styles of text that distinguish between different kinds of information. Here are some examples of these styles, and an explanation of their meaning.

Package and function names are shown as follows: "You can then install and load the package `RCurl`."

A block of code is set as follows:

```
> install.packages("RCurl")
> library(RCurl)
```

Any URL is written as follows:

`http://data.worldbank.org/topic/economy-and-growth`

Variable name, **argument name**, **new terms** and **important words** are shown in bold. Words that you see on the screen, in menus or dialog boxes for example, appear in the text like this: "In R, a missing value is noted with the symbol **NA (not available)**, and an impossible value is **NaN (not a number)**."

Reader feedback

Feedback from our readers is always welcome. Let us know what you think about this book—what you liked or disliked. Reader feedback is important for us as it helps us develop titles that you will really get the most out of.

To send us general feedback, simply e-mail `feedback@packtpub.com`, and mention the book's title in the subject of your message.

If there is a topic that you have expertise in and you are interested in either writing or contributing to a book, see our author guide at `www.packtpub.com/authors`.

Customer support

Now that you are the proud owner of a Packt book, we have a number of things to help you to get the most from your purchase.

Downloading the example code

You can download the example code files for this book from your account at `http://www.packtpub.com`. If you purchased this book elsewhere, you can visit `http://www.packtpub.com/support` and register to have the files e-mailed directly to you.

You can download the code files by following these steps:

1. Log in or register to our website using your e-mail address and password.
2. Hover the mouse pointer on the **SUPPORT** tab at the top.
3. Click on **Code Downloads & Errata**.
4. Enter the name of the book in the **Search** box.
5. Select the book for which you're looking to download the code files.
6. Choose from the drop-down menu where you purchased this book from.
7. Click on **Code Download**.

Once the file is downloaded, please make sure that you unzip or extract the folder using the latest version of:

- WinRAR / 7-Zip for Windows
- Zipeg / iZip / UnRarX for Mac
- 7-Zip / PeaZip for Linux

The code bundle for the book is also hosted on GitHub at `https://github.com/PacktPublishing/R-for-Data-Science-Cookbook`. We also have other code bundles from our rich catalog of books and videos available at `https://github.com/PacktPublishing/`. Check them out!

Downloading the color images of this book

We also provide you with a PDF file that has color images of the screenshots/diagrams used in this book. The color images will help you better understand the changes in the output. You can download this file from `https://www.packtpub.com/sites/default/files/downloads/RforDataScienceCookbook_ColorImages.pdf`.

Errata

Although we have taken every care to ensure the accuracy of our content, mistakes do happen. If you find a mistake in one of our books—maybe a mistake in the text or the code—we would be grateful if you could report this to us. By doing so, you can save other readers from frustration and help us improve subsequent versions of this book. If you find any errata, please report them by visiting `http://www.packtpub.com/submit-errata`, selecting your book, clicking on the **Errata Submission Form** link, and entering the details of your errata. Once your errata are verified, your submission will be accepted and the errata will be uploaded to our website or added to any list of existing errata under the Errata section of that title.

To view the previously submitted errata, go to `https://www.packtpub.com/books/content/support` and enter the name of the book in the search field. The required information will appear under the **Errata** section.

Piracy

Piracy of copyrighted material on the Internet is an ongoing problem across all media. At Packt, we take the protection of our copyright and licenses very seriously. If you come across any illegal copies of our works in any form on the Internet, please provide us with the location address or website name immediately so that we can pursue a remedy.

Please contact us at `copyright@packtpub.com` with a link to the suspected pirated material.

We appreciate your help in protecting our authors and our ability to bring you valuable content.

Questions

If you have a problem with any aspect of this book, you can contact us at `questions@packtpub.com`, and we will do our best to address the problem.

1
Functions in R

This chapter covers the following topics:

- Creating R functions
- Matching arguments
- Understanding environments
- Working with lexical scope
- Understanding closure
- Performing lazy evaluation
- Creating infix operators
- Using the replacement function
- Handling errors in a function
- The debugging function

Introduction

R is the mainstream programming language of choice for data scientists. According to polls conducted by KDnuggets, a leading data analysis website, R ranked as the most popular language for analytics, data mining, and data science in the three most recent surveys (2012 to 2014). For many data scientists, R is more than just a programming language because the software also provides an interactive environment that can perform all types of data analysis.

R has many advantages in data manipulation and analysis, and the three most well-known are as follows:

- **Open Source and free**: Using SAS or SPSS requires the purchase of a usage license. One can use R for free, allowing users to easily learn how to implement statistical algorithms from the source code of each function.
- **Powerful data analysis functions**: R is famous in the data science community. Many biologists, statisticians, and programmers have wrapped their models into R packages before distributing these packages worldwide through CRAN (Comprehensive R Archive Network). This allows any user to start their analysis project by downloading and installing an R package from CRAN.
- **Easy to use**: As R is a self-explanatory, high-level language, programming in R is fairly easy. R users only need to know how to use the R functions and how each parameter works through its powerful documentation. We can easily conduct high-level data analysis without having knowledge of the complex underlying mathematics.

R users will most likely agree that these advantages make complicated data analysis easier and more approachable. Notably, R also allows us to take the role of just a basic user or a developer. For an R user, we only need to know how a function works without requiring detailed knowledge of how it is implemented. Similarly to SPSS, we can perform various types of data analysis through R's interactive shell. On the other hand, as an R developer, we can write their function to create a new model, or they can even wrap implemented functions into a package.

Instead of explaining how to write an R program from scratch, the aim of this book is to cover how to become a developer in R. The main purpose of this chapter is to show users how to define their function to accelerate the analysis procedure. Starting with creating a function, this chapter covers the environment of R, and it explains how to create matching arguments. There is also content on how to perform functional programming in R, how to create advanced functions, such as infix operator and replacement, and how to handle errors and debug functions.

Creating R functions

The R language is a collection of functions; a user can apply built-in functions from various packages to their project, or they can define a function for a particular purpose. In this recipe, we will show you how to create an R function.

Getting ready

If you are new to the R language, you can find a detailed introduction, language history, and functionality on the official R site (`http://www.r-project.org/`). When you are ready to download and install R, please connect to the comprehensive R archive network (`http://cran.r-project.org/`).

How to do it...

Perform the following steps in order to create your first R function:

1. Type the following code on your R console to create your first function:

```
>addnum<- function(x, y){
+ s <- x+y
+ return(s)
+ }
```

2. Execute the addnum user-defined function with the following command:

```
>addnum (3,7)
[1] 10
```

   Or, you can define your function without a return statement:

```
>addnum2<- function(x, y){
+ x+y
+ }
```

3. Execute the addnum2 user-defined function with the following command:

```
>addnum2(3,7)
[1] 10
```

4. You can view the definition of a function by typing its function name:

```
>addnum2
function(x, y){
x+y
}
```

5. Finally, you can use body and formals to examine the body and formal arguments of a function:

```
>body(addnum2)
{
x + y
}
>formals(addnum2)
$x
$y
>args(addnum2)
function (x, y)
NULL
```

How it works...

R functions are a block of organized and reusable statements, which makes programming less repetitive by allowing you to reuse code. Additionally, by modularizing statements within a function, your R code will become more readable and maintainable.

By following these steps, you can now create two `addnum` and `addnum2` R functions, and you can successfully add two input arguments with either function. In R, the function usually takes the following form:

```
FunctionName<- function (arg1, arg2) {
body
return(expression)
}
```

`FunctionName` is the name of the function, and `arg1` and `arg2` are arguments. Inside the curly braces, we can see the function body, where a body is a collection of a valid statement, expression, or assignment. At the bottom of the function, we can find the `return` statement, which passes expression back to the caller and exits the function.

The `addnum` function is in standard function syntax, which contains both `body` and `return` statement. However, you do not necessarily need to put a `return` statement at the end of the function. Similar to the `addnum2` function, the function itself will return the last expression back to the caller.

If you want to view the composition of the function, simply type the function name on the interactive shell. You can also examine the `body` and `formal` arguments of the function further using the `body` and `formal` functions. Alternatively, you can use the `args` function to obtain the argument list of the function.

There's more...

If you want to see the documentation of a function in R, you can use the `help` function or simply type `?` in front of the function name. For example, if you want to examine the documentation of the `sum` function, you would do the following:

```
>help(sum)
> ?sum
```

Matching arguments

In R functions, the arguments are the input variables supplied when you invoke the function. We can pass the argument, named argument, argument with default variable, or unspecific numbers of argument into functions. In this recipe, we will demonstrate how to pass different kinds of arguments to our defined function.

Getting ready

Ensure that you completed the previous recipe by installing R on your operating system.

How to do it...

Perform the following steps to create a function with different types of argument lists:

1. Type the following code to your R console to create a function with a default value:

   ```
   >defaultarg<- function(x, y = 5){
   + y <- y * 2
   + s <- x+y
   + return(s)
   + }
   ```

2. Then, execute the `defaultarg` user-defined function by passing `3` as the input argument:

   ```
   >defaultarg(3)
   [1] 13
   ```

3. Alternatively, you can pass different types of input argument to the function:

   ```
   >defaultarg(1:3)
   [1] 11 12 13
   ```

4. You can also pass two arguments into the function:

   ```
   >defaultarg(3,6)
   [1] 15
   ```

5. Or, you can pass a named argument list into the function:

   ```
   >defaultarg(y = 6, x = 3)
   [1] 15
   ```

6. Moving on, you can use if-else-condition together with the function of the named argument:

```
>funcarg<- function(x, y, type= "sum"){
+ if (type == "sum"){
+ sum(x,y)
+ }else if (type == "mean"){
+ mean(x,y)
+ }else{
+ x * y
+ }
+ }
>funcarg(3,5)
[1] 8
>funcarg(3,5, type = 'mean')
[1] 3
>funcarg(3,5, type = 'unknown')
[1] 15
```

7. Additionally, one can pass an unspecified number of parameters to the function:

```
>unspecarg<- function(x, y, ...){
+ x <- x + 2
+ y <- y * 2
+ sum(x,y, ...)
+ }
>unspecarg(3,5)
[1] 15
>unspecarg(3,5,7,9,11)
[1] 42
```

How it works...

R provides a flexible argument binding mechanism when creating functions. In this recipe, we first create a function called `defaultag` with two formal arguments: `x` and `y`. Here, the `y` argument has a default value, defined as `5`. Then, when we make a function call by passing `3` to `defaultarg`, it passes `3` to `x` and `5` to `y` in the function and returns `13`. Besides passing a scalar as the function input, we can pass a vector (or any other data type) to the function. In this example, if we pass a `1:3` vector to `defaultarg`, it returns a vector.

Moving on, we can see how arguments bind to the function. When calling a function by passing an argument without a parameter name, the function binds the passing value by position. Take step 4 as an example; the first `3` argument matches to `x`, and `6` matches to `y`, and it returns `15`. On the other hand, you can pass arguments by name. In step 5, we can pass named arguments to the function in any order. Thus, if we pass `y=6` and `x=3` to `defaultarg`, the function returns `15`.

Furthermore, we can use an argument as a control statement. In step 6, we specify three formal arguments: `x`, `y`, and `type`, in which the `type` argument has the default value defined as `sum`. Next, we can specify the value for the `type` argument as a condition in the if-else control flow. That is, when we pass `sum` to `type`, it returns the summation of `x` and `y`. When we pass `mean` to `type`, it returns the average of `x` and `y`. When we pass any value other than `sum` and `mean` to the `type` argument, it returns the product of `x` and `y`.

Lastly, we can pass an unspecified number of arguments to the function using the `...` notation. In the final step of this example, if we pass only `3` and `5` to the function, the function first passes `3` to `x` and `5` to `y`. Then, the function adds `2` to `x`, multiplies `y` by `2`, and sums the value of both `x` and `y`. However, if we pass more than two arguments to the function, the function will also sum the additional parameters.

There's more...

In addition to giving a full argument name, we can abbreviate the argument name when making a function call:

```
>funcarg(3,5, t = 'unknown')
[1] 15
```

Here, though we do not specify the argument's name, `type`, correctly, the function passes a value of `unknown` to the argument `type`, and returns `15` as the output.

Understanding environments

Besides the function name, body, and formal arguments, the environment is another basic component of a function. In a nutshell, the environment is where R manages and stores different types of variables. Besides the global environment, each function activates its environment whenever a new function is created. In this recipe, we will show you how the environment of each function works.

Getting ready

Ensure that you completed the previous recipes by installing R on your operating system.

How to do it...

Perform the following steps to work with the environment:

1. First, you can examine the current environment with the `environment` function:

```
>environment()
<environment: R_GlobalEnv>
```

2. You can also browse the global environment with `.GlobalEnv` and `globalenv`:

```
> .GlobalEnv
<environment: R_GlobalEnv>
>globalenv()
<environment: R_GlobalEnv>
```

3. You can compare the environment with the `identical` function:

```
>identical(globalenv(), environment())
[1] TRUE
```

4. Furthermore, you can create a new environment as follows:

```
>myenv<- new.env()
>myenv
<environment: 0x0000000017e3bb78>
```

5. Next, you can find the variables of different environments:

```
>myenv$x<- 3
>ls(myenv)
[1] "x"
>ls()
[1] "myenv"
>x
Error: object 'x' not found
```

6. At this point, you can create an `addnum` function and use `environment` to get the environment of the function:

```
>addnum<- function(x, y){
+ x+y
+ }
>environment(addnum)
<environment: R_GlobalEnv>
```

7. You can also determine that the environment of a function belongs to the package:

```
>environment(lm)
<environment: namespace:stats>
```

8. Moving on, you can print the environment within a function:

```
>addnum2<- function(x, y){
+ print(environment())
+ x+y
+ }
>addnum2(2,3)
<environment: 0x0000000018468710>
[1] 5
```

9. Furthermore, you can compare the environment inside and outside a function:

```
>addnum3<- function(x, y){
+ func1<- function(x){
+ print(environment())
+ }
+ func1(x)
+ print(environment())
+ x + y
+ }
>addnum3(2,5)
<environment: 0x000000001899beb0>
<environment: 0x000000001899cc50>
[1] 7
```

How it works...

We can regard an R environment as a place to store and manage variables. That is, whenever we create an object or a function in R, we add an entry to the environment. By default, the top-level environment is the `R_GlobalEnv` global environment, and we can determine the current environment using the environment function. Then, we can use either `.GlobalEnv` or `globalenv` to print the global environment, and we can compare the environment with the `identical` function.

Besides the global environment, we can actually create our environment and assign variables into the new environment. In the example, we created the `myenv` environment and then assigned `x <- 3` to `myenv` by placing a dollar sign after the environment name. This allows us to use the `ls` function to list all variables in `myenv` and global environment. At this point, we find `x` in `myenv`, but we can only find `myenv` in the global environment.

Moving on, we can determine the environment of a function. By creating a function called `addnum`, we can use `environment` to get its environment. As we created the function under global environment, the function obviously belongs to the global environment. On the other hand, when we get the environment of the `lm` function, we get the package name instead. That means that the `lm` function is in the namespace of the `stat` package.

Furthermore, we can print out the current environment inside a function. By invoking the `addnum2` function, we can determine that the `environment` function outputs a different environment name from the global environment. That is, when we create a function, we also create a new environment for the global environment and link a pointer to its parent environment. To further examine this characteristic, we create another `addnum3` function with a `func1` nested function inside. At this point, we can print out the environment inside `func1` and `addnum3`, and it is possible that they have completely different environments.

There's more...

To get the parent environment, we can use the `parent.env` function. In the following example, we can see that the parent environment of `parentenv` is `R_GlobalEnv`:

```
>parentenv<- function(){
+ e <- environment()
+ print(e)
+ print(parent.env(e))
+ }
>parentenv()
<environment: 0x0000000019456ed0>
<environment: R_GlobalEnv>
```

Working with lexical scoping

Lexical scoping, also known as **static binding**, determines how a value binds to a free variable in a function. This is a key feature that originated from the **scheme** functional programming language, and it makes R different from S. In the following recipe, we will show you how lexical scoping works in R.

Getting ready

Ensure that you completed the previous recipes by installing R on your operating system.

How to do it...

Perform the following steps to understand how the scoping rule works:

1. First, we create an `x` variable, and we then create a `tmpfunc` function with `x+3` as the return:

```
>x<- 5
>tmpfunc<- function(){
+ x + 3
+ }
>tmpfunc()
[1] 8
```

2. We then create a function named `parentfunc` with a `childfunc` nested function and see what returns when we call the `parentfunc` function:

```
>x<- 5
>parentfunc<- function(){
+ x<- 3
+ childfunc<- function(){
+ x
+ }
+ childfunc()
+ }
>parentfunc()
[1] 3
```

3. Next, we create an `x` string, and then we create a `localassign` function to modify `x` within the function:

```
> x <- 'string'
>localassign<- function(x){
+ x <- 5
+ x
+ }
>localassign(x)
```

```
[1] 5
>x
[1] "string"
```

4. We can also create another `globalassign` function but reassign the `x` variable to `5` using the `<<-` notation:

```
> x <- 'string'
>gobalassign<- function(x){
+ x <<- 5
+ x
+ }
>gobalassign(x)
[1] 5
>x
[1] 5
```

How it works...

There are two different types of variable binding methods: one is lexical binding, and the other is dynamic binding. Lexical binding is also called static binding in which every binding scope manages variable names and values in the lexical environment. That is, if a variable is lexically bound, it will search the binding of the nearest lexical environment. In contrast to this, dynamic binding keeps all variables and values in the global state. That is, if a variable is dynamically bound, it will bind to the most recently created variable.

To demonstrate how lexical binding works, we first create an `x` variable and assign `5` to `x` in the global environment. Then, we can create a function named `tmpfunc`. The function outputs `x + 3` as the return value. Even though we do not assign any value to `x` within the `tmpfunc` function, `x` can still find the value of `x` as `5` in the global environment.

Next, we create another function named `parentfunc`. In this function, we assign `x` to `3` and create a `childfunc` nested function (a function defined within a function). At the bottom of the `parentfunc` body, we invoke `childfunc` as the function return. Here, we find that the function uses the `x` defined in `parentfunc` instead of the one defined outside `parentfunc`. This is because R searches the global environment for a matched symbol name, and then subsequently searches the namespace of packages on the search list.

Moving on, let's take a look at what will return if we create an `x` variable as a string in the global state and assign an `x` local variable to `5` within the function. When we invoke the `localassign` function, we discover that the function returns `5` instead of the string value. On the other hand, if we print out the value of `x`, we still see string in return. While the local variable and global variable have the same name, the assignment of the function does not alter the value of `x` in global state. If you need to revise the value of `x` in the global state, you can use the `<<-` notation instead.

There's more...

In order to examine the search list (or path) of R, you can type `search()` to list the search list:

```
>search()
[1] ".GlobalEnv""tools:rstudio"
[3] "package:stats" "package:graphics"
[5] "package:grDevices" "package:utils"
[7] "package:datasets" "package:methods"
[9] "Autoloads" "package:base"
```

Understanding closure

Functions are the first-class citizens of R. In other words, you can pass a function as the input to an other function. In previous recipes, we illustrated how to create a named function. However, we can also create a function without a name, known as closure (that is, an anonymous function). In this recipe, we will show you how to use closure in a standard function.

Getting ready

Ensure that you completed the previous recipes by installing R on your operating system.

How to do it...

Perform the following steps to create a closure in function:

1. First, let's review how a named function works:

```
>addnum<- function(a,b){
+ a + b
+ }
>addnum(2,3)
[1] 5
```

2. Now, let's perform the same task to sum up two variables with closure:

```
> (function(a,b){
+ a + b
+ })(2,3)
[1] 5
```

3. We can also invoke a closure function within another function:

```
>maxval<- function(a,b){
+ (function(a,b){
+ return(max(a,b))
+ }
+ )(a, b)
+ }
>maxval(c(1,10,5),c(2,11))
[1] 11
```

4. In a similar manner to the apply family function, you can use the vectorization calculation:

```
> x <- c(1,10,100)
> y <- c(2,4,6)
> z <- c(30,60,90)
> a <- list(x,y,z)
>lapply(a, function(e){e[1] * 10})
[[1]]
[1] 10
[[2]]
[1] 20
[[3]]
[1] 300
```

5. Finally, we can add functions into a list and apply the function to a given vector:

```
> x <- c(1,10,100)
>func<- list(min1 = function(e){min(e)},
 max1 = function(e){max(e)} )
>func$min1(x)
[1] 1
>lapply(func, function(f){f(x)})
```

```
$min1
[1] 1
$max1
[1] 100
```

How it works...

In R, you do not have to create a function with the actual name. Instead, you can use closure to integrate methods within objects. Thus, you can create a smaller and simpler function within another object to accomplish complicated tasks.

In our first example, we illustrated how a normally-named function is created. We can simply invoke the function by passing values into the function. On the other hand, we demonstrate how closure works in our second example. In this case, we do not need to assign a name to the function, but we can still pass the value to the anonymous function and obtain the return value.

Next, we demonstrate how to add a closure within a `maxval` named function. This function simply returns the maximum value of two passed parameters. However, it is possible to use closure within any other function. Moreover, we can use closure as an argument in higher order functions, such as `lapply` and `sapply`. Here, we can input an anonymous function as a function argument to return the multiplication of 10 and the first value of any vector within a given list.

Furthermore, we can specify a single function, or we can store functions in a list. Therefore, when we want to apply multiple functions to a given vector, we can pass the function calls as an argument list to the `lapply` function.

There's more...

Besides using closure within a `lapply` function, we can also pass a closure to other functions of the apply function family. Here, we demonstrate how we can pass the same closure to the `sapply` function:

```
> x <- c(1,10,100)
> y <- c(2,4,6)
> z <- c(30,60,90)
> a <- list(x,y,z)
>sapply(a, function(e){e[1] * 10})
[1] 10 20 300
```

Performing lazy evaluation

R functions evaluate arguments lazily; the arguments are evaluated as they are needed. Thus, lazy evaluation reduces the time needed for computation. In the following recipe, we will demonstrate how lazy evaluation works.

Getting ready

Ensure that you completed the previous recipes by installing R on your operating system.

How to do it...

Perform the following steps to see how lazy evaluation works:

1. First, we create a `lazyfunc` function with `x` and `y` as the argument, but only return `x`:

```
>lazyfunc<- function(x, y){
+ x
+ }
>lazyfunc(3)
[1] 3
```

2. On the other hand, if the function returns the summation of `x` and `y` but we do not pass `y` into the function, an error occurs:

```
>lazyfunc2<- function(x, y){
+ x + y
+ }
>lazyfunc2(3)
Error in lazyfunc2(3) : argument "y" is missing, with no default
```

3. We can also specify a default value to the `y` argument in the function but pass the `x` argument only to the function:

```
>lazyfunc4<- function(x, y=2){
+ x + y
+ }
>lazyfunc4(3)
[1] 5
```

4. In addition to this, we can use lazy evaluation to perform Fibonacci computation in a function:

```
>fibonacci<- function(n){
+ if (n==0)
+ return(0)
+ if (n==1)
+ return(1)
+ return(fibonacci(n-1) + fibonacci(n-2))
+ }
>fibonacci(10)
[1] 55
```

How it works...

R performs a lazy evaluation to evaluate an expression if its value is needed. This type of evaluation strategy has the following three advantages:

- It increases performance due to the avoidance of repeated evaluation
- It recursively constructs an infinite data structure
- It inherently includes iteration in its data structure

In this recipe, we demonstrate some lazy evaluation examples in the R code. In our first example, we create a function with two arguments, `x` and `y`, but return only `x`. Due to the characteristics of lazy evaluation, we can successfully obtain function returns even though we pass the value of `x` to the function. However, if the function return includes both `x` and `y`, as step 2 shows, we will get an error message because we only passed one value to the function. If we set a default value to `y`, then we do not necessarily need to pass both `x` and `y` to the function.

As lazy evaluation has the advantage of creating an infinite data structure without an infinite loop, we use a Fibonacci number generator as the example. Here, this function first creates an infinite list of Fibonacci numbers and then extracts the nth element from the list.

There's more...

Additionally, we can use the `force` function to check whether `y` exists:

```
>lazyfunc3<- function(x, y){
+ force(y)
+ x
+ }
```

```
>lazyfunc3(3)
Error in force(y) : argument "y" is missing, with no default
>input_function<- function(x, func){
+ func(x)
+ }
>input_function(1:10, sum)
[1] 55
```

Creating infix operators

In the previous recipe, we learned how to create a user-defined function. Most of the functions that we mentioned so far are `prefix` functions, where the arguments are in between the parenthesis after the function name. However, this type of syntax makes a simple binary operation of two variables harder to read as we are more familiar with placing an operator in between two variables. To solve the concern, we will show you how to create an infix operator in this recipe.

Getting ready

Ensure that you completed the previous recipes by installing R on your operating system.

How to do it...

Perform the following steps to create an infix operator in R:

1. First, let's take a look at how to transform infix operation to prefix operation:

   ```
   > 3 + 5
   [1] 8
   > '+'(3,5)
   [1] 8
   ```

2. Furthermore, we can look at a more advanced example of the transformation:

   ```
   > 3:5 * 2 - 1
   [1] 5 7 9
   > '-'('*'(3:5, 2), 1)
   [1] 5 7 9
   ```

3. Moving on, we can create our infix function that finds the intersection between two vectors:

```
>x <-c(1,2,3,3, 2)
>y <-c(2,5)
> '%match%' <- function(a,b){
+ intersect(a, b)
+ }
>x %match% y
[1] 3
```

4. Let's also create a `%diff%` infix to extract the set difference between two vectors:

```
> '%diff%' <- function(a,b){
+ setdiff(a, b)
+ }
>x %diff% y
[1] 1 2
```

5. Lastly, we can use the infix operator to extract the intersection of three vectors. Or, we can use the `Reduce` function to apply the operation to the list:

```
>x %match% y %match% z
[1] 3
> s <- list(x,y,z)
>Reduce('%match%',s)
[1] 3
```

How it works...

In a standard function, if we want to perform some operations on the `a` and `b` variables, we would probably create a function in the form of `func(a,b)`. While this form is the standard function syntax, it is harder to read than regular mathematical notation, (that is, `a * b`). However, we can create an infix operator to simplify the function syntax.

Before creating our infix operator, we examine different syntax when we apply a binary operator on two variables. In the first step, we demonstrate how to perform arithmetic operations with binary operators. Similar to a standard mathematical formula, all we need to do is to place a binary operator between two variables. On the other hand, we can transform the representation from infix form to prefix form. Like a standard function, we can use the binary operator as the function name, and then we can place the variables in between the parentheses.

In addition to using a predefined infix operator in R, the user can define the infix operator. To create an operator, we need to name the function that starts and ends with `%`, and surround the name with a single quote (`'`) or back tick (`` ` ``). Here, we create an infix operator named `%match%` to extract the interaction between two vectors. We can also create another infix function named `%diff%` to extract the set difference between two vectors. Lastly, though we can apply the created infix function to more than two vectors, we can use the reduce function to apply the `%match%` operation on the list.

There's more...

We can also overwrite the existing operator by creating an infix operator with the same name:

```
>'+' <- function(x,y) paste(x,y, sep = "|")
> x = '123'
> y = '456'
>x+y
[1] "123|456"
```

Here, we can concatenate two strings with the + operator.

Using the replacement function

On some occasions in R, we may discover that we can assign a value to a function call, which is what the replacement function does. Here, we will show you how the replacement function works, and how to create your own.

Getting ready

Ensure that you completed the previous recipes by installing R on your operating system.

How to do it...

Perform the following steps to create a replacement function in R:

1. First, we assign names to data with the `names` function:

   ```
   > x <- c(1,2,3)
   >names(x) <- c('a','b','c')
   >x
   a b c
   1 2 3
   ```

What the `names` function actually does is similar to the following commands:

```
> x <- 'names<-'(x,value=c('a','b','c'))
>x
a b c
1 2 3
```

2. Here, we can also make our replacement function:

```
> x<-c(1,2,3)
> "erase<-" <- function(x, value){
+ x[!x %in% value]
+ }
>erase(x) <- 2
>x
[1] 1 3
```

3. We can invoke the `erase` function in the same way that we invoke the normal function:

```
>x <- c(1,2,3)
> x <- 'erase<-'(x,value=c(2))
>x
[1] 1 3
```

We can also remove multiple values with the `erase` function:

```
> x <- c(1,2,3)
>erase(x) = c(1,3)
>x
[1] 2
```

4. Finally, we can create a replacement function that can remove values of certain positions:

```
> x <- c(1,2,3)
> y <- c(2,2,3)
> z <- c(3,3,1)
> a = list(x,y,z)
> "erase<-" <- function(x, pos, value){
+ x[[pos]] <- x[[pos]][!x[[pos]] %in% value]
+ x
+ }
>erase(a, 2) = c(2)
```

```
>a
[[1]]
[1] 1 2 3
[[2]]
[1] 3
[[3]]
[1] 3 3 1
```

How it works...

In this recipe, we first demonstrated how we could use the `names` function to assign argument names for each value. This type of function method may appear confusing, but it is actually what the replacement function does: assigning the value to the function call. We then illustrated how this function works in a standard function form, which we achieved by placing an assignment arrow (`<-`) after the function name and placing the `x` object and value between the parentheses.

Next, we learned how to create our replacement function. We made a function named `erase`, which removed certain values from a given object. We invoked the function by wrapping the vector to replace within the `erase` function and assigning the value to remove on the right-hand side of the assignment notation. Alternatively, we can still call the replacement function by placing an assignment arrow after `erase` as the function name. In addition to removing a single value from a given vector object, we can also remove multiple values by placing a vector on the right-hand side of the assignment function.

Furthermore, we can remove the values of certain positions with the replacement function. Here, we only needed to add a position argument between the object and value within the parentheses. As our last step shows, we removed `2` from the second value of the list with our newly-created replacement function.

There's more...

As mentioned earlier, `names<-` is a replacement function. To examine whether a function is a replacement function or not, use the `get` function:

```
>get("names<-")
function (x, value) .Primitive("names<-")
```

Handling errors in a function

If you are familiar with modern programming languages, you may have experience with how to use `try`, `catch`, and finally, `block`, to handle possible errors during development. Likewise, R provides similar error-handling operations in its functions. Thus, you can add error-handling mechanisms into R code to make programs more robust. In this recipe, we will introduce some basic error-handling functions in R.

Getting ready

Ensure that you completed the previous recipes by installing R on your operating system.

How to do it...

Perform the following steps to handle errors in an R function:

1. First, let's observe what an error message looks like:

```
> 'hello world' + 3
Error in "hello world" + 3 : non-numeric argument to binary
operator
```

2. In a user-defined function, we can also print out the error message using `stop` if something beyond our expectation happens:

```
>addnum<- function(a,b){
+ if(!is.numeric(a) | !is.numeric(b)){
+ stop("Either a or b is not numeric")
+ }
+ a + b
+ }
>addnum(2,3)
[1] 5
>addnum("hello world",3)
Error in addnum("hello world", 3) : Either a or b is not numeric
```

3. Now, let's see what happens if we replace the `stop` function with a `warning` function:

```
>addnum2<- function(a,b){
+ if(!is.numeric(a) | !is.numeric(b)){
+ warning("Either a or b is not numeric")
+ }
```

```
+ a + b
+ }
>addnum2("hello world",3)
Error in a + b : non-numeric argument to binary operator
In addition: Warning message:
In addnum2("hello world", 3) : Either a or b is not numeric
```

4. We can also see what happens if we replace the `stop` function with a `warning` function:

```
>options(warn=2)
>addnum2("hello world", 3)
Error in addnum2("hello world", 3) :
(converted from warning) Either a or b is not numeric
```

5. To suppress warnings, we can wrap the function to invoke with a `suppressWarnings` function:

```
>suppressWarnings(addnum2("hello world",3))
Error in a + b : non-numeric argument to binary operator
```

6. We can also use the `try` function to catch the error message:

```
>errormsg<- try(addnum("hello world",3))
Error in addnum("hello world", 3) : Either a or b is not numeric
>errormsg
[1] "Error in addnum("hello world", 3) : Either a or b is not
numeric\n"
attr(,"class")
[1] "try-error"
attr(,"condition")
<simpleError in addnum("hello world", 3): Either a or b is not
numeric>
```

7. By setting the `silent` option, we can suppress the error message displayed on the console:

```
>errormsg<- try(addnum("hello world",3), silent=TRUE)
```

8. Furthermore, we can use the `try` function to prevent interrupting the for-loop. Here, we show a for-loop without using the `try` function:

```
>iter<- c(1,2,3,'O',5)
>res<- rep(NA, length(iter))
>for (i in 1:length(iter)) {
```

```
+ res[i] = as.integer(iter[i])
+ }
Error: (converted from warning) NAs introduced by coercion
>res
[1] 1 2 3 NA NA
```

9. Now, let's see what happens if we insert the `try` function into the code:

```
>iter<- c(1,2,3,'O',5)
>res<- rep(NA, length(iter))
>for (i in 1:length(iter)) {
+ res[i] = try(as.integer(iter[i]), silent=TRUE)
+ }
>res
[1] "1"
[2] "2"
[3] "3"
[4] "Error in try(as.integer(iter[i]), silent = TRUE) : \n
(converted from warning) NAs introduced by coercion\n"
[5] "5"
```

10. For arguments, we can use the `stopifnot` function to check the argument:

```
>addnum3<- function(a,b){
+ stopifnot(is.numeric(a), !is.numeric(b))
+ a + b
+ }
>addnum3("hello", "world")
Error: is.numeric(a) is not TRUE
```

11. To handle all kinds of errors, we can use the `tryCatch` function for error handling:

```
>dividenum<- function(a,b){
+ result<- tryCatch({
+ print(a/b)
+ }, error = function(e) {
+ if(!is.numeric(a) | !is.numeric(b)){
+ print("Either a or b is not numeric")
+ }
+ }, finally = {
+ rm(a)
```

```
+ rm(b)
+ print("clean variable")
+ }
+ )
+ }
>dividenum(2,4)
[1] 0.5
[1] "clean variable"
>dividenum("hello", "world")
[1] "Either a or b is not numeric"
[1] "clean variable"
>dividenum(1)
Error in value[[3L]](cond) : argument "b" is missing, with no
default
[1] "clean variable"
```

How it works...

Similar to other programming languages, R provides developers with an error-handling mechanism. However, the error-handling mechanism in R is implemented in the function instead of a pure code block. This is due to the fact that all operations are pure function calls.

In the first step, we demonstrate what will output if we add an integer to a string. If the operation is invalid, the system will print an error message on the console. There are three basic types of error handling messages in R, which are `error`, `warning`, and `interrupt`.

Next, we create a function named `addnum`, which is designed to return the addition of two arguments. However, sometimes you will pass an unexpected type of input (for example, string) into a function. For this condition, we can add an argument type check condition before the `return` statement. If none of the input data types is numeric, the `stop` function will print an error message quoted in the `stop` function.

Besides using the `stop` function, we can use a `warning` function instead to handle an error. However, only using a warning function, the function process will not terminate but proceed to return `a + b`. Thus, we might find both an error and warning message displayed on the console. To suppress the warning message, we can set `warn=2` in the `options` function, or we can use `suppressWarnings` instead to mute the warning message. On the other hand, we can also use the `stopifnot` function to check whether the argument is valid or not. If the input argument is invalid, we can stop the program and print an error message on the screen.

Moving on, we can catch the error using the `try` function. Here, we store the error message into `errormsg` in the operation of adding a character string to an integer. However, the function will still print the error message on the screen. We can mute the message by setting a `silent` argument to `TRUE`. Furthermore, the `try` function is very helpful if don't want a for-loop being interrupted by unexpected errors. Therefore, we first demonstrate how an error may unexpectedly interrupt the loop execution. In that step, we may find that the loop execution stops, and we have successfully assigned only three variables to `res`. However, we can actually proceed with the for-loop execution by wrapping the code into a `try` function.

Besides the `try` function, we can use a more advanced error-handling function, `tryCatch`, to handle errors including `warning` and `error`. We use the `tryCatch` function in the following manner:

```
tryCatch({
result<- expr
}, warning = function(w) {
# handling warning
}, error = function(e) {
# handling error
}, finally = {
#Cleanup
})
```

In this function, we can catch `warning` and `error` messages in different function code blocks. By following the function form, we can create a function named `dividenum`. The function first performs numeric division; if any error occurs, we can catch the error and print an error message in the `error` function. At the end of the block, we remove any defined value within the function and print the message of `clean variable`. At this point, we can test how this function works in three different situations: performing a normal division, dividing a string from a string, and passing only one parameter into the function. We can now observe the output message under different conditions. In the first condition, the function prints out the division result, followed by `clean variable` because it is coded in the block of `finally`. For the second condition, the function first catches the error of missing value in the `error` block and then outputs `clean variable` at the end. For the last condition, while we do not catch the error of not passing a value to the `b` parameter, the function still returns an error message first and then prints `clean variable` on the console.

There's more...

If you want to catch the error message while using the `tryCatch` function, you can put a `conditionMessage` to the error argument of the `tryCatch` function:

```
>dividenum<- function(a,b){
+ result<- tryCatch({
+ a/b
+ }, error = function(e) {
+ conditionMessage(e)
+ }
+ )
+ result
+ }
>dividenum(3,5)
[1] 0.6
>dividenum(3,"hello")
[1] "non-numeric argument to binary operator"
```

In this example, if you pass two valid numeric arguments to the `dividenum` function, the function returns a computation of 3/5 as output. On the other hand, if you pass a non-numeric value to the function, the function catches the error with the `conditionMessage` function and returns the error as the function output.

The debugging function

As a programmer, debugging is the most common task faced on a daily basis. The simplest debugging method is to insert a `print` statement at every desired location; however, this method is rather inefficient. Here, we will illustrate how to use some R debugging tools to help accelerate the debugging process.

Getting ready

Make sure that you know how a function and works and how to create a new function.

How to do it...

Perform the following steps to debug an R function:

1. First, we create a `debugfunc` function with `x` and `y` as argument, but we only return `x`:

```
>debugfunc<- function(x, y){
+ x <- y + 2
+ x
+ }
>debug(2)
```

2. We then pass only `2` to `dubugfunc`:

```
>debugfunc(2)
Error in debugfunc(2) : argument "y" is missing, with no default
```

3. Next, we apply the `debug` function onto `debugfunc`:

```
>debug(debugfunc)
```

4. At this point, we pass `2` to `debugfunc` again:

```
>debugfunc(2)
debugging in: debugfunc(2)
debug at #1: {
x <- y + 2
x
}
```

5. You can type `help` to list all possible commands:

```
Browse[2]> help
n next
s step into
f finish
c or cont continue
Q quit
where show stack
help show help
<expr> evaluate expression
```

6. Then, you can type `n` to move on to the next debugging step:

```
Browse[2]> n
debug at #2: x <- y + 2
```

7. At this point, you can use `objects` or `ls` to list variables:

```
Browse[2]> objects()
[1] "x" "y"
Browse[2]>ls()
[1] "x" "y"
```

8. At each step, you can type the variable name to obtain the current value:

```
Browse[2]> x
[1] 2
Browse[2]> y
Error: argument "y" is missing, with no default
```

9. At the last step, you can quit the debug mode by typing the `Q` command:

```
Browse[2]> Q
```

10. You can then leave the debug mode using the `undebug` function:

```
>undebug(debugfunc)
```

11. Moving on, let's debug the function using the `browser` function:

```
debugfunc2<- function(x, y){
x <- 3
browser()
x <- y + 2
x
}
```

12. The debugger will then step right into where the `browser` function is located:

```
>debugfunc2(2)
Called from: debugfunc2(2)
Browse[1]> n
debug at #4: x <- y + 2
```

13. To recover the debug process, type `recover` during the browsing process:

```
Browse[2]> recover()
Enter a frame number, or 0 to exit
1: debugfunc2(2)
```

```
Selection: 1
Browse[4]> Q
```

14. On the other hand, you can use the `trace` function to insert code into the `debug` function at step 4:

```
>trace(debugfunc2, quote(if(missing(y)){browser()}), at=4)
[1] "debugfunc2"
```

15. You can then track the debugging process from step 4, and determine the inserted code:

```
>debugfunc2(3)
Called from: debugfunc2(3)
Browse[1]> n
debug at #4: {
.doTrace(if (missing(y)) {
browser()
}, "step 4")
x <- y + 2
}
Browse[2]> n
debug: .doTrace(if (missing(y)) {
browser()
}, "step 4")
Browse[2]> Q
```

16. On the other hand, you can track the usage of certain functions with the `trace` function:

```
>debugfunc3<- function(x, y){
+ x <- 3
+ sum(x)
+ x <- y + 2
+ sum(x,y)
+ x
+ }
>trace(sum)
>debugfunc3(2,3)
trace: sum(x)
trace: sum(x, y)
[1] 5
```

17. You can also print the calling stack of a function with the `traceback` function:

```
>lm(y~x)
Error in eval(expr, envir, enclos) : object 'y' not found
>traceback()
7: eval(expr, envir, enclos)
6: eval(predvars, data, env)
5: model.frame.default(formula = y ~ x, drop.unused.levels = TRUE)
4: stats::model.frame(formula = y ~ x, drop.unused.levels = TRUE)
3: eval(expr, envir, enclos)
2: eval(mf, parent.frame())
1: lm(y ~ x)
```

How it works...

As it is inevitable for all code to include bugs, an R programmer has to be well prepared for them with a good debugging toolset. In this recipe, we showed you how to debug a function with the `debug`, `browser`, `trace`, and `traceback` functions.

In the first section, we explained how to debug a function by applying `debug` to an existing function. We first made a function named `debugfunc`, with two input arguments: `x` and `y`. Then, we applied a `debug` function onto `debugfunc`. Here, we applied the `debug` function on the name, argument, or function. At this point, whenever we invoke `debugfunc`, our R console will enter into a browser mode with `Browse` as the prompt at the start of each line.

Browser mode enables us to make a single step through the execution of the function. We list the single-letter commands that one can use while debugging here:

Command	Meaning
`c` or `cont` (continue)	This executes all the code of the current function
`n` (next)	This evaluates the next statement, stepping over function calls
`s` (step into)	This evaluates the next statement, stepping into function calls
`objects` or `ls`	This lists all current objects
`help`	This lists all possible commands
`where`	This prints the stack trace of active function calls
`f` (finish)	This finishes the execution of current function
`Q` (quit)	This terminates the debugging mode

In the following operations, we first use `help` to list all possible commands. Then, we type `n` to step to the next line. Next, we type `objects` and `ls` to list all current objects. At this point, we can type the variable name to find out the current value of each object. Finally, we can type `Q` to quit debugging mode, and use `undebug` to unflag the function.

Besides using the `debug` function, we can insert the browser function within the code to debug it. After we have inserted `browser()` into `debugfunc2`, whenever we invoke the function, the R function will step right into the next line below the `browser` function. Here, we can perform any command mentioned in the previous command table. If you want to move among frames or return to the top level of debugging mode, we can use the `recover` function. Additionally, we can use the `trace` function to insert debugging code into the function. Here, we assign what to `trace` as `debugfunc2`, and set the tracer to examine whether `y` is missing. If `y` is missing, it will execute the `browser()` function. At that argument, we set `4` to the argument so that the tracer code will be inserted at line 4 of `debugfunc2`. Then, when we call the `debugfunc2` function, the function enters right into where the tracer is located and executes the browser function as the `y` argument is missing.

Finally, we introduce the `traceback` function, which prints the calling stack of the function. At this step, we pass two unassigned parameters, `x` and `y`, into an `lm` linear model fitting function. As we do not assign any value to these two parameters, the function returns an error message in the console output. To understand the calling stack sequence, we can use the `traceback` function to print out the stack.

There's more...

Besides using the command line, we can use RStudio to debug functions:

1. First, you select `Toggle Breakpoint` from the dropdown menu of `Debug`:

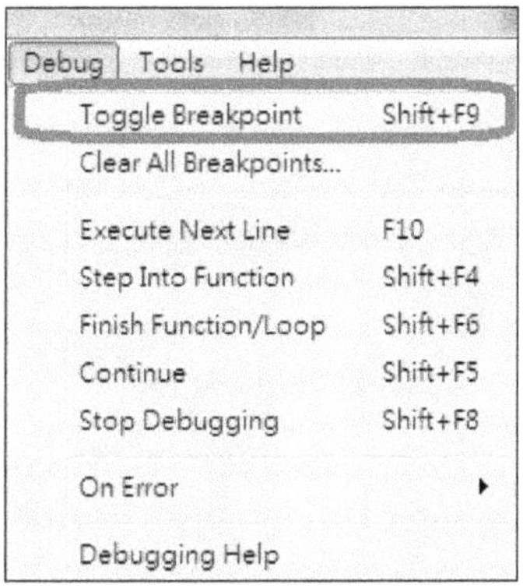

Figure 1: Toggle Breakpoint

2. Then, you set breakpoint on the left of line number:

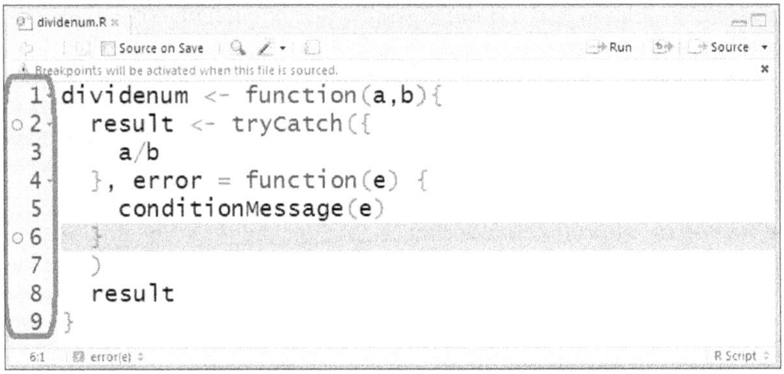

Figure 2: Set breakpoint

3. Next, you save the code file and click on `Source` to activate the debugging process:

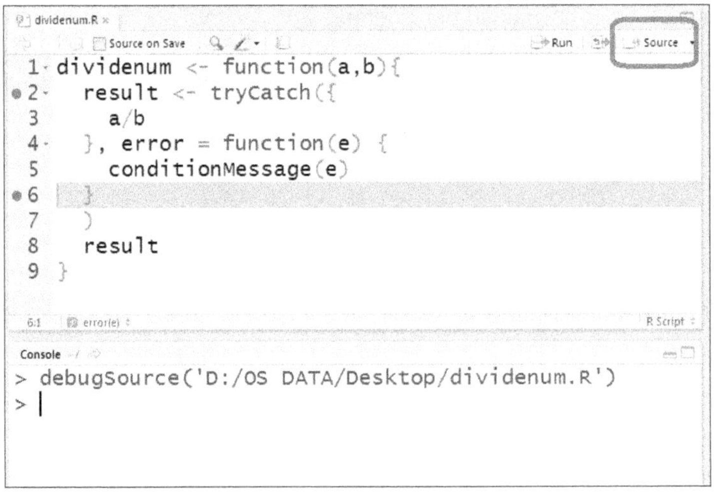

Figure 3: Activate debugging process

4. Finally, when you invoke the function, your R console will then enter into `Browse` mode:

Figure 4: Browse the function

You can now use the command line or dropdown menu of Debug to debug the function:

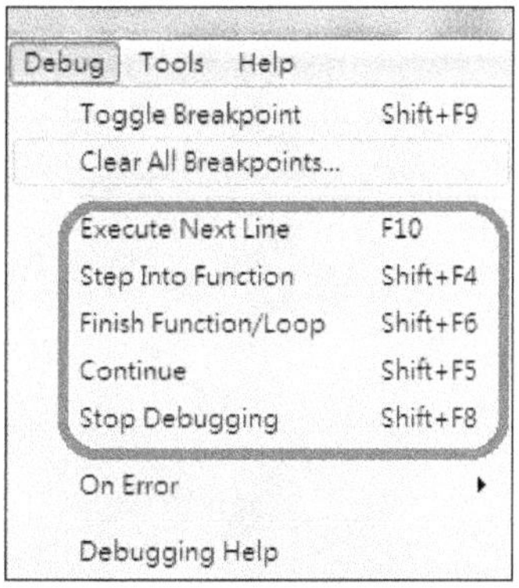

Figure 5: Use debug functions

2

Data Extracting, Transforming, and Loading

This chapter covers the following topics:

- Downloading open data
- Reading and writing CSV files
- Scanning text files
- Working with Excel files
- Reading data from databases
- Scraping web data
- Accessing Facebook data
- Working with twitteR

Introduction

Before using data to answer critical business questions, the most important thing is to prepare it. Data is normally archived in files, and using Excel or text editors allows it to be easily obtained. However, data can be located in a range of different sources, such as databases, websites, and various file formats. Being able to import data from these sources is crucial.

There are four main types of data. Data recorded in text format is the simplest. As some users require storing data in a structured format, files with a `.tab` or `.csv` extension can be used to arrange data in a fixed number of columns. For many years, Excel has had a leading role in the field of data processing, and this software uses the `.xls` and `.xlsx` formats. Knowing how to read and manipulate data from databases is another crucial skill. Moreover, as most data is not stored in a database, one must know how to use the web scraping technique to obtain data from the Internet. As part of this chapter, we introduce how to scrape data from the Internet using the `rvest` package.

Many experienced developers have already created packages to allow beginners to obtain data more easily, and we focus on leveraging these packages to perform data extraction, transformation, and loading. In this chapter, we first learn how to utilize R packages to read data from a text format and scan files line by line. We then move to the topic of reading structured data from databases and Excel. Last, we learn how to scrape Internet and social network data by using the R web scraper.

Downloading open data

Before conducting any data analysis, an essential step is to collect high-quality, meaningful data. One important data source is open data, which is selected, organized, and freely available to the public. Most open data is published online in either text format or as APIs. Here, we introduce how to download the text format of an open data file with the `download.file` function.

Getting ready

In this recipe, you need to prepare your environment with R installed and a computer that can access the Internet.

How to do it...

Please perform the following steps to download open data from the Internet:

1. First, visit the `http://finance.yahoo.com/q/hp?s=%5EGSPC+Historical+Prices` link to view the historical price of the S&P 500 in Yahoo Finance:

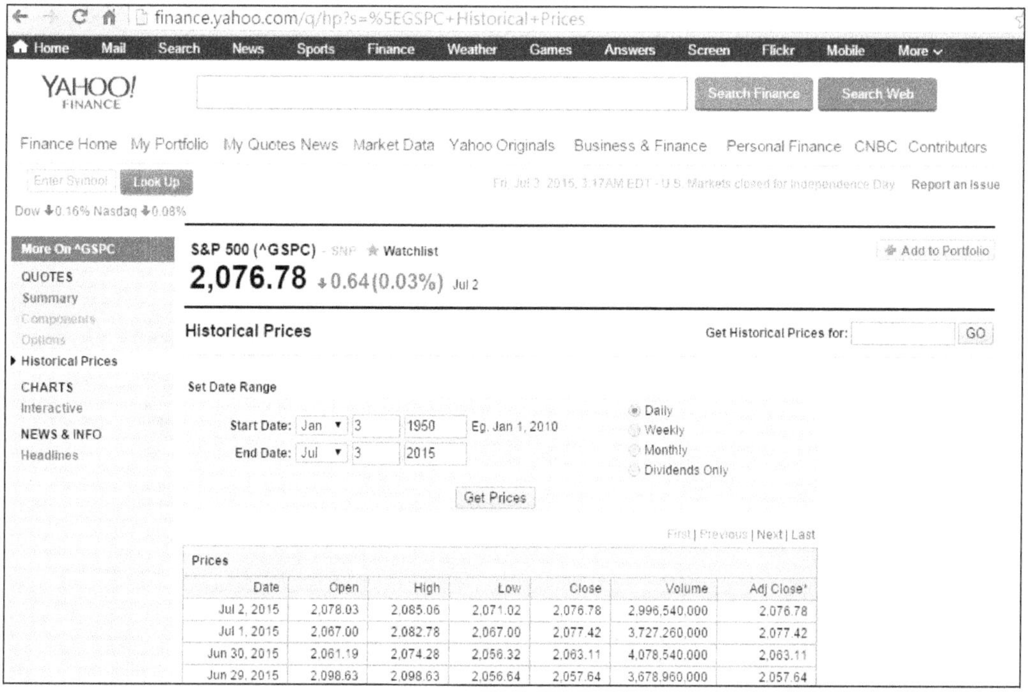

Figure 1: Historical price of S&P 500

2. Scroll down to the bottom of the page, right-click and copy the link in **Download to Spreadsheet** (the link should appear similar to `http://real-chart.finance.yahoo.com/table.csv?s=%5EGSPC&d=6&e=3&f=2015&g=d&a=0&b=3&c=1950&ignore=.csv`):

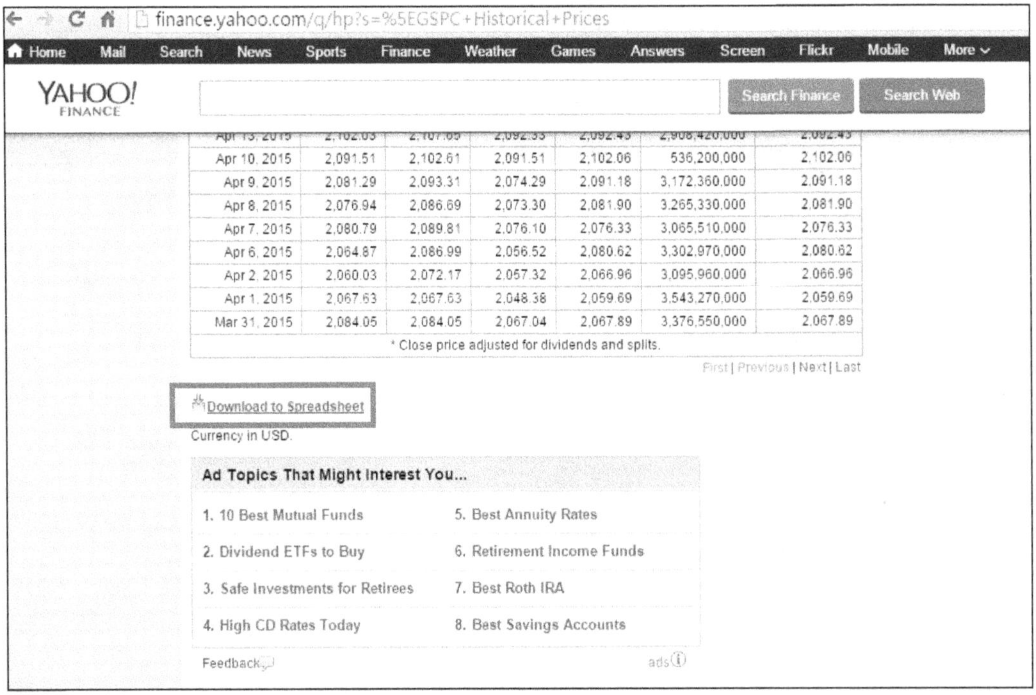

Figure 2: Download to Spreadsheet

3. Download this file with the `download.file` function:

```
> download.file('http://real-chart.finance.yahoo.com/table.
csv?s=%5EGSPC&d=6&e=3&f=2015&g=d&a=0&b=3&c=1950&ignore=.csv',
'snp500.csv')
```

4. You can now use the `getwd` function to determine the current directory, and then use `list.files` to search for the downloaded file:

```
> getwd()
> list.files('./')
```

How it works...

In this recipe, we demonstrated how to download a file using `download.file` in R. First, we used Yahoo Finance to view historical prices of the S&P 500. At the bottom of the page, we found a link with a `http://` URL prefix. The `http://` URL prefix stands for **Hypertext Transfer Protocol** (**HTTP**), which serves the purpose of transmitting and receiving information over the Internet. Therefore, we can request the remote server with the link address through the use of `download.file`. Last, we can make the request for the link and save the remote file into our local directory.

There's more...

Apart from using the `download.file` function to download the file, you can use `RCurl` to download a file with either a HTTP URL prefix or HTTPS URL prefix:

1. First, go to the `https://nycopendata.socrata.com/Social-Services/NYC-Wi-Fi-Hotspot-Locations/a9we-mtpn?` link to explore the Wi-Fi hotspot location file in the NYC open data:

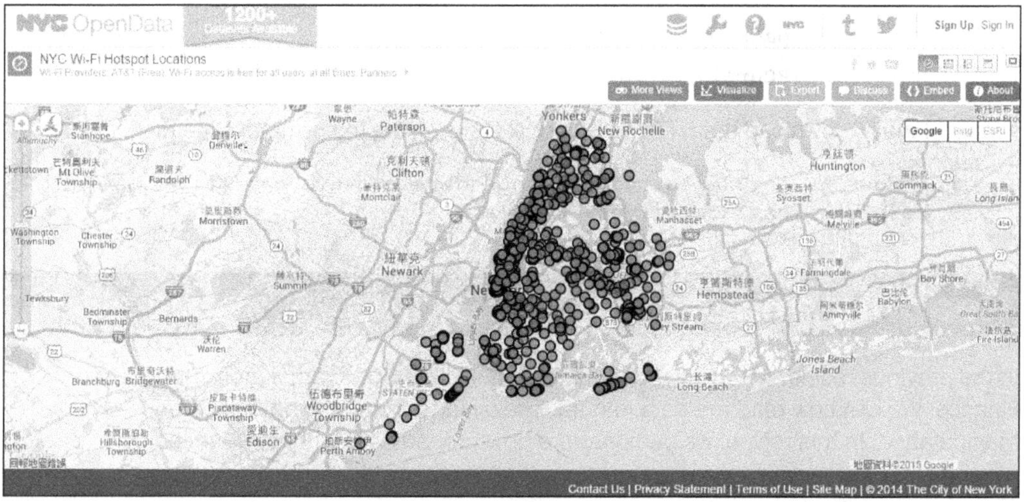

Figure 3: Wi-Fi hotspot location of NYC

2. Next, click on **Export** and find the **CSV** download link:

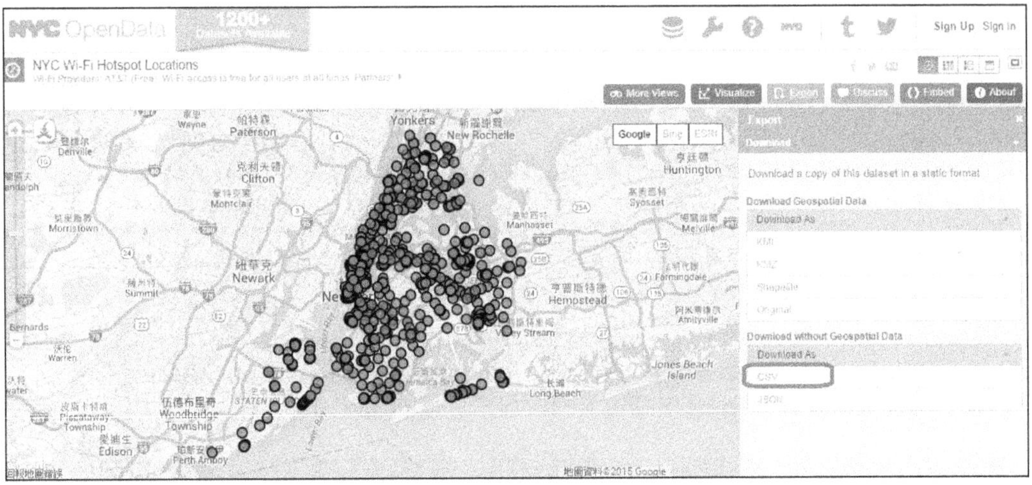

Figure 4: Downloading the CSV format of the Wi-Fi hotspot location

3. You can then install and load the `RCurl` package:

```
> install.packages("RCurl")
> library(RCurl)
```

4. Finally, download the HTTPS URL prefix file by using the `getURL` function:

```
> rows <- getURL("https://nycopendata.socrata.com/api/views/jd4g-
ks2z/rows.csv?accessType=DOWNLOAD")
```

Reading and writing CSV files

In the previous recipe, we downloaded the historical S&P 500 market index from Yahoo Finance. We can now read the data into an R session for further examination and manipulation. In this recipe, we demonstrate how to read a file with an R function.

Getting ready

In this recipe, you need to have followed the previous recipe by downloading the S&P 500 market index text file to the current directory.

How to do it...

Please perform the following steps to read text data from the CSV file.

1. First, determine the current directory with `getwd`, and use `list.files` to check where the file is, as follows:

```
> getwd()
> list.files('./')
```

2. You can then use the `read.table` function to read data by specifying the comma as the separator:

```
> stock_data <- read.table('snp500.csv', sep=',' , header=TRUE)
```

3. Next, filter data by selecting the first six rows with column `Date`, `Open`, `High`, `Low`, and `Close`:

```
> subset_data <- stock_data[1:6, c("Date", "Open", "High", "Low",
"Close")]
```

4. Examine the first six rows of loaded data with the `head` function:

```
> head(stock_data)
```

5. As the file to be loaded is in CSV format, you can also use `read.csv` to read the file:

```
> stock_data2 <- read.csv('snp500.csv', header=TRUE)
> head(stock_data2)
```

How it works...

By following the previous recipe, you should now have Yahoo Finance data downloaded in the current directory. As the downloaded file is organized in a table, you can use the `read.table` function to read data from the file into an R data frame.

As the downloaded data is separated with a comma and contains a column header, you can set `header` equal to `TRUE` and `','` as the field separator in function parameters. After you have read `snp500.csv` into the `stock_data` data frame, you can then select the first six rows from the data for further examination with the `head` function.

Similar to the `read.table` function, you can also use `read.csv` to read the text file. The only difference between `read.csv` and `read.table` is that `read.csv` uses commas as the default separator to read the file, while `read.table` uses white space as the default separator. You can also use the `head` function to examine the loaded data frame.

There's more...

In the previous section, we demonstrated how to use RCurl to obtain Wi-Fi hotspot data from the NYC open data site. As the downloaded data is in character vector, we can use read.csv to read text into an R session by setting text equal to character vector rows in the function argument:

```
> wifi_hotspot <- read.csv(text = rows)
```

Scanning text files

In previous recipes, we introduced how to use read.table and read.csv to load data into an R session. However, read.table and read.csv only work if the number of columns is fixed and the data size is small. To be more flexible in data processing, we will demonstrate how to use the scan function to read data from the file.

Getting ready

In this recipe, you need to have completed the previous recipes and have snp500.csv downloaded in the current directory.

How to do it...

Please perform the following steps to scan data from the CSV file:

1. First, you can use the scan function to read data from snp500.csv:

   ```
   > stock_data3 <- scan('snp500.csv',sep=',', what=list(Date = '',
   Open = 0, High = 0, Low = 0,Close = 0, Volume = 0, Adj_Close = 0),
   skip=1, fill=T)
   Read 16481 records
   ```

2. You can then examine loaded data with mode and str:

   ```
   > mode(stock_data3)
   [1] "list"
   > str(stock_data3)
   List of 7
    $ Date      : chr [1:16481] "2015-07-02" "2015-07-01" "2015-06-30"
   "2015-06-29" ...
    $ Open      : num [1:16481] 2078 2067 2061 2099 2103 ...
    $ High      : num [1:16481] 2085 2083 2074 2099 2109 ...
   ```

```
 $ Low      : num [1:16481] 2071 2067 2056 2057 2095 ...
 $ Close    : num [1:16481] 2077 2077 2063 2058 2102 ...
 $ Volume   : num [1:16481] 3.00e+09 3.73e+09 4.08e+09 3.68e+09
5.03e+09 ...
 $ Adj_Close: num [1:16481] 2077 2077 2063 2058 2102 ...
```

How it works...

When comparing `read.csv` and `read.table`, the `scan` function is more flexible and efficient in data reading. Here, we specify the field name and support type of each field within a list in the `what` parameter. In this case, the first field is of character type, and the rest of the fields are of numeric type. Therefore, we can set two single (or double) quotes for the `Date` column, and `0` for the rest of the fields. Then, as we need to skip the header row and automatically add empty fields to any lines with fewer fields than the number of columns, we set `skip` to `1` and `fill` to `True`.

At this point, we can now examine the data with some built-in functions. Here, we use `mode` to obtain the type of the object and use `str` to display the structure of the data.

There's more...

On some occasions, the data is separated by fixed width rather than fixed delimiter. To specify the width of each column, you can use the `read.fwf` function:

1. First, you can use `download.file` to download `weather.op` from the author's GitHub page:

```
> download.file("https://github.com/ywchiu/rcookbook/raw/
master/chapter2/weather.op", "weather.op")
```

2. You can then examine the data with the file editor:

1	STN---	WBAN	YEARMODA	TEMP		MAX	MIN
2	008403	99999	20140101	85.8	24	102.7*	69.3*
3	008403	99999	20140102	86.3	24	102.9*	71.1*
4	008403	99999	20140103	85.9	24	101.1*	72.0*
5	008403	99999	20140104	85.6	24	102.7*	70.5*
6	008403	99999	20140105	84.8	23	102.0*	66.6*

Figure 5: Using the file editor to examine the file

3. Read the data by specifying the width of each column in `widths`, the column name in `col.names`, and skip the first row by setting `skip` to `1`:

```
> weather <- read.fwf("weather.op", widths = c(6,6,10,11,9,8),
col.names = c("STN","WBAN","YEARMODA","TEMP","MAX","MIN"), skip=1)
```

4. Lastly, you can examine the data using the `head` and `names` functions:

```
> head(weather)
   STN  WBAN YEARMODA       TEMP       MAX       MIN
1 8403 99999 20140101     85.8 24     102.7*     69.3*
2 8403 99999 20140102     86.3 24     102.9*     71.1*
3 8403 99999 20140103     85.9 24     101.1*     72.0*
4 8403 99999 20140104     85.6 24     102.7*     70.5*
5 8403 99999 20140105     84.8 23     102.0*     66.6*
6 8403 99999 20140106     86.8 23     102.0*     70.9*

> names(weather)
[1] "STN"        "WBAN"       "YEARMODA" "TEMP"       "MAX"
[6] "MIN"
```

Working with Excel files

Excel is another popular tool used to store and analyze data. Of course, one can convert Excel files to CSV files or other text formats by using Excel. Alternatively, to simplify the process, you can use `install` and load the `xlsx` package to read and process Excel data in R.

Getting ready

In this recipe, you need to prepare your environment with R installed and a computer that can access the Internet.

How to do it...

Please perform the following steps to read Excel documents:

1. First, install and load the `xlsx` package:

```
> install.packages("xlsx")
> library(xlsx)
```

2. Access `www.data.worldbank.org/topic/economy-and-growth` to find world economy indicator data in Excel:

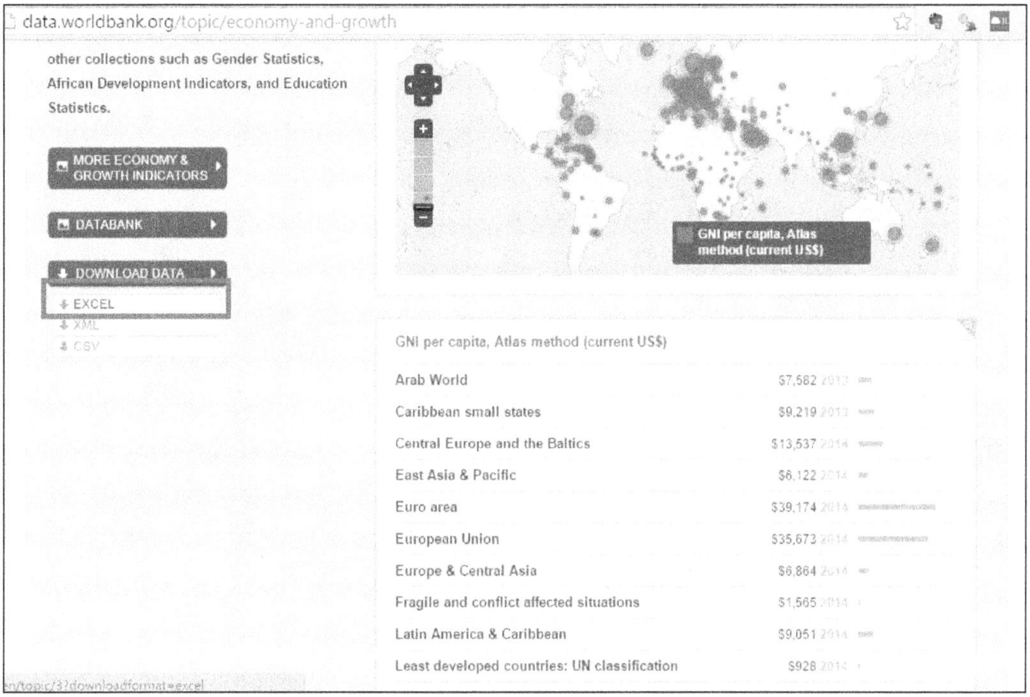

Figure 6: World economy indicator

3. Download world economy indicator data from the following URL using `download.file`:

```
> download.file("http://api.worldbank.org/v2/en/
topic/3?downloadformat=excel", "worldbank.xls", mode="wb")
```

4. Examine the downloaded file with Excel (or Open Office):

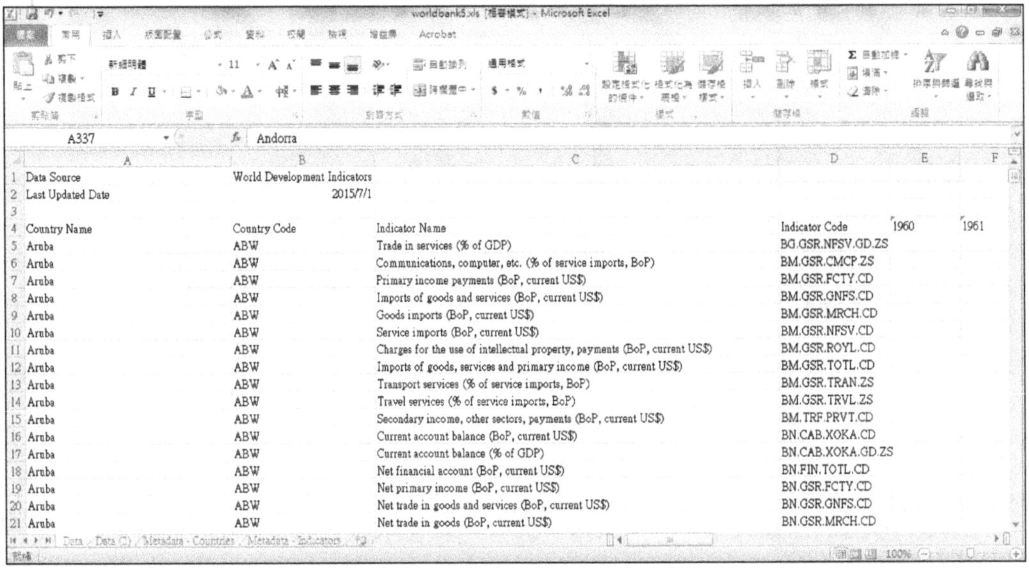

	A	B	C	D	E	F
1	Data Source	World Development Indicators				
2	Last Updated Date	2015/7/1				
3						
4	Country Name	Country Code	Indicator Name	Indicator Code	1960	1961
5	Aruba	ABW	Trade in services (% of GDP)	BG.GSR.NFSV.GD.ZS		
6	Aruba	ABW	Communications, computer, etc. (% of service imports, BoP)	BM.GSR.CMCP.ZS		
7	Aruba	ABW	Primary income payments (BoP, current US$)	BM.GSR.FCTY.CD		
8	Aruba	ABW	Imports of goods and services (BoP, current US$)	BM.GSR.GNFS.CD		
9	Aruba	ABW	Goods imports (BoP, current US$)	BM.GSR.MRCH.CD		
10	Aruba	ABW	Service imports (BoP, current US$)	BM.GSR.NFSV.CD		
11	Aruba	ABW	Charges for the use of intellectual property, payments (BoP, current US$)	BM.GSR.ROYL.CD		
12	Aruba	ABW	Imports of goods, services and primary income (BoP, current US$)	BM.GSR.TOTL.CD		
13	Aruba	ABW	Transport services (% of service imports, BoP)	BM.GSR.TRAN.ZS		
14	Aruba	ABW	Travel services (% of service imports, BoP)	BM.GSR.TRVL.ZS		
15	Aruba	ABW	Secondary income, other sectors, payments (BoP, current US$)	BM.TRF.PRVT.CD		
16	Aruba	ABW	Current account balance (BoP, current US$)	BN.CAB.XOKA.CD		
17	Aruba	ABW	Current account balance (% of GDP)	BN.CAB.XOKA.GD.ZS		
18	Aruba	ABW	Net financial account (BoP, current US$)	BN.FIN.TOTL.CD		
19	Aruba	ABW	Net primary income (BoP, current US$)	BN.GSR.FCTY.CD		
20	Aruba	ABW	Net trade in goods and services (BoP, current US$)	BN.GSR.GNFS.CD		
21	Aruba	ABW	Net trade in goods (BoP, current US$)	BN.GSR.MRCH.CD		

Figure 7: Using Excel to examine the downloaded file

5. You can use `read.xlsx2` to read data from the downloaded Excel file:

```
> options(java.parameters = "-Xmx2000m")
> wb <- read.xlsx2("worldbank.xls", sheetIndex = 1,
startRow = 4)
```

6. Select the country name, country code, indicator name, indicator code, and data for 2014 out of the read data:

```
> wb2 <- wb[c("Country.Name","Country.Code","Indicator.Name",
"Indicator.Code", "X2014")]
```

7. Subsequently, you can examine the dimension of the file using the `dim` function:

```
> dim(wb2)
```

8. Finally, you can write filtered data back to a file named `2014wbdata.xlsx`:

```
> write.xlsx2(wb2, "2014wbdata.xlsx", sheetName = "Sheet1")
```

How it works...

In this recipe, we covered how to read and write an Excel file containing world development indicators with the `xlsx` package. In the first step, we needed to install and load the `xlsx` package, which enables the user to read and write Excel files in the R command prompt through the use of a Java POI package. Thus, to utilize the Java POI package, the installation process will also install `rJava` and `xlsxjars` at the same time. You can find the Java POI `.jar` file under `<R Installed Path>\library\xlsx]jars\java`. Using the author's computer as an example, which has the Windows 7 operating system installed, the `.jar` files are located at the `C:\Program Files\R\R-3.2.1\library\xlsxjars\java` path.

Next, we downloaded world economy indicator data from the link (`http://data.worldbank.org/topic/economy-and-growth`) with the `download.file` function. By default, `download.file` downloads the file in ASCII mode. To download the file in binary mode, we need to set download mode to `wb`.

After the Excel file is downloaded, we can examine it with Excel. The screenshot of the Excel file shows that the economy indicator starts from row **4** in **Sheet 1**. Therefore, we can use the `read.xlsx2` function to read world economy indicators from this location. The `xlsx` package provides two functions to read data from Excel: `read.xlsx` and `read.xlsx2`. As the `read.xlsx2` function mainly processes data in Java, `read.xlsx2` performs better (in particular, `read.xlsx2` processes data considerably faster on sheets with more than 100,000 cells).

After we have read the contents of the worksheet into an R data frame, we can select variables `Country.Name`, `Country.Code`, `Indicator.Name`, `Indicator.Code`, and `X2014` out of the extracted R data frame. Next, we can use the `dim` function to examine the dimensions of the data frame. Finally, we can use `write.xlsx2` to write transformed data to an Excel file, `2014wbdata.xlsx`.

Reading data from databases

As R reads data into memory, it is perfect for processing and analyzing small datasets. However, as an enterprise accumulates much more data than individuals in their daily lives, database documents are becoming more common for the purpose of storing and analyzing bigger data. To access databases with R, one can use `RJDBC`, `RODBC`, or `RMySQL` as the communications bridge. In this section, we will demonstrate how to use `RJDBC` to connect data stored in the database.

Getting ready

In this section, we need to prepare a MySQL environment first. If you have a MySQL environment installed on your machine (Windows), you can inspect server status from MySQL Notifier. If the local server is running, the server status should prompt **localhost (Online)**, as shown in the following screenshot:

Figure 8: MySQL Notifier

Once we have our database server online, we need to validate whether we are authorized to access the database with a given username and password by using any database connection client. For example, you can use the MySQL command line client to connect to the database.

How to do it...

Please perform the following steps to connect R to MySQL with `RJDBC`:

1. First, we need to install and load the `RJDBC` package:

```
> install.packages("RJDBC")
> library(RJDBC)
```

2. We can then download the JDBC driver for MySQL from the `https://dev.mysql.com/get/Downloads/Connector-J/mysql-connector-java-5.0.8.zip` link.
3. Unzip the downloaded `mysql-connector-java-5.0.8.zip` and place the unzipped `.jar` file `mysql-connector-java-5.0.8-bin.jar` in the relevant location. For example, on the author's computer the extracted `.jar` file was placed in the `C:\Program Files\MySQL\` path.
4. Next, we can load the MySQL driver for the purpose of connecting to MySQL:

```
> drv <- JDBC("com.mysql.jdbc.Driver",
+        "C:\\Program Files\\MySQL\\mysql-connector-java-5.0.8-bin.jar"
+        )
```

5. We can then establish a connection to MySQL using a registered MySQL driver:

```
> conn <- dbConnect(drv,
+                    "jdbc:mysql://localhost:3306/finance",
+                    "root",
+                    "test")
```

6. Use some basic operations to retrieve the table list from the connection:

```
> dbListTables(conn)
[1] "majortrade"
```

7. Obtain the data by using the `SELECT` operation:

```
> trade_data <- dbGetQuery(conn, "SELECT * FROM majortrade")
```

8. Finally, we can disconnect from MySQL:

```
> dbDisconnect(conn)
[1] TRUE
```

How it works...

Two major standards that R can use to access databases are ODBC and JDBC. **JDBC** (also known as **Java Database Connectivity**) is made up of a set of Java implemented classes and interfaces, which enables communication between Java and the database. On the other hand, **ODBC** (also known as **Open Database Connectivity**) is a standard interface developed by Microsoft.

To compare these two standards, ODBC performs better in importing and exporting data; however, it is also platform dependent. In other words, before making your program work, you must configure the connectivity for different operating systems. In contrast, JDBC is platform independent, which means that your written program can work on any operating system.

To connect R to MySQL with RJDBC, we first need to install and load the RJDBC package from CRAN. RJDBC provides the ability to connect to a database through the JDBC interface. As JDBC is implemented in Java, you should install `rJava` prior to RJDBC.

To proceed, we downloaded MySQL Connector/J, which is the official JDBC driver for MySQL, from the MySQL official download site. After extracting the `.zip` file (or `.tar` file), we placed the file in the proper file path (or you can add `.jar` files to the classpath). We can now begin writing an R program to access the database.

In our R script, we first need to register and initialize the MySQL driver before issuing any query to the database. Here, we need to specify the driver's class name, `com.mysql.jdbc.Driver` (different databases have different class names), and the `.jar` file, `mysql-connector-java-5.0.8-bin.jar`, where we can find the class. Next, we can establish the connection to the database with a registered driver. Here, we have to provide the connection string (`"jdbc:mysql://localhost:3306/finance"`), the username (`"root"`), and password (`"test"`) to access the database. As our MySQL server is installed and running on localhost, we can craft the connection string into `"jdbc:mysql://localhost:3306/finance"`. `3306` is the default port of MySQL and `finance` is our target database.

After the connection is established, we can now issue some SQL queries to the database. We first list the table within the `finance` database with the `dbListTables` command. Next, we create the table named `majortrade`, and insert records loaded from `snp500.csv` to the `majortrade` table with the `insert` statement. We then obtain the data from the database using the `select` statement. Finally, to release the connection, we need to disconnect by using the `dbDisconnect` command.

There's more...

In R, you can also use `RODBC` and `RMySQL` to connect to a database. In this section, we illustrate how to access the database through `RMySQL`. Follow the next few steps to install and load the `RMySQL` package, and then issue some queries to the `MySQL` database:

1. First, we need to install and load the `RMySQL` package:

```
> install.packages("RMySQL")
> library(RMySQL)
```

2. Next, we can access MySQL with a valid username and password:

```
> mydb <-dbConnect(MySQL(), user='root',
password='test', host='localhost')
```

3. At this point, we can send the query to the database and select trading data from the finance database:

```
> dbSendQuery(mydb, "USE finance")
> fetch(dbSendQuery(mydb, "SELECT * FROM majortrade;"))
```

Scraping web data

In most cases, the majority of data will not exist in your database, but will instead be published in different forms on the Internet. To dig up more valuable information from these data sources, we need to know how to access and scrape data from the Web. Here, we will illustrate how to use the `rvest` package to harvest finance data from `http://www.bloomberg.com/`.

Getting ready

In this recipe, you need to prepare your environment with R installed and a computer that can access the Internet.

How to do it...

Perform the following steps to scrape data from `http://www.bloomberg.com/`:

1. First, access the following link to browse the S&P 500 index on the Bloomberg Business website `http://www.bloomberg.com/quote/SPX:IND`:

Figure 9: S&P 500 index

2. Once the page appears, as shown in the preceding screenshot, we can begin installing and loading the `rvest` package:

```
> install.packages("rvest")
> library(rvest)
```

3. Next, you can use the HTML function from the `rvest` package to scrape and parse the HTML page of the link to the S&P 500 index at `http://www.bloomberg.com/`:

```
> spx_quote <- html("http://www.bloomberg.com/quote/SPX:IND")
```

4. Use the browser's built-in web inspector to inspect the location of the detail quote (marked with a red rectangle) below the index chart:

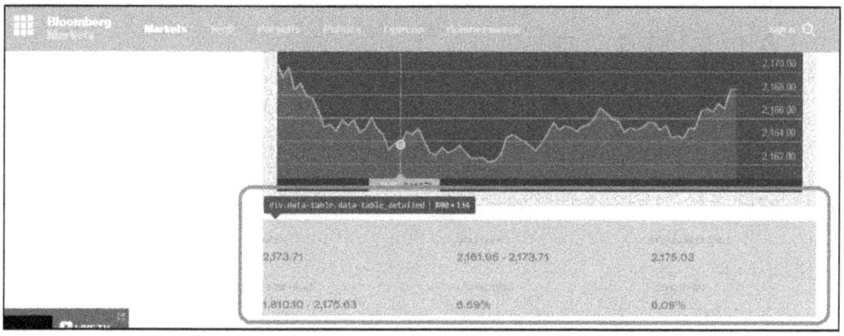

Figure 10: Inspecting the DOM location of S&P 500 index

5. You can then move the mouse over the detail quote and click on the target element that you wish to scrape down. As the following screenshot shows, the `<div class="cell">` section holds all the information we need:

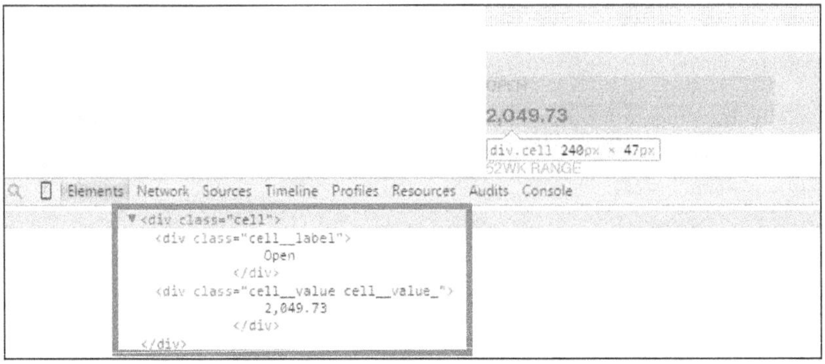

Figure 11: Inspecting the DOM location of the detail quote

6. Extract elements with the class of `cell` with the `html_nodes` function:

```
> cell  <- spx_quote %>% html_nodes(".cell")
```

7. Furthermore, we can parse the label of the detailed quote from elements with the class of `cell__label`, extract text from scraped HTML, and eventually clean spaces and newline characters from the extracted text:

```
> label <- cell %>%
+       html_nodes(".cell__label")  %>%
+       html_text() %>%
+       lapply(function(e) gsub("\n|\\s+", "", e))
```

8. Also, we can extract the value of a detailed quote from the element with the class of `cell__value`, extract text from scraped HTML, as well as clean spaces and newline characters:

```
> value <- cell %>%
+       html_nodes(".cell__value") %>%
+       html_text() %>%
+       lapply(function(e)gsub("\n|\\s+", "", e))
```

9. Finally, we can set the extracted `label` as the name to `value`:

```
> names(value) <- title
```

10. Next, we can access the energy and oil market index page at the following link (`http://www.bloomberg.com/energy`):

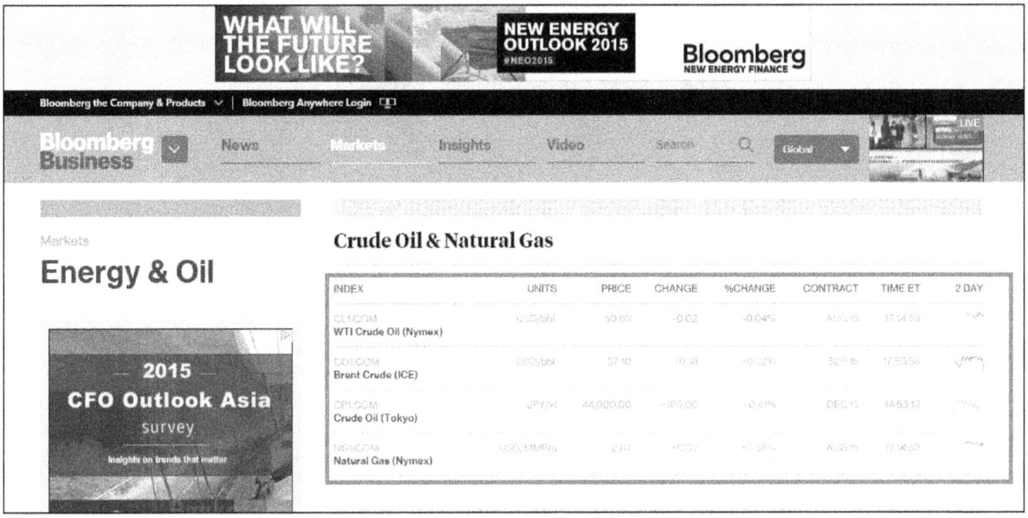

Figure 12: Inspecting the DOM location of crude oil and natural gas

11. We can then use the web inspector to inspect the location of the table element:

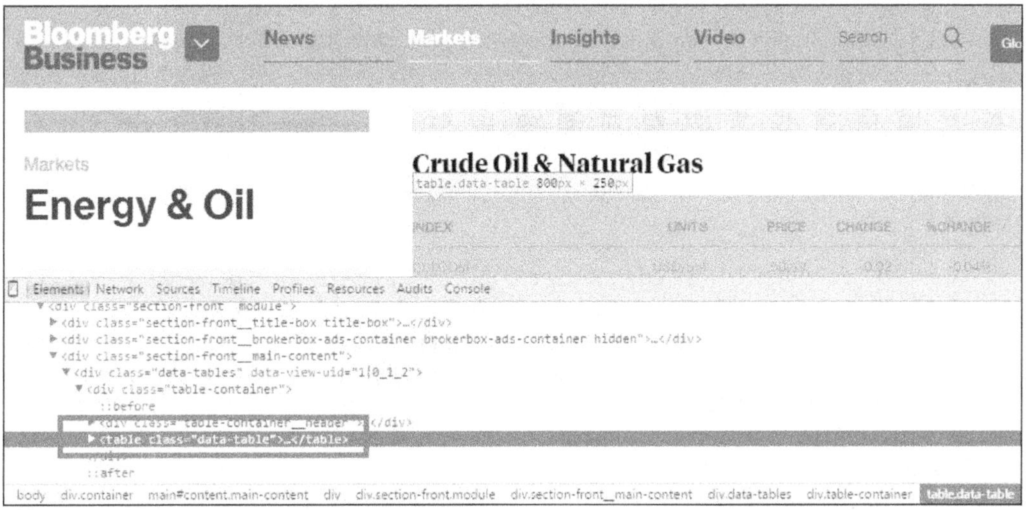

Figure 13: Inspecting the DOM location of table element

12. Finally, we can use `html_table` to extract the table element with the class of `data-table`:

```
> energy <- html("http://www.bloomberg.com/energy")
> energy.table <- energy %>%
  html_node(".data-table") %>% html_table()
```

How it works...

The most difficult step in scraping data from a website is that web data is published and structured in different formats. You have to fully understand how data is structured within the HTML tag before continuing.

As **HTML** (**Hypertext Markup Language**) is a language that has similar syntax to XML, we can use the XML package to read and parse HTML pages. However, the XML package only provides the `XPath` method, which has two main shortcomings, as follows:

- Inconsistent behavior in different browsers
- It is hard to read and maintain

For these reasons, we recommend using CSS selector over XPath when parsing HTML.

Python users may be familiar with how to scrape data quickly by using requests and `BeautifulSoup` packages. The `rvest` package is the counterpart package in R, which provides the same ability to simply and efficiently harvest data from HTML pages.

In this recipe, our target is to scrape the finance data of the S&P 500 detail quote from `http://www.bloomberg.com/`. Our first step is to make sure that we can access our target webpage via the Internet, which is followed by installing and loading the `rvest` package. After installation and loading is complete, we can then use the HTML function to read the source code of the page to `spx_quote`.

Once we have confirmed that we can read the HTML page, we can start parsing the detail quote from the scraped HTML. However, we first need to inspect the CSS path of the detail quote. There are many ways to inspect the CSS path of a specific element. The most popular method is to use the development tool built into each browser (press *F12* or *FN* + *F12*) to inspect the CSS path. Using Google Chrome as an example, you can open the development tool by pressing *F12*. A DevTools window may show up somewhere in the visual area (refer to the following document page: `https://developer.chrome.com/devtools/docs/dom-and-styles#inspecting-elements`).

Then, you can move the mouse cursor to the upper left of the **DevTools** window and select the Inspect Element icon (a magnifier icon similar to). Next, click on the target element, and the **DevTools** window will highlight the source code of the selected area. You can then move the mouse cursor to the highlighted area and right-click on it. From the pop-up menu, click on **Copy CSS Path** to extract the CSS path. Or, you can examine the source code and find that the selected element is structured in HTML code with the class of `cell`.

One highlight of `rvest` is that it is designed to work with `magrittr`, so that we can use a pipelines operator `%>%` to chain output parsed at each stage. Thus, we can first obtain the output source by calling `spx_quote` and then pipe the output to `html_nodes`. As the `html_nodes` function uses CSS selector to parse elements, the function takes basic selectors with type (for example, `div`), ID (for example, `#header`), and class (for example, `.cell`). As the elements to be extracted have the class of `cell`, you should place a period (`.`) in front of `cell`.

Lastly, we should extract both label and value from previously parsed nodes. Here, we first extract the element of class `cell__label`, and then use `html_text` to extract text. We can then use the `gsub` function to clean spaces and newline characters from the parsed text. Likewise, we apply the same pipeline to extract the element of class `class__value`. As we have extracted both label and value from the detail quote, we can apply the label as the name to the extracted values. We now have organized data from the Web to structured data.

Alternatively, we can also use `rvest` to harvest tabular data. Similar to the process used to harvest the S&P 500 index, we can first access the energy and oil market index page. We can then use the web element inspector to find the element location of table data. As we have found the element located in the class of `data-table`, we can use the `html_table` function to read the table content into an R data frame.

There's more...

Instead of using the web inspector built into each browser, one can consider using **SelectorGadget** (`http://selectorgadget.com/`) to search for the CSS path. SelectorGadget is a very powerful and simple-to-use extension for Google Chrome, which enables the user to extract the CSS path of the target element with only a few clicks:

1. To begin using SelectorGadget, access the following link: `https://chrome.google.com/webstore/detail/selectorgadget/mhjhnkcfbdhnjickkkdbjoemdmbfginb`. Then, click on the green button (shown in the red rectangle in the following screenshot) to install the plugin in Chrome:

Figure 14: Adding SelectorGadget to Chrome

2. Next, click the upper-right icon to open SelectorGadget, and then select the area that needs to be scraped down. The selected area will be colored green, and the gadget will display the CSS path of the area and the number of elements matched to the path:

Figure 15: Using SelecorGadget to inspect the DOM location of the table element

3. Finally, you can paste the extracted CSS path to `html_nodes` as an input argument to parse the data.

Besides `rvest`, you can connect R with Selenium via `Rselenium` to scrape the web page. Selenium was originally designed as an automated web application that enables the user to command a web browser to automate processes through simple scripts. However, you can also use Selenium to scrape data from the Internet. The following instruction presents a sample demo on how to scrape `http://www.bloomberg.com/` using `Rselenium`:

1. First, access the following link to download the Selenium standalone server (`http://www.seleniumhq.org/download/`):

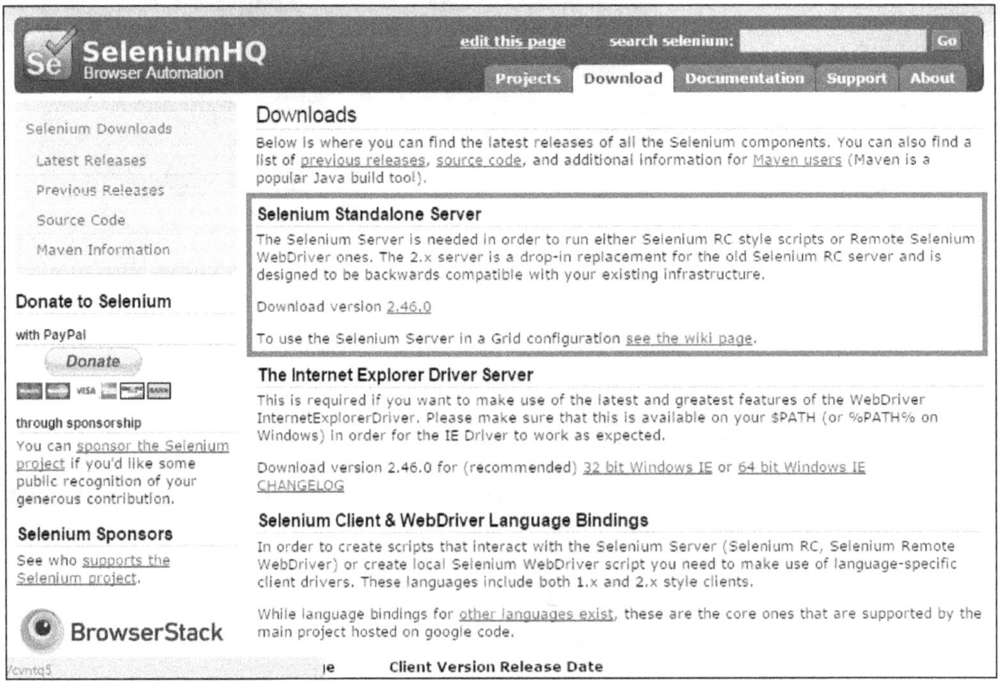

Figure 16: Downloading the Selenium standalone server driver

2. Next, start the Selenium standalone server using the following command:

```
$ java -jar selenium-server-standalone-2.46.0.jar
```

3. If you can successfully launch the standalone server, you should see the following message, which means you can connect to the server that binds to port `4444`:

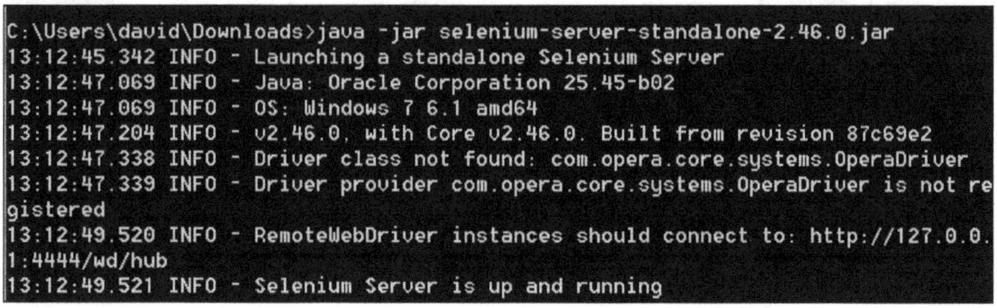
C:\Users\david\Downloads>java -jar selenium-server-standalone-2.46.0.jar
13:12:45.342 INFO - Launching a standalone Selenium Server
13:12:47.069 INFO - Java: Oracle Corporation 25.45-b02
13:12:47.069 INFO - OS: Windows 7 6.1 amd64
13:12:47.204 INFO - v2.46.0, with Core v2.46.0. Built from revision 87c69e2
13:12:47.338 INFO - Driver class not found: com.opera.core.systems.OperaDriver
13:12:47.339 INFO - Driver provider com.opera.core.systems.OperaDriver is not re
gistered
13:12:49.520 INFO - RemoteWebDriver instances should connect to: http://127.0.0.
1:4444/wd/hub
13:12:49.521 INFO - Selenium Server is up and running

Figure 17: Initiating the Selenium standalone server

4. At this point, you can begin installing and loading `RSelenium` with the following command:

```
> install.packages("RSelenium")
> library(RSelenium)
```

5. After `RSelenium` is installed, register the driver and connect to the Selenium server:

```
> remDr <- remoteDriver(remoteServerAddr = "localhost"
+                       , port = 4444
+                       , browserName = "firefox"
+)
```

6. Examine the status of the registered driver:

```
> remDr$getStatus()
```

7. Next, we navigate to `http://www.bloomberg.com/`:

```
> remDr$open()
> remDr$navigate("http://www.bloomberg.com/quote/SPX:IND ")
```

8. Finally, we can scrape the data by using CSS selector:

```
> webElem <- remDr$findElements('css selector', ".cell")
> webData <- sapply(webElem, function(x){
+   label <- x$findChildElement('css selector', '.cell__label')
+   value <- x$findChildElement('css selector', '.cell__value')
+   cbind(c("label" = label$getElementText(), "value" =
value$getElementText()))
+ }
+ )
```

Accessing Facebook data

Social network data is another great source for the user who is interested in exploring and analyzing social interactions. The main difference between social network data and web data is that social network platforms often provide a semi-structured data format (mostly JSON). Thus, one can easily access the data without the need to inspect how the data is structured. In this recipe, we will illustrate how to use `rvest` and `rson` to read and parse data from Facebook.

Getting ready

In this recipe, you need to prepare your environment with R installed and a computer that can access the Internet.

How to do it...

Perform the following steps to access data from Facebook:

1. First, we need to log in to Facebook and access the developer page (`https://developers.facebook.com/`):

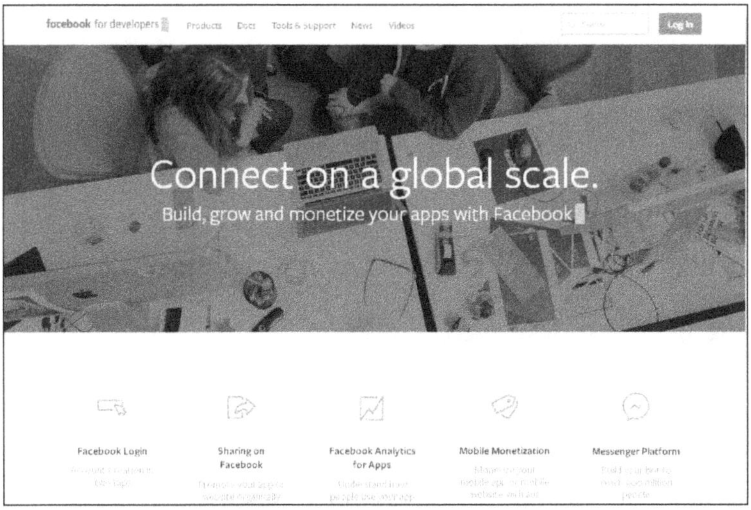

Figure 18: Accessing the Facebook developer page

2. Click on **Tools & Support**, and select **Graph API Explorer**:

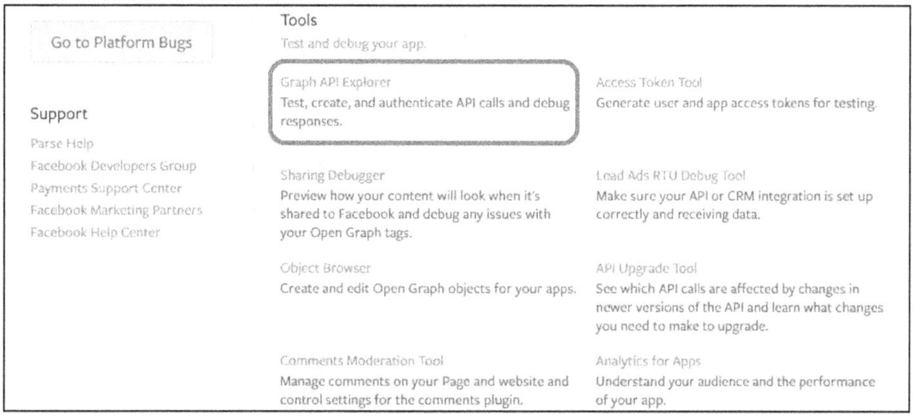

Figure 19: Selecting the Graph API Explorer

3. Next, click on **Get Token**, and choose **Get Access Token**:

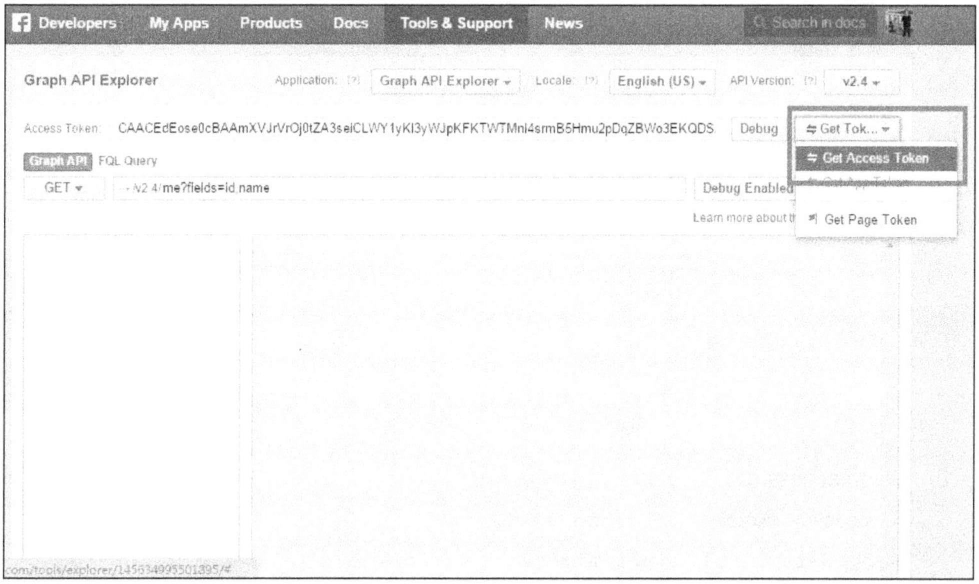

Figure 20: Selecting the Get Access Token

4. On the **User Data Permissions** pane, select **user_tagged_places** and then click **Get Access Token**:

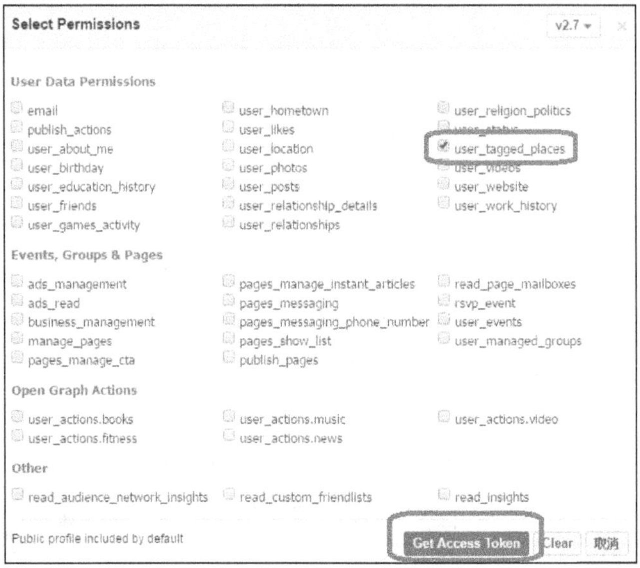

Figure 21: Selecting permissions

5. Copy the generated access token to the clipboard:

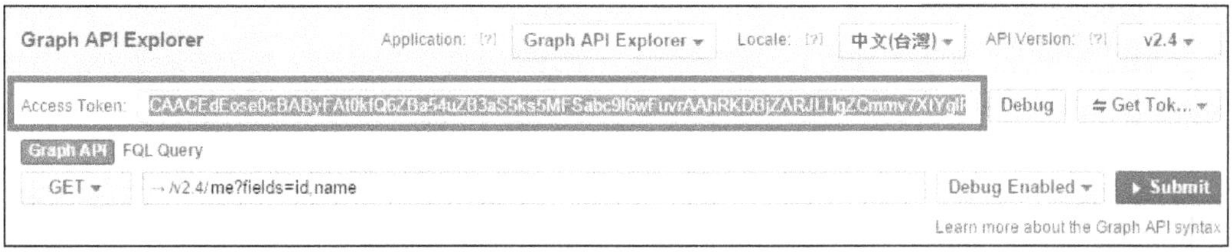

Figure 22: Copying the access token

6. Try to access the Facebook API by using `rvest`:

```
> access_token <- '<access_token>'
> fb_data <- html(sprintf("https://graph.facebook.com/me/tagged_places?access_token=%s",access_token))
```

7. Install and load the `rjson` package:

```
> install.packages("rjson")
> library(rjson)
```

8. Extract the text from `fb_data` and then use `fromJSON` to read the JSON data:

```
> fb_json <-  fromJSON(fb_data %>% html_text())
```

9. Use `sapply` to extract the name and ID of the place from `fb_json`:

```
> fb_place <- sapply(fb_json$data, function(e){e$place$name})
> fb_id <- sapply(fb_json$data, function(e){e$place$id})
```

10. Last, use `data.frame` to wrap the data:

```
> data.frame(place = fb_place, id = fb_id)
```

How it works...

In this recipe, we cover how to retrieve social network data through Facebook's Graph API. Unlike scraping web pages, you need to obtain a Facebook access token before making any request for insight information. There are two ways to retrieve the access token: one is to use Facebook's Graph API Explorer, and the other is to create a Facebook application. In this recipe, we illustrate how to use the Graph API Explorer to obtain the access token.

Facebook's Graph API Explorer is where you can craft your requests URL to access Facebook data on your behalf. To access the Explorer page, we first visit Facebook's developer page (`https://developers.facebook.com/`). The Graph API Explorer page is under the drop-down menu for **Tools & Support**. After going to the Explorer page, we select **Get Access Token** from the drop-down menu for **Get Token**. Subsequently, a tabbed window will appear; you can check access permission to various levels of the application. For example, we can check **tagged_places** to access the locations we have previously tagged. After we have selected the permissions we require, we can click on **Get Access Token** to allow Graph API Explorer to access our insight data. After completing these steps, you will see an access token, which is a temporary, short-lived token that you can use to access the Facebook API.

With the access token, we can then access the Facebook API with R. First, we need a HTTP request package. Similar to the web scraping recipe, we can use the `rvest` package to make the request. We craft a request URL with the addition of the `access_token` (copied from Graph API Explorer) to the Facebook API. From the response, we should receive JSON formatted data. To read the attributes of JSON formatted data, we install and load the `RJSON` package. We can then use the `fromJSON` function to read the JSON format string extracted from the response.

Lastly, we read places and ID information through the use of the `sapply` function, and we can then use data frame to transform extracted information into the data frame. At the end of the recipe, we should see data formatted in the data frame.

There's more...

To learn more about Graph API, read the following official documentation from Facebook: `https://developers.facebook.com/docs/reference/api/field_expansion/`.

1. First, we need to install and load the `Rfacebook` package:

```
> install.packages("Rfacebook")
> library(Rfacebook)
```

2. We can then use built-in functions to retrieve data from the user or access similar information with the provision of an access token:

```
> getUsers("me", "<access_token>")
```

If you would like to scrape public fan pages without logging in to Facebook every time, you can create a Facebook app to access insight information on behalf of the app:

1. To create an authorized app token, log in to the Facebook developer page and click on **Add a New Page**:

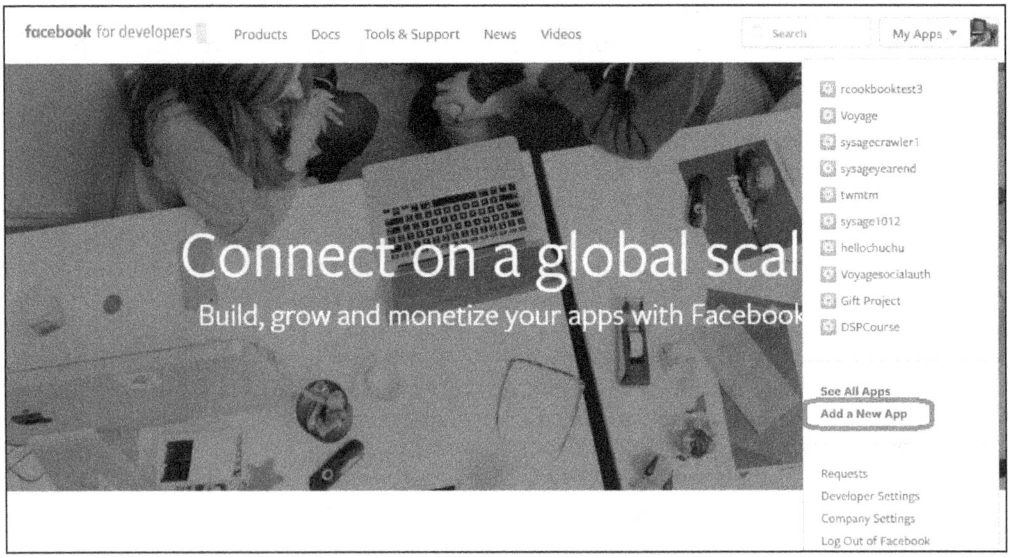

Figure 23: Creating a new app

2. You can create a new Facebook app with any name and a valid e-mail ID, providing that it has not already been registered:

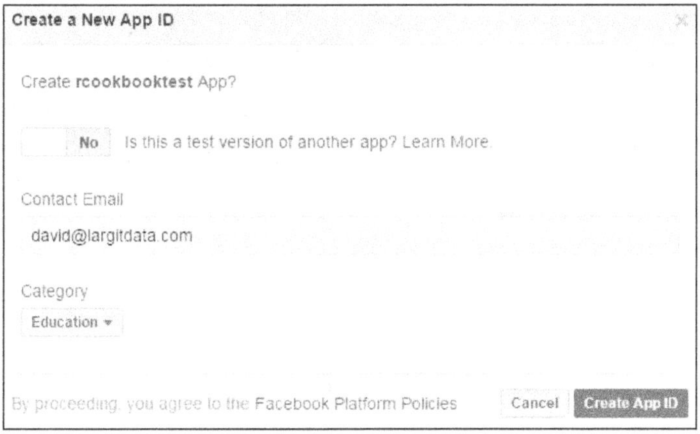

Figure 24: Creating a new app ID

3. Next, you can copy both the app ID and app secret and craft the access token to `<APP ID>|<APP SECRET>`. You can now use this token to scrape public fan page information with Graph API:

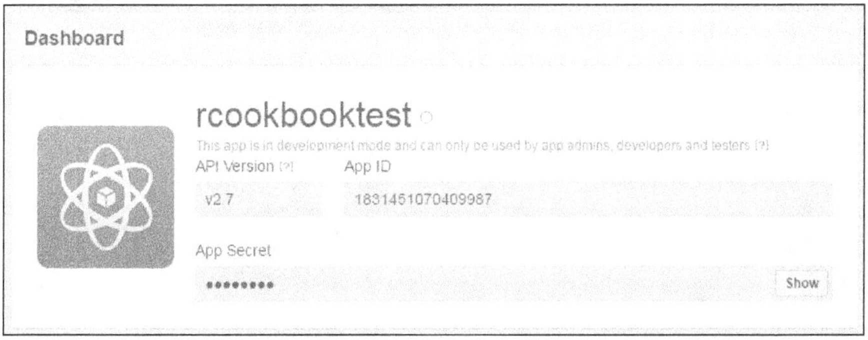

Figure 25: Obtaining the app ID and secret

4. Similar to Rfacebook, we can then replace the `access_token` with `<APP ID>|<APP SECRET>`:

```
> getUsers("me", "<access_token>")
```

Working with twitteR

In addition to obtaining social network interaction data, one can collect millions of tweets from Twitter for further text mining tasks. The method for retrieving data from Twitter is very similar to Facebook. For both social platforms, all we need is an access token to access insight data. After we have retrieved the access token, we can then use `twitteR` to access millions of tweets.

Getting ready

In this recipe, you need to prepare your environment with R installed and a computer that can access the Internet.

How to do it...

Perform the following steps to read data from Twitter:

1. First, you need to log in to Twitter and access the page of **Twitter Apps** at `https://apps.twitter.com/`. Click on **Create New App**:

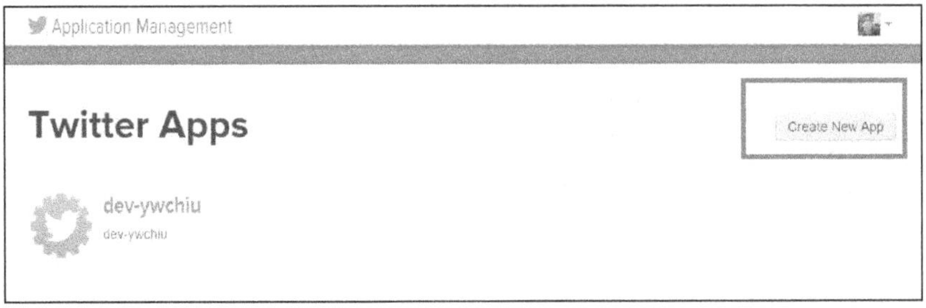

Figure 26: Creating a new Twitter app

2. Fill in all required application details to create a new application:

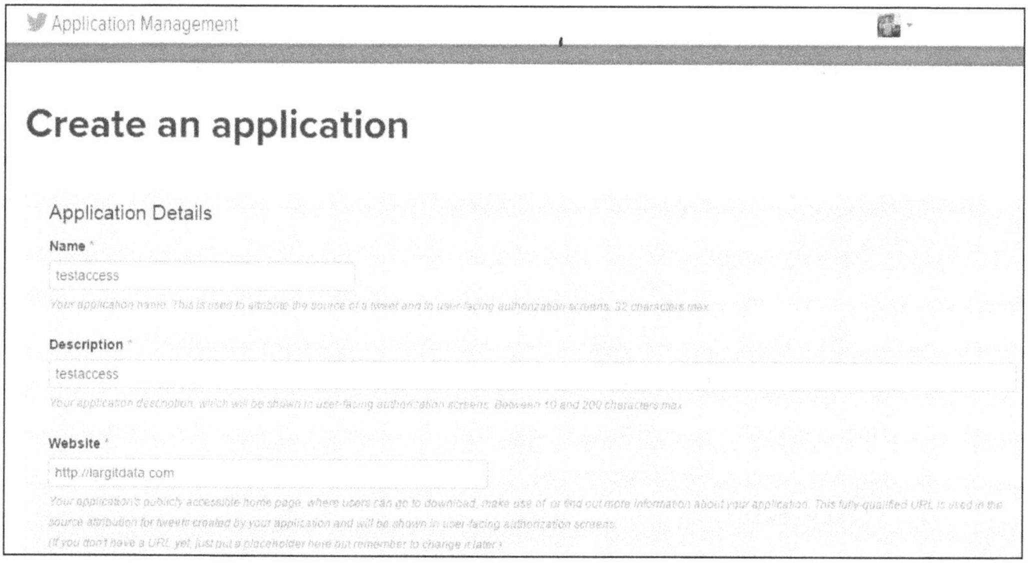

Figure 27: Filling in the required details

3. Next, you can select **Keys and Access Tokens** and then access **Application Settings**:

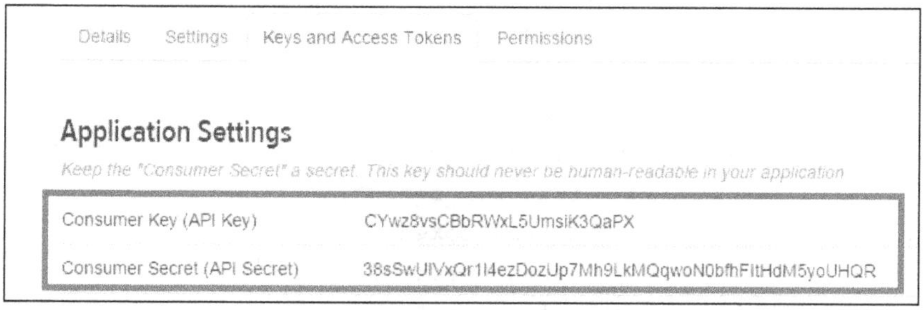

Figure 28: Copying API key and secret

4. Click on the **Create my access token** button, and the explorer will generate an authorized access token and the secret:

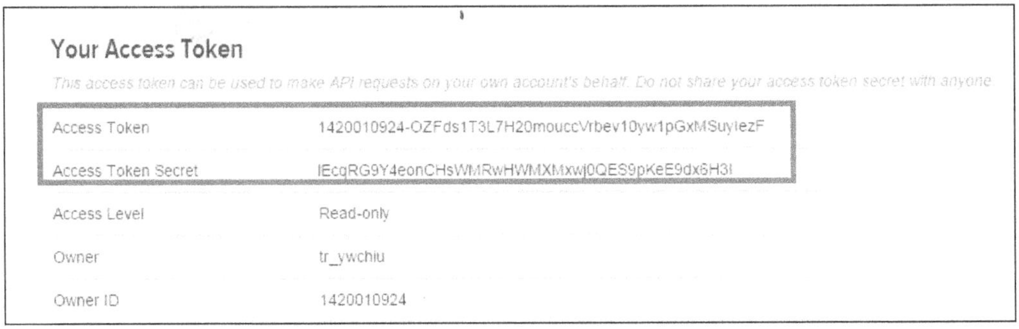

Figure 29: Creating the access token and secret

5. Install and load the `twitteR` package:

```
> install.packages("twitteR")
> library(twitteR)
```

6. Set up Twitter OAuth with the copied consumer key and consumer secret from **Application Settings**, and the copied access token and access secret from **Your Access Token**:

```
> consumer_key <- '<consumer_key>'
> consumer_secret <- '<consumer_secret>'
> access_token <- '<access_token>'
> access_secret <- '<access_secret>'
> setup_twitter_oauth(consumer_key,
+                     consumer_secret,
+                     access_token,
+                     access_secret)
[1] "Using direct authentication"
Use a local file to cache OAuth access credentials between R
sessions?
1: Yes
2: No

Selection: 1
Adding .httr-oauth to .gitignore
```

7. At this point, you can use the `searchTwitter` function to extract the top 100 search results with `world cup` as the search key:

```
> res <- searchTwitter("world cup", n=100)
```

How it works...

In this recipe, we use `twitteR` to obtain tweets from Twitter. Similar to Facebook, we need an access token before accessing any tweets. To apply an access token, one must first create an application with one login account and then fill in all the required information to create a new app.

After the app is created, we select the **Keys and Access Tokens** tab and find both consumer keys and secrets under a section labeled **Application Settings**. Scroll down further to the button named **Create my access token**. Clicking on this provides an access token and access token secrets under a section labeled **Your Access Token**.

At this point, we can now connect Twitter with `twitteR`. First, install and load the `twitteR` package. You can then copy the consumer key and consumer secret from **Application Settings**, and the access token and access secret from **Your Access Token**. This copied information can be used to set up Twitter OAuth. Finally, we can use the `searchTwitter` function to find the top 100 search results for the `world cup` keyword.

There's more...

Similar to Facebook, Twitter also provides its users a console for API testing. You can access the API console from `https://dev.twitter.com/rest/tools/console`:

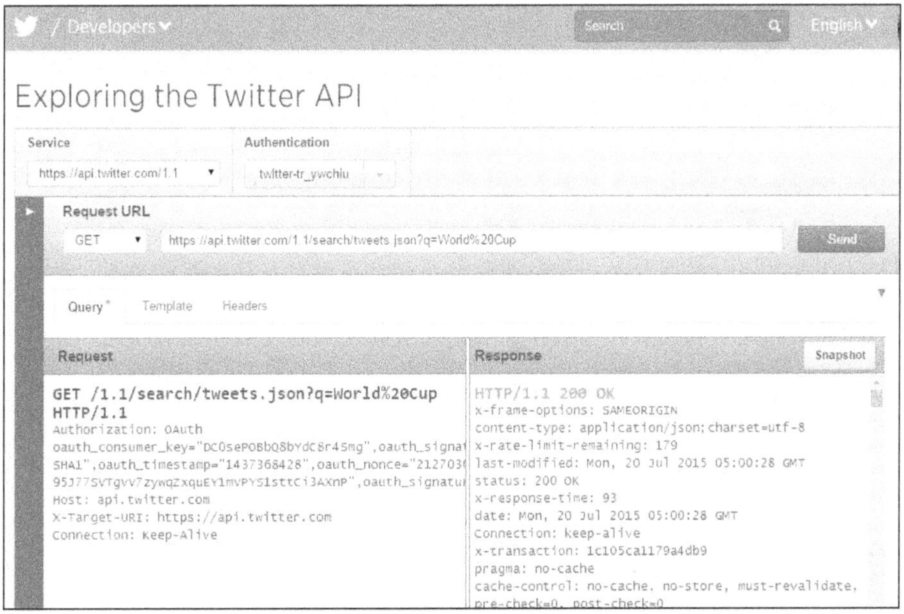

Figure 30: Exploring the Twitter API

3
Data Preprocessing and Preparation

This chapter covers the following topics:

- Renaming the data variable
- Converting data types
- Working with the date format
- Adding new records
- Filtering data
- Dropping data
- Merging data
- Sorting data
- Reshaping data
- Detecting missing data
- Imputing missing data

Introduction

In the previous chapter, we covered how to integrate data from various data sources. However, simply collecting data is not enough; you also have to ensure the quality of the collected data. If the quality of data used is insufficient, the results of the analysis may be misleading due to biased samples or missing values. Moreover, if the collected data is not well structured and shaped, you may find it hard to correlate and investigate the data. Therefore, data preprocessing and preparation is an essential task that you must perform prior to data analysis.

Those of you familiar with how SQL operates may already understand how to use databases to process data. For example, SQL allows users to add new records with the `insert` operation, modify data with the `update` operation, and remove records with the `delete` operation. However, we do not need to move collected data back to the database; R already provides more powerful and convenient preprocessing functions and packages. In this chapter, we will cover how simple it is to perform data preprocessing in R.

Renaming the data variable

The use of a data frame enables the user to select and filter data by row names and column names. As not all imported datasets contain row names and column names, we need to rename this dataset with a built-in naming function.

Getting ready

In this recipe, you need to prepare your environment with R installed and a computer that can access the Internet.

How to do it...

Perform the following steps to rename data:

1. First, download `employees.csv` from the GitHub link `https://github.com/ywchiu/rcookbook/raw/master/chapter3/employees.csv`:

   ```
   > download.file("https://github.com/ywchiu/rcookbook/raw/master/chapter3/employees.csv", " employees.csv")
   ```

2. Additionally, download `salaries.csv` from the GitHub link `https://github.com/ywchiu/rcookbook/raw/master/chapter3/salaries.csv`:

   ```
   > download.file("https://github.com/ywchiu/rcookbook/raw/master/chapter3/salaries.csv", "salaries.csv")
   ```

3. Next, read the file into an R session with the `read.csv` function:

   ```
   > employees <- read.csv('employees.csv', head=FALSE)
   > salaries <- read.csv('salaries.csv', head=FALSE)
   ```

4. Use the `names` function to view the column names of the dataset:

   ```
   > names(employees)
   [1] "V1" "V2" "V3" "V4" "V5" "V6"
   > names(salaries)
   [1] "V1" "V2" "V3" "V4"
   ```

5. Next, rename columns with a given `names` vector:

```
> names(employees) <- c("emp_no", "birth_date", "first_name",
"last_name", "gender", "hire_date")
> names(employees)
[1] "emp_no"      "birth_date" "first_name" "last_name"
[5] "gender"      "hire_date"
```

6. Besides using `names`, you can also rename columns with the `colnames` function:

```
> colnames (salaries) <- c("emp_no", "salary",  "from_date",
"to_date")
> colnames (salaries)
[1] "emp_no"    "salary"    "from_date" "to_date"
```

7. In addition to revising the column names, we can also revise row names with the `rownames` function:

```
> rownames (salaries) <- salaries$emp_no
```

How it works...

In this recipe, we demonstrated how to rename datasets with the `names` function. First, we used the `download.file` function to download both `salaries.csv` and `employees.csv` from GitHub. Then, we used the `names` function to examine the column names of these two datasets. To revise the column names of these two datasets, we simply assigned a character vector to the name of the dataset. We can also revise column names with the `colnames` function. Finally, we can revise the row names of the dataset to `emp_no` with the `rownames` function.

There's more...

To avoid having to specify column names and row names separately with the `colnames` and `rownames` functions, we can use the `dimnames` function to specify both column names and row names in one operation:

```
> dimnames(employees) <- list(c(1,2,3,4,5,6,7,8,9,10),
c("emp_no", "birth_date", "first_name", "last_name",
"gender", "hire_date"))
```

In this code, the first input vector within the list indicates the row names, and the second input vector points to the column names.

Converting data types

If we do not specify a data type during the import phase, R will automatically assign a type to the imported dataset. However, if the data type assigned is different to the actual type, we may face difficulties in further data manipulation. Thus, data type conversion is an essential step during the preprocessing phase.

Getting ready

Complete the previous recipe and import both `employees.csv` and `salaries.csv` into an R session. You must also specify column names for these two datasets to be able to perform the following steps.

How to do it...

Perform the following steps to convert the data type:

1. First, examine the data type of each attribute using the `class` function:

```
> class(employees$birth_date)
[1] "factor"
```

2. You can also examine types of all attributes using the `str` function:

```
> str(employees)

'data.frame':  10 obs. of  6 variables:
 $ emp_no    : int  10001 10002 10003 10004 10005 10006 10007 10008 10009 10010
 $ birth_date: Factor w/ 10 levels "1952-04-19","1953-04-20",..: 3 10 8 4 5 2 6 7 1 9
 $ first_name: Factor w/ 10 levels "Anneke","Bezalel",..: 5 2 7 3 6 1 10 8 9 4
 $ last_name : Factor w/ 10 levels "Bamford","Facello",..: 2 9 1 4 5 8 10 3 6 7
 $ gender    : Factor w/ 2 levels "F","M": 2 1 2 2 2 1 1 2 1 1
 $ hire_date : Factor w/ 10 levels "1985-02-18","1985-11-21",..: 3 2 4 5 9 7 6 10 1 8
```

3. Then, you need to convert both `birth_date` and `hired_date` to the date format:

```
> employees$birth_date <- as.Date(employees$birth_date)
> employees$hire_date <- as.Date(employees$hire_date)
```

4. You also need to convert both `first_name` and `last_name` into character type:

```
> employees$first_name <- as.character(employees$first_name)
> employees$last_name <- as.character(employees$last_name)
```

5. Again, you can use `str` to examine the dataset:

```
> str(employees)

'data.frame': 10 obs. of  6 variables:
 $ emp_no     : int  10001 10002 10003 10004 10005 10006 10007 
10008 10009 10010
 $ birth_date: Date, format: "1953-09-02" ...
 $ first_name: chr  "Georgi" "Bezalel" "Parto" "Chirstian" ...
 $ last_name : chr  "Facello" "Simmel" "Bamford" "Koblick" ...
 $ gender    : Factor w/ 2 levels "F","M": 2 1 2 2 2 1 1 2 1 1
 $ hire_date : Date, format: "1986-06-26" ...
```

6. Furthermore, you can convert the data type of `from_date` and `to_date` to date type within `salaries`:

```
> salaries$from_date <- as.Date(salaries$from_date)
> salaries$to_date <- as.Date(salaries$to_date)
```

How it works...

In this recipe, we demonstrated how to convert the data type of each attribute within the dataset. Before conducting further conversion on any attribute, you must first examine the current type of each attribute. To identify the data type, you can use the `class` function to determine the data-selecting attribute. Furthermore, to inspect all data types, you can use the `str` function.

From the output of applying the `str` function to the `employees` data frame, we can see that both `birth_date` and `hire_date` are in factor type. However, if we need to calculate one's age with the `birth_date` attribute, we need to convert it to date format. Thus, we change both `birth_date` and `hire_date` to date format using the `as.Date` function.

Also, as the factor type limits the choice of values in one attribute, we may not freely add a record to the dataset. As it is hard to find exactly the same last name and first name from the dataset, we need to convert `last_name` and `first_name` to the character type. We can then proceed to append a new record to the `employees` dataset in the next recipe. Finally, we should also convert `from_date` and `to_date` of the salaries dataset to date type, and we can then perform date calculations in the next recipe.

There's more...

Besides using an as function to convert data type, you can specify the data type during the data import phase. Using the `read.csv` function as an example, you can specify the data type in the `colClasses` argument. If you want R to automatically select the data type (that is, automatically convert `emp_no` to integer type), simply specify `NA` within `colClasses`:

```
> employees <- read.csv('~/Desktop/employees.csv', colClasses =
c(NA,"Date", "character", "character", "factor", "Date"), head=FALSE)
> str(employees)
'data.frame': 10 obs. of 6 variables:
 $ V1: int 10001 10002 10003 10004 10005 10006 10007 10008 10009 10010
 $ V2: Date, format: "1953-09-02" ...
 $ V3: chr "Georgi" "Bezalel" "Parto" "Chirstian" ...
 $ V4: chr "Facello" "Simmel" "Bamford" "Koblick" ...
 $ V5: Factor w/ 2 levels "F","M": 2 1 2 2 2 1 1 2 1 1
 $ V6: Date, format: "1986-06-26" ...
```

By specifying the `colClasses` argument, `emp_no`, `birth_date`, `first_name`, `last_name`, `gender`, and `hire_date` will be converted into integer type, date type, character type, character type, factor type, and date type respectively.

Working with the date format

After we have converted each data attribute to the proper data type, we may determine that some attributes in `employees` and `salaries` are in the date format. Thus, we can calculate the number of years between the employees' date of birth and current year to estimate the age of each employee. Here, we will show you how to use some built-in date functions and the `lubridate` package to manipulate date format data.

Getting ready

Refer to the previous recipe and convert each attribute of imported data into the correct data type. Also, you have to rename the columns of the `employees` and `salaries` datasets by following the steps from the *Renaming the data variable* recipe.

How to do it...

Perform the following steps to work with the date format in `employees` and `salaries`:

1. We can add or subtract days on the date format attribute using the following:

```
> employees$hire_date + 30
```

2. We can obtain time differences in days between `hire_date` and `birth_date` using the following:

```
> employees$hire_date - employees$birth_date
Time differences in days
 [1] 11985  7842  9765 11902 12653 13192 11586 13357
 [9] 11993  9581
```

3. Besides getting time differences in days, we can obtain differences in weeks using the `difftime` function:

```
> difftime(employees$hire_date, employees$birth_date,
unit="weeks")
Time differences in weeks
 [1] 1712.143 1120.286 1395.000 1700.286 1807.571
 [6] 1884.571 1655.143 1908.143 1713.286 1368.714
```

4. In addition to built-in date operation functions, we can install and load the `lubridate` package to manipulate dates:

```
> install.packages("lubridate")
> library(lubridate)
```

5. Next, we can convert date data to POSIX format using the `ymd` function:

```
> ymd(employees$hire_date)
```

6. Then, we can examine the period between `hire_date` and `birth_date` using the `as.period` function:

```
> span <- interval(ymd(employees$birth_date), ymd(employees$hire_
date))
> time_period <- as.period(span)
```

7. Furthermore, we can obtain time difference in the using with the `year` function:

```
> year(time_period)
```

8. Moreover, we can retrieve the current date using the `now` function:

```
> now()
```

9. Finally, we can now calculate the age of each employee using the following:

```
> span2 <- interval(now() , ymd(employees$birth_date))
> year(as.period(span2))
```

How it works...

After following the steps in the previous section, both `employees` data and `salaries` data should now be renamed, and the data type of each attribute should have already been converted to the proper data type. As some of the attributes are in the date format, we can then use some date functions to calculate the time difference in days between these attributes.

Date type data allows arithmetic operations; we can add or subtract some days from its value. Thus, we first demonstrate that we can add 30 to `hire_date`. Then, we can check whether 30 more days have been added to all hire dates. Next, we can calculate the time difference in days between the `birth_date` and `hire_date` attributes in order to find out the age at which each employee started working at that company. However, the minus operation can only show us the time differences in days; we need to perform more calculations to change the differences in time from days to a different measurement. Thus, we can use the `difftime` function to determine time differences in a different unit (for example, hours, days, and weeks). While `difftime` provides more measurement choices, we still need to make some further calculations to obtain the difference in months and years.

To simplify date computation, we can use a convenient `lubridate` date operation package. As the data is in *year-month-date* format, we can use the `ymd` function to convert the data to POSIX format first. Then, we can use an interval function to calculate the time span between `hire_date` and `birth_date`. Subsequently, we can use the `as.period` function to compute the period of the time span. This allows us to use the year function to obtain the number of years between each employee's birthday and hire date.

Finally, to calculate the age of the employee, we can use the `now` function to obtain the current time. We then use `interval` to obtain the time interval between the birth date of the employee and the current date. With this information, we can finally use the `year` function to obtain the actual age of the employee.

There's more...

When using the `lubridate` package (version 1.3.3), you might find the following error message:

```
Error in (function (..., deparse.level = 1)  :
  (converted from warning) number of columns of result is not a multiple
of vector length (arg 3)
```

This error message occurs due to a locale configuration bug. You can fix the problem by setting locale to `English_United States.1252`:

```
> Sys.setlocale(category = "LC_ALL", locale = "English_United States.1252")
[1] "LC_COLLATE=English_United States.1252;LC_CTYPE=English_United States.1252;LC_MONETARY=English_United States.1252;LC_NUMERIC=C;LC_TIME=English_United States.1252"
```

Adding new records

For those of you familiar with databases, you may already know how to perform an `insert` operation to append a new record to the dataset. Alternatively, you can use an `alter` operation to add a new column (attribute) into a table. In R, you can also perform `insert` and `alter` operations but much more easily. We will introduce the `rbind` and `cbind` function in this recipe so that you can easily append a new record or new attribute to the current dataset with R.

Getting ready

Refer to the *Converting data types* recipe and convert each attribute of imported data into the proper data type. Also, rename the columns of the `employees` and `salaries` datasets by following the steps from the *Renaming the data variable* recipe.

How to do it...

Perform the following steps to add a new record or new variable into the dataset:

1. First, use `rbind` to insert a new record to `employees`:

   ```
   > employees <- rbind(employees, c(10011, '1960-01-01', 'Jhon', 'Doe', 'M', '1988-01-01'))
   ```

2. We can then reassign the combined results of the data frame `employees` and new records back to `employees`:

   ```
   > employees <- rbind(employees, c(10011, '1960-01-01', 'Jhon', 'Doe', 'M', '1988-01-01'))
   ```

3. Besides adding a new record to the original dataset, we can add a new `position` attribute with `NA` as the default value:

   ```
   > cbind(employees, position = NA)
   ```

4. Furthermore, we can add a new age attribute, based on a calculation using the current date and birth_date of each employee:

```
> span <- interval(ymd(employees$birth_date), now())
> time_period <- as.period(span)
> employees$age <- year(time_period)
```

5. Alternatively, we can use the transform function to add multiple variables:

```
> transform(employees, age = year(time_period), position = "RD",
marrital = NA)
```

How it works...

Similar to database operations, we can add a new record to the data frame by the schema of the dataset (the number of attributes and data type of each attribute). Here, we first introduced how to use the rbind function to add a new record to a data frame. As the employees dataset consists of six columns, we can add a record with six values to the employees dataset with the rbind function. In the first column, emp_no is in integer format. Thus, we do not have to wrap the input value with single quotes. For the first_name and last_name attributes, we can freely input any character string as a value because we already converted their type to character type. For the last gender attribute, which is in factor type, we can only input either M or F as a value.

In addition to adding a new record to a target dataset, we can add a new variable with the cbind function. To add a new variable, we can assign a variable with a default value while calling cbind. Here, we use NA as the default value for a new position variable. We can also assign the calculated results from other columns as the value of the new variable. In this demonstration, we first computed each employee's age from the current date to their birthday. Then, we used the dollar sign to assign the computed value to a new attribute, age. Besides using the dollar sign to assign a new variable, we can use the transform function to create age, position, and marital variables in the employees dataset.

There's more...

Besides using the dollar sign and transform function, we can use the with function to create new variables:

```
> with(employees, year(birth_date))
 [1] 1953 1964 1959 1954 1955 1953 1957 1958 1952 1963
> employees $birth_year <- with(employees, year(birth_date))
```

Filtering data

Data filtering is the most common requirement for users who want to analyze partial data of interest rather than the whole dataset. In database operations, we can use a SQL command with a `where` clause to subset the data. In R, we can simply use the square bracket to perform filtering.

Getting ready

Refer to the *Converting data types* recipe and convert each attribute of imported data into the proper data type. Also, rename the columns of the `employees` and `salaries` datasets by following the steps from the *Renaming the data variable* recipe.

How to do it...

Perform the following steps to filter data:

1. First, use `head` and `tail` to subset the first three rows and last three rows from the `employees` dataset:

```
> head(employees, 3)
  emp_no birth_date first_name last_name gender  hire_date
1  10001 1953-09-02     Georgi   Facello      M 1986-06-26
2  10002 1964-06-02   Bezalel    Simmel      M 1985-11-21
3  10003 1959-12-03      Parto   Bamford      M 1986-08-28

> tail(employees, 3)
   emp_no birth_date first_name last_name gender  hire_date
8   10008 1958-02-19     Saniya  Kalloufi      M 1994-09-15
9   10009 1952-04-19     Sumant      Peac      F 1985-02-18
10  10010 1963-06-01  Duangkaew  Piveteau      F 1989-08-24
```

2. You can also use the square bracket to subset the first three rows of the data with a given sequence from `1` to `3`:

```
> employees[1:3,]
  emp_no birth_date first_name last_name gender  hire_date
1  10001 1953-09-02     Georgi   Facello      M 1986-06-26
2  10002 1964-06-02    Bezalel    Simmel      F 1985-11-21
3  10003 1959-12-03      Parto   Bamford      M 1986-08-28
```

3. Then, you can also specify the sequence of columns that you should select:

```
> employees[1:3, 2:4]
  birth_date first_name last_name
1 1953-09-02     Georgi   Facello
2 1964-06-02    Bezalel    Simmel
3 1959-12-03      Parto   Bamford
```

4. Besides subsetting a sequence of columns and rows from the dataset, you can specify certain rows and columns to subset with index vectors:

```
> employees[c(2,5), c(1,3)]
  emp_no first_name
2  10002    Bezalel
5  10005    Kyoichi
```

5. If you know the name of the column, you can also select columns with a given name vector:

```
> employees[1:3, c("first_name","last_name")]
  first_name last_name
1     Georgi   Facello
2    Bezalel    Simmel
3      Parto   Bamford
```

6. On the other hand, you can exclude columns with negative index:

```
> employees[1:3,-6]
  emp_no birth_date first_name last_name gender
1  10001 1953-09-02     Georgi   Facello      M
2  10002 1964-06-02    Bezalel    Simmel      F
3  10003 1959-12-03      Parto   Bamford      M
```

7. You can also exclude some attributes using the `in` and `!` operators:

```
> employees[1:3, !names(employees) %in% c("last_name", "first_
name")]
  emp_no birth_date gender  hire_date
1  10001 1953-09-02      M 1986-06-26
2  10002 1964-06-02      F 1985-11-21
3  10003 1959-12-03      M 1986-08-28
```

8. Furthermore, you can set the equal condition to subset data:

```
> employees[employees$gender == 'M',]
  emp_no birth_date first_name last_name gender  hire_date
1  10001 1953-09-02     Georgi   Facello      M 1986-06-26
3  10003 1959-12-03      Parto   Bamford      M 1986-08-28
4  10004 1954-05-01  Chirstian   Koblick      M 1986-12-01
5  10005 1955-01-21    Kyoichi Maliniak      M 1989-09-12
8  10008 1958-02-19     Saniya Kalloufi      M 1994-09-15
```

9. You can use a comparison operator to subset data:

```
> salaries[salaries$salary >= 60000 & salaries$salary < 70000,]
```

10. Additionally, the `substr` function can extract partial records:

```
> employees[substr(employees$first_name,0,2)=="Ge",]
  emp_no birth_date first_name last_name gender  hire_date
1  10001 1953-09-02     Georgi   Facello      M 1986-06-26
```

11. The regular expression is another useful and powerful tool for a user to subset data that they are interested in:

```
> employees[grep('[aeiou]$', employees$first_name),]
  emp_no birth_date first_name last_name gender  hire_date
1  10001 1953-09-02     Georgi   Facello      M 1986-06-26
3  10003 1959-12-03      Parto   Bamford      M 1986-08-28
5  10005 1955-01-21    Kyoichi  Maliniak      M 1989-09-12
6  10006 1953-04-20     Anneke   Preusig      F 1989-06-02
8  10008 1958-02-19     Saniya  Kalloufi      M 1994-09-15
```

How it works...

In this recipe, we demonstrated how to filter data with R. In the first case, we used the `head` and `tail` functions to examine the first few rows. By default, the `head` and `tail` functions will return the first six rows and last six rows of the dataset. We can still specify the number of records to subset in the second input parameter of the function.

Besides using the `head` and `tail` functions, we can use a square bracket to subset data. Using the square bracket, the value on the left-hand side of the comma assigns the rows to subset, and the value on the right-hand side of the comma indicates the columns to select. In the second step, we demonstrated that we can subset the first three records from the dataset by assigning a sequence of 1 to 3 on the left-hand side of the comma. If we do not specify anything on the right-hand side of the comma, this means that we will select all variables in our subset. Otherwise, we can specify the columns to select on the right-hand side of the comma. Similarly to step 3, we can select the second to fourth column by specifying a sequence on the right-hand side of the comma; select columns with a given `c(3,5)` index vector. Furthermore, we can select data with a given `c("first_name","last_name")` attribute name vector.

In addition to selecting the variables that we require, we can exclude the columns that we do not need with a negative index. Thus, we can place `-6` on the right-hand side of the comma to exclude the sixth column from the dataset. We can also use both the `!` and `in` operators to exclude data with a certain column name. In the seventh case, we can exclude attributes with the name as `first_name` and `last_name`.

Furthermore, we can filter data similar to SQL operation with a given condition. Here, as we need to use conditions to filter records from the data, we should place the criteria on the left-hand side of the comma. Thus, in cases 8 to 10, we demonstrate that we can filter male employee data with an equality condition, extract salary data ranging between 60,000 and 70,000, and retrieve employees with the first two letters matching *Ge* with the `substr` function. Finally, we can also employ the `grep` function along with a regular expression to subset employees with a vowel as the final letter of their first name.

There's more...

Besides using square brackets, we can also use the `subset` function to subset data:

1. We can select `first_name` and `last_name` of the first three rows of the `employees` data:

```
> subset(employees, rownames(employees) %in% 1:3, select=c("first_
name","last_name"))
  first_name last_name
1     Georgi   Facello
2    Bezalel    Simmel
3      Parto   Bamford
```

2. We can also set the condition to filter employee data by gender:

```
>subset(employees, employees$gender == 'M')
  emp_no birth_date first_name last_name gender  hire_date
1  10001 1953-09-02     Georgi   Facello      M 1986-06-26
3  10003 1959-12-03      Parto   Bamford      M 1986-08-28
4  10004 1954-05-01 Chirstian   Koblick      M 1986-12-01
5  10005 1955-01-21    Kyoichi  Maliniak      M 1989-09-12
8  10008 1958-02-19     Saniya  Kalloufi      M 1994-09-15
```

Dropping data

In the previous recipes, we introduced how to revise and filter datasets. Following these steps almost concludes the data preprocessing and preparation phase. However, we may still find some bad data within our dataset. Thus, we should discard this bad data or unwanted records to prevent it from generating misleading results. Here, we introduce some practical methods to remove this unnecessary data.

Getting ready

Refer to the *Converting data types* recipe and convert each attribute of imported data into the proper data type. Also, rename the columns of the `employees` and `salaries` datasets by following the steps from the *Renaming the data variable* recipe.

How to do it...

Perform the following steps to drop an attribute from the current dataset:

1. First, you can drop the `last_name` column by excluding `last_name` in our filtered subset:

```
> employees <- employees[,-5]
```

2. Or, you can assign `NULL` to the attribute you wish to drop:

```
> employees$hire_date <- NULL
```

3. To drop rows, you can specify the index of the row that you want to drop by assigning a negative index:

```
> employees <- employees[c(-2,-4,-6),]
```

How it works...

The idea of dropping rows is very similar to data filtering; you only need to specify the negative index of rows (or columns) that you want to drop during the filtering. Then, you can replace the original dataset with the filtered subset. Thus, as the `last_name` column is at the fifth index, you can remove the attribute by specifying `-5` at the right-hand side of the comma within the square bracket. In addition to reassignment, you can also assign `NULL` to the attribute that you want to drop. As for removing rows, you can place negative indexes on the left-hand side of comma within the square bracket, and then replace the original dataset with the filtered subset.

There's more...

In addition to data filtering or assigning the specific attribute to `NULL`, you can use the `within` function to remove unwanted attributes. All you need to do is place the unwanted attribute names inside the `rm` function:

```
> within(employees, rm(birth_date, hire_date))
   emp_no first_name last_name gender
1   10001     Georgi   Facello      M
2   10002    Bezalel    Simmel      F
3   10003      Parto   Bamford      M
4   10004  Chirstian   Koblick      M
5   10005    Kyoichi  Maliniak      M
6   10006     Anneke   Preusig      F
7   10007    Tzvetan Zielinski      F
8   10008     Saniya  Kalloufi      M
9   10009     Sumant      Peac      F
10  10010  Duangkaew  Piveteau      F
```

Merging data

Merging data enables us to understand how different data sources relate to each other. The `merge` operation in R is similar to the `join` operation in a database, which combines fields from two datasets using values that are common to each.

Getting ready

Refer to the *Converting data types* recipe and convert each attribute of imported data into the proper data type. Also, rename the columns of the `employees` and `salaries` datasets by following the steps from the *Renaming the data variable* recipe.

How to do it...

Perform the following steps to merge `salaries` and `employees`:

1. As `employees` and `salaries` are common in `emp_no`, we can merge these two datasets using `emp_no` as the join key:

```
> employees_salary  <- merge(employees, salaries, by="emp_no")
> head(employees_salary,3)
  emp_no birth_date first_name last_name salary  from_date    to_
date
1  10001 1953-09-02     Georgi    Facello  60117 1986-06-26 1987-
06-26
2  10001 1953-09-02     Georgi    Facello  62102 1987-06-26 1988-
06-25
3  10001 1953-09-02     Georgi    Facello  66596 1989-06-25 1990-
06-25
```

2. Or, we can assign `NULL` to the attribute that we want to drop:

```
> merge(employees, salaries, by="emp_no", all.x =TRUE)
```

3. In addition to the `merge` function, we can install and load the `plyr` package to manipulate data:

```
> install.packages("plyr")
> library(plyr)
```

4. Besides the standard `merge` function, we can use the `join` function in `plyr` to merge data:

```
> join(employees, salaries, by="emp_no")
```

How it works...

Similarly to data tables in a database, we sometimes need to combine two datasets for the purpose of correlating data. In R, we can simply combine two different data frames with common values using the `merge` function.

In the `merge` function, we use both `salaries` and `employees` as our input data frame. For the `by` parameter, we can specify `emp_no` as the key to join these two tables. We will then see that the data with the same `emp_no` value has now merged into a new data frame. However, sometimes we want to perform either a left join or a right join for the purpose of preserving every data value from either employees or salaries. To perform the left join, we can set `all.x` to `TRUE`. Then, we can find every row from the `employees` dataset preserved in the merged dataset. On the other hand, if one wants to preserve all rows from the `salaries` dataset, we can set `all.y` to `TRUE`.

In addition to using the built-in `merge` function, we can install the `plyr` package to merge datasets. The usage of `join` is very similar to `merge`; we only have to specify the data to join and the columns with the common values within the `by` parameter.

There's more...

In the `plyr` package, we can use the `join_all` function to join recursive datasets within a list. Here, we can use `join_all` to join the `employees` and `salaries` datasets by `emp_no`:

```
> join_all(list(employees, salaries), "emp_no")
```

Sorting data

The power of sorting enables us to view data in an arrangement so that we can analyze the data more efficiently. In a database, we can use an `order by` clause to sort data with appointed columns. In R, we can use the `order` and `sort` functions to place data in an arrangement.

Getting ready

Refer to the *Converting data types* recipe and convert each attribute of imported data into the proper data type. Also, rename the columns of the `employees` and `salaries` datasets by following the steps from the *Renaming the data variable* recipe.

How to do it...

Perform the following steps to sort the `salaries` dataset:

1. First, we can use the `sort` function to sort data:

```
> a <- c(5,1,4,3,2,6,3)
> sort(a)
[1] 1 2 3 3 4 5 6
> sort(a, decreasing=TRUE)
[1] 6 5 4 3 3 2 1
```

2. Next, we can determine how the `order` function works on the same input vector:

```
> order(a)
[1] 2 5 4 7 3 1 6
> order(a, decreasing = TRUE)
[1] 6 1 3 4 7 5 2
```

3. To sort a data frame by a specific column, we first obtain the ordered index and then employ the index to retrieve the sorted dataset:

```
> sorted_salaries <- salaries[order(salaries$salary, decreasing =
TRUE),]
> head(sorted_salaries)
    emp_no salary  from_date    to_date
684  10068 113229 2001-08-03 9999-01-01
683  10068 112470 2000-08-03 2001-08-03
682  10068 111623 1999-08-04 2000-08-03
681  10068 108345 1998-08-04 1999-08-04
680  10068 106204 1997-08-04 1998-08-04
679  10068 105533 1996-08-04 1997-08-04
```

4. Besides sorting data by a single column, we can sort data by multiple columns:

```
> sorted_salaries2 <-salaries[order(salaries$salary,
salaries$from_date, decreasing = TRUE),]
> head(sorted_salaries2)
    emp_no salary  from_date    to_date
684  10068 113229 2001-08-03 9999-01-01
683  10068 112470 2000-08-03 2001-08-03
682  10068 111623 1999-08-04 2000-08-03
681  10068 108345 1998-08-04 1999-08-04
680  10068 106204 1997-08-04 1998-08-04
679  10068 105533 1996-08-04 1997-08-04
```

How it works...

R provides two methods to sort data: one is `sort` and the other is `order`. For the sort function, the function returns sorted vector as output. In our first case, we set up an `a` integer vector with seven integer elements. We then applied the `sort` function to sort the `a` vector, which yielded a sorted vector as the output. By default, the sorted vector is in ascending order. However, we can change the order sequence by specifying decreasing to `TRUE`. On the other hand, the `order` function returns an ordering index vector as output. Still, we can specify whether the returned index vector is in ascending or descending order.

To arrange elements in the vector in ascending or descending order, we can simply use the `sort` function. However, to arrange records in a specific column, we should use the `order` function. In our example, we first obtained the ordering index in descending order from the `salary` attribute and then retrieved the record from `salaries` with an ordering index. As a result, we found records in `salaries` arranged by salary. Besides sorting records by a single attribute, we can sort records by multiple attributes. All we need to do is to place the `salary` and `from_date` attributes one by one in the `order` function.

There's more...

You can use the arrange function in `plyr` to sort salary data with `salary` in ascending order and `from_date` in descending order:

```
> arranged_salaries <- arrange(salaries, salary, desc(from_date))
> head(arranged_salaries)
  emp_no salary  from_date    to_date
1  10048  39507 1986-02-24 1987-01-27
2  10027  39520 1996-04-01 1997-04-01
3  10064  39551 1986-11-20 1987-11-20
4  10072  39567 1990-05-21 1991-05-21
5  10072  39724 1991-05-21 1992-05-20
6  10049  39735 1993-05-04 1994-05-04
```

Reshaping data

Reshaping data is similar to creating a contingency table, which enables the user to aggregate data of specific values. The `reshape2` package is designed for this specific purpose. Here, we introduce how to use the `reshape2` package to transform our dataset from long to wide format with the `dcast` function. We also cover how to transform it from wide format back to long format with the `melt` function.

Getting ready

Refer to the *Merging data* recipe and merge `employees` and `salaries` into `employees_salary`.

How to do it...

Perform the following steps to reshape data:

1. First, we can use the `dcast` function to transform data from long to wide:

```
> wide_salaries <- dcast(salaries, emp_no ~ year(ymd(from_date)),
value.var="salary")
> wide_salaries[1:3, 1:7]
  emp_no 1985  1986  1987  1988  1989  1990
1  10001   NA 60117 62102 66074 66596 66961
2  10002   NA    NA    NA    NA    NA    NA
3  10003   NA    NA    NA    NA    NA    NA
```

2. We can also transform the data by keeping `emp_no` and the formatted name string as two of the attributes. Then, we can shape the year of salary payment as the column name and shape salary as its value:

```
> wide_employees_salary <- dcast(employees_salary, emp_no +
paste(first_name, last_name) ~ year(ymd(from_date)), value.
var="salary", variable.name="condition")
> wide_employees_salary[1:3,1:7]
  emp_no paste(first_name, last_name) 1985  1986  1987  1988  1989
1  10001               Georgi Facello   NA 60117 62102
66074 66596
2  10002               Bezalel Simmel   NA    NA    NA
NA    NA
3  10003                Parto Bamford   NA    NA    NA
NA    NA
```

3. On the other hand, we can transform the data from wide back to long format using the `melt` function:

```
> long_salaries <- melt(wide_salaries, id.vars=c("emp_no"))
> head(long_salaries)
  emp_no variable value
1  10001     1985    NA
2  10002     1985    NA
3  10003     1985    NA
4  10004     1985    NA
5  10005     1985    NA
6  10006     1985    NA
```

4. To remove data with missing values in `long_salaries`, we can use `na.omit` to remove this data:

```
> head(na.omit(long_salaries))
   emp_no variable value
9   10009     1985 60929
13  10013     1985 40000
48  10048     1985 40000
64  10064     1985 40000
70  10070     1985 55999
98  10098     1985 40000
```

How it works...

In this recipe, we demonstrated how to reshape data with the `reshape2` package. First, we used the `dcast` function to transform data from the long to wide format. Using this function, we specified that we wanted the data to transform as `salaries` in the first parameter. Then, we specified the shaping formula by setting `emp_no` as row and the year of salary paid as the column. Finally, we set `salary` as presented value in wide format layout.

We can also shape the data with multiple columns; all that we need to do is add another column information on the left-hand side of the formula with a + operator. Thus, in our second case, we set the shaping formula as `emp_no + paste(first_name, last_name) ~ year(ymd(from_date))` on the `employees_salary` merged data. Then, we saw that the output data was presented with `emp_no` and the formatted name on the left-hand side, and the year of salary paid on the upper side and the presented salary as the value.

Besides shaping the data from long format to wide format, we can also reshape the data back to long format using the `melt` function. Therefore, we can transform the `wide_salaries` data back to long format using `emp_no` as the basis. As there are many missing values (presented as `NA`), we can use the `na.omit` function to remove these records.

There's more...

In addition to the `dcast` and `melt` functions in the `plyr` package, we can use `stack` and `unstack` to ungroup and group values:

1. We can group data by value using the `unstack` function:

```
> un_salaries <- unstack(long_salaries[,c(3,1)])
> head(un_salaries, 3)
$`10001`
 [1] 60117 62102 66074 66596 66961 71046 74333 75286 75994 76884
80013
```

```
[12] 81025 81097 84917 85112 85097 88958

$`10002`
[1] 65828 65909 67534 69366 71963 72527

$`10003`
[1] 40006 43616 43466 43636 43478 43699 43311
```

2. In contrast, we can concatenate multiple data frames or lists using the `stack` function:

```
> stack_salaries <- stack(un_salaries )
> head(stack_salaries)
  values   ind
1  60117 10001
2  62102 10001
3  66074 10001
4  66596 10001
5  66961 10001
6  71046 10001
```

Detecting missing data

There are numerous causes behind missing data. For example, it could be the result of typos or data process flaws. However, if there is missing data in our analysis process, the results of the analysis may be misleading. Thus, it is important to detect missing values before proceeding with further analysis.

Getting ready

Refer to the *Converting data types* recipe and convert each attribute of imported data into the proper data type. Also, rename the columns of the `employees` and `salaries` datasets by following the steps from the *Renaming the data variable* recipe.

How to do it...

Perform the following steps to detect missing values:

1. First, we set the `to_date` attribute with a date over `2100-01-01`:

```
> salaries[salaries$to_date > "2100-01-01",]
```

2. We then change the data with a date over `2100-01-01` to a missing value:

```
> salaries[salaries$to_date > "2100-01-01","to_date"] = NA
```

3. Next, we can use the `is.na` function to find which rows contain missing values:

```
> is.na(salaries$to_date)
```

4. We can also count the number of missing values in `to_date` with the `sum` function:

```
> sum(is.na(salaries$to_date))
[1] 81
```

5. Moreover, we can calculate the ratio of missing values:

```
> sum(is.na(salaries$to_date) == TRIE)/length(salaries$to_date)
[1] 0.081
```

6. However, if we want to know the percentage of missing values in each column, we can use the `sapply` function:

```
> wide_salaries <- dcast(salaries, emp_no ~ year(ymd(from_date)), value.var="salary")
> wide_salaries[1:3, 1:7]
  emp_no 1985  1986  1987  1988  1989  1990
1  10001   NA 60117 62102 66074 66596 66961
2  10002   NA    NA    NA    NA    NA    NA
3  10003   NA    NA    NA    NA    NA    NA
```

7. To determine the percentage of missing values of attributes, we can use `sapply` to calculate the percentage of all attributes:

```
> sapply(wide_salaries, function(df) {
+                     sum(is.na(df)==TRUE)/ length(df);
+                 })
   emp_no      1985      1986      1987      1988      1989
0.0000000 0.9405941 0.8910891 0.8217822 0.7524752 0.6534653
     1990      1991      1992      1993      1994      1995
0.5544554 0.5148515 0.4158416 0.3960396 0.3267327 0.2475248
     1996      1997      1998      1999      2000      2001
0.2079208 0.1782178 0.1485149 0.1287129 0.1485149 0.1881188
     2002
0.5841584
```

8. Furthermore, we can install and load the `Amelia` package:

```
> install.packages("Amelia")
> library(Amelia)
```

9. Then, we can use the `missmap` function to plot the missing value map:

```
> missmap(wide_salaries, main="Missingness Map of Salary")
```

You'll see the following plot:

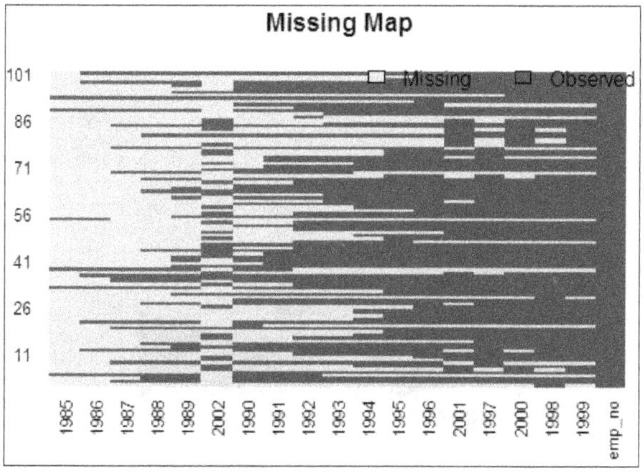

Figure 1: The visualization of a missing value

How it works...

In R, a missing value is often noted using the `NA` symbol, which stands for **not available**. Most functions (such as `mean` or `sum`) may output `NA` when encountering an `NA` value in the dataset. Though we can assign an argument, such as `na.rm`, to remove the effect of `NA`, it is better to impute or remove the missing data in the dataset to prevent the propagation effect of the missing value.

In this recipe, we first searched records for dates after `2100-01-01`. As it is very unlikely that one's payroll is paid after `2100-01-01`, we can regard these date values as typos or the result of system errors. Thus, we can first assign these values to missing values (denoted with `NA`). Next, we can start searching for missing values within the data using some built-in functions.

To find missing values in the dataset, we first sum all of the `NA` values, divide by the number of values within each attribute, and then apply the calculation to all attributes with `sapply`.

Also, to display the calculation results using a table, we can utilize the `Amelia` package to plot the missing value map of every attribute in one chart. The visualization of missing values enables users to get a better understanding of the missing percentage within each dataset. From the plot, we can determine that `1985` contains the most missing values.

There's more...

For missing-value handling, we introduced Amelia to visualize the missing value. Apart from typing console commands, we can also use the interactive GUI of Amelia, **AmeliaView**, to run the application.

To start running AmeliaView, simply type `AmeliaView()` in the R console:

```
>AmeliaView()
```

You'll then see the following window:

Figure 2: The GUI of AmeliaView

Imputing missing data

The previous recipe showed us how to detect missing values within the dataset. Though the data with missing values is rather incomplete, we can still adapt a heuristic approach to complete our dataset. Here, we introduce some techniques one can employ to impute missing values.

Getting ready

Refer to the *Converting data types* recipe and convert each attribute of imported data into the proper data type. Also, rename the columns of the `employees` and `salaries` datasets by following the steps in the *Renaming the data variable* recipe.

How to do it...

Perform the following steps to impute missing values:

1. First, we subset user data with `emp_no` equal to `10001`:

```
> test.emp <- salaries[salaries$emp_no == 10001,]
```

2. Then, we purposely assign `salary` as the missing value of row `8`:

```
> test.emp[8,c("salary")]
[1] 75286
> test.emp[8,c("salary")] = NA
```

3. For the first imputing method, we can remove records with missing values using the `na.omit` function:

```
> na.omit(test.emp)
```

4. On the other hand, we can calculate the mean salary of employee `10001`:

```
> mean_salary <- mean(salaries$salary[salaries$emp_no == 10001],
na.rm=TRUE)
> mean_salary
[1] 75388.94
```

5. Then, we can impute the missing value with mean salary of employee `10001`:

```
> salaries$salary[salaries$emp_no == 10001 &
is.na(salaries$salary)] = mean_salary
```

6. In addition to these approaches, we can use the `mice` package to impute data:

```
> install.packages("mice")
> library(mice)
```

7. The following steps allow us to use the `mice` function to impute data:

```
> test.emp$from_date = year(ymd(test.emp$from_date))
> test.emp$to_date = year(ymd(test.emp$to_date))
> imp = mice(test.emp, meth=c('norm'), set.seed=7)
> complete(imp)
```

How it works...

In general, there are three basic approaches to impute missing values: first, we remove missing values, then we perform mean imputation, and finally we perform multiple imputations. In this recipe, we used data of one of the employee's as an example. First, we extracted the subset with `emp_no` equal to `10001`. Then, we purposely replaced one `salary` value with an `NA` missing value. Next, we used one of the imputation methods to impute the missing value.

First, we demonstrated how to remove missing values using the `na.omit` function, which automatically removes records containing missing values. Missing value removal is straightforward and simple, and it is best suited for cases where missing values occupy only a small proportion of the dataset. In contrast, if missing values represent a large proportion of the dataset, removing these missing values may bias the analysis result and lead to a false conclusion.

Secondly, we imputed missing values with mean value. Thus, we first compute the mean salary value with the `mean` function by ignoring missing values (`na.rm` indicates that the calculation process will not take missing values into account). This type of imputation will not affect the estimation of the mean value, but it may bias the standard deviation and square error of the dataset, and further lead to bias estimation. Therefore, this is not a recommended approach.

Thirdly, we can use the `mice` function to perform multiple imputations. Here, we first installed and load the `mice` package to an R session. We then extracted the integer year from the `from_date` and `to_date` attributes. Next, as employee salary increases each year, we can choose to use Bayesian linear regression as the estimation method (by configuring `norm` to `meth` parameter) to predict missing values. Finally, we can obtain the completed data with the complete function. The multiple imputation method employs the relation between data to predict missing values, and it uses a Monte Carlo method to generate a possible missing value, which is the most common approach toward multiple missing value imputation problems.

There's more...

Besides using the `mice` package to predict missing values within a dataset, we can use the regression method to predict missing values:

1. First, make every attribute of `test.emp` accessible:

```
> attach(test.emp)
```

2. Next, fit a regression line regarding `salary` against `from_date`:

```
> fit = lm(salary ~ from_date)
```

3. Finally, we can predict the salary of the missing value:

```
> predict(fit, data.frame(from_date = 1993))
       1
73575.91
```

The `lm` function will generate a fitted regression line regarding `salary` against `from_date`, and we can then use the fitted model to predict the possible salary in `1993`.

4

Data Manipulation

This chapter covers the following topics:

- Enhancing a `data.frame` with a `data.table`
- Managing data with a `data.table`
- Performing fast aggregation with a `data.table`
- Merging large datasets with a `data.table`
- Subsetting and slicing data with `dplyr`
- Sampling data with `dplyr`
- Selecting columns with `dplyr`
- Chaining operations in `dplyr`
- Arranging rows with `dplyr`
- Eliminating duplicated rows with `dplyr`
- Adding new columns with `dplyr`
- Summarizing data with `dplyr`
- Merging data with `dplyr`

Introduction

Most R users will agree that data frames provide a flexible and expressive structure for tabular data. While data frames are effective for small datasets, they are not ideal to use when processing data that is larger than a Gigabyte in size. Additionally, it is not easy to summarize data within the data frame itself; we need to load an additional package, such as `plyr` or `reshape2`, to perform advanced aggregation. Therefore, we would like to introduce how to use `data.table` and `dplyr` to perform descriptive statistics.

We first illustrate what these two packages do:

- `data.table`: This is an extension of `data.frame`; it provides the ability to quickly aggregate and process large datasets. Additionally, it provides a much more readable and less confusing syntax compared to data frames.
- `dplyr`: This provides users with SQL-like functions so that we can quickly aggregate and summarize data from various sources.

These two packages can help users quickly and easily generate descriptive statistics from any given tabular dataset.

In this chapter, we first cover how to create a `data.table`. We then introduce operations to manipulate, join, and process data. Moving on, we present how to use `dplyr` to perform fast data aggregation. In advance, with the help of the `%>%` chaining operator, we can quickly perform descriptive statistics with SQL-like functions on a given tabular dataset.

Enhancing a data.frame with a data.table

When you process a dataset that is a Gigabyte or larger in size, you may find that a `data.frame` is rather inefficient. To address this issue, you can use the enhanced extension of `data.frame`—`data.table`. In this recipe, we will show how to create a `data.table` in R.

Getting ready

Download the `purchase_view.tab` and `purchase_order.tab` datasets from the following GitHub links, respectively:

- `https://github.com/ywchiu/rcookbook/blob/master/chapter6/purchase_view.tab`
- `https://github.com/ywchiu/rcookbook/blob/master/chapter6/purchase_order.tab`

How to do it...

Perform the following steps to create a `data.table`:

1. First, install and load the `data.table` package using the following commands:

```
> install.packages("data.table")
> library(data.table)
```

2. Next, we can create an R data frame using `read.table`:

```
> purchase <- read.table("purchase_view.tab", header=TRUE,
sep='\t')
 [1] "data.frame"
> dim(purchase)
[1] 1191486       4

> order <- read.table("purchase_order.tab", header=TRUE, sep='\t')
> dim(order)
[1] 54772     6
```

3. Convert the `data.frame` into an R `data.table`:

```
> purchase.transform = as.data.table(purchase)
> class(purchase.transform)
[1] "data.table" "data.frame"
```

4. Alternatively, we can create a new `data.table`:

```
> dt <- data.table(product=c("p1", "p2", "p3"),
price=c(100,200,300), category="beverage" )
> dt
   product price category
1:      p1   100 beverage
2:      p2   200 beverage
3:      p3   300 beverage
```

5. Instead of using `colnames` to change column names, we can use the `setnames` function:

```
> setnames(dt, c("Product", "Price", "Drink"))
> head(dt)
   Product Price    Drink
1:      p1   100 beverage
2:      p2   200 beverage
3:      p3   300 beverage
```

6. Additionally, we can use the `fread` function from the `data.table` package to read data:

```
> purchase.dt <- fread("purchase_view.tab", header=TRUE, sep='\t')
> order.dt <- fread("purchase_order.tab", header=TRUE, sep='\t')
```

7. Moving on, let's examine the data loading speed of the data frame:

```
> system.time(purchase <- read.table("purchase_view.tab",
header=TRUE, sep='\t'))
   user  system elapsed
  31.74    0.19   32.34
```

8. Now, let's compare the read performance when using the `fread` function:

```
> system.time(purchase.dt <- fread("purchase_view.tab",
header=TRUE, sep='\t'))
   user  system elapsed
   0.84    0.00    0.86
```

9. Similar to how we examine data from `data.frame`, we can also use the `dim` and `str` functions to examine the structure of the `data.table`:

```
> dim(purchase.dt)
[1] 1191486       4


> str(purchase.dt)
Classes 'data.table' and 'data.frame': 1191486 obs. of  4
variables:
 $ Time   : chr  "2015-07-01 00:00:01" "2015-07-01 00:00:03"
"2015-07-01 00:00:05" "2015-07-01 00:00:05" ...
 $ Action : chr  "view" "view" "view" "view" ...
 $ User   : chr  "U129297265" "U321001337" "U10070718237"
"U393805241" ...
 $ Product: chr  "P0023468384" "P0018926456" "P0000063593"
"P0005541535" ...
 - attr(*, ".internal.selfref")=<externalptr>
```

How it works...

At the start of this recipe, we demonstrated how to create a data frame by reading the `purchase_view.tab` and `purchase_order.tab` files with the `read.table` function. We used the `as.data.table` function to convert the data type from `data.frame` into `data.table`. This allows us to list the object type of saved data with the `class` function, and the output returned both `data.frame` and `data.table`. The output reveals that a `data.table` is also a data frame; we can also perform any `data.frame` compatible function on a `data.table`.

Moving on, we illustrated that we can initialize a new `data.table` or create a `data.table` from a file with the `fread` function. In a `data.frame`, we need to use the `colnames` function to set the column name, but the `colnames` syntax would inherently copy the whole table, which will cause performance issues when processing larger datasets. However, `data.table` provides the `setnames` function, which changes the input by reference and avoids copying the table.

At this point, we can compare the read performance between the `read.table` function and the `fread` function. Here, we measure the reading time with the `system.time` function. Surprisingly, when reading the `purchase_view.tab` file that has over a million rows (11,91,486 obs. of four variables), it takes the `read.table` function `31.74` seconds to complete the reading task. However, `fread` takes only `0.84` seconds. This is mainly due to the fact that the `fread` function is implemented in pure C. Besides having better performance, `fread` can automatically guess the delimiter of the input file, which makes reading data more efficient and simple.

As mentioned previously, a `data.table` is also a `data.frame`. Therefore, we can still examine the data overview and structure with the `dim` and `str` functions. Moreover, we can also perform data visualization and model fitting on a `data.table` object.

There's more...

Besides using the `fread` function from `data.table` to read datasets that contain millions of rows, we can consider using another `readr` package to read flat files into R. The `readr` function is implemented in C++ and Rcpp. As a result, the performance is slightly slower than `fread`; however, it is still a better choice than `read.csv` or `read.table` when reading a large amount of data from flat files. You can use the following command to read `purchase_order.tab` into an R session:

```
> install.packages("readr")
> library("readr")

> order.readr = read_tsv("purchase_order.tab")
> head(order.readr)
Source: local data frame [6 x 6]

                 Time Action        User      Product Quantity Price
               (time)  (chr)       (chr)        (chr)    (int) (int)
1 2015-07-01 00:00:01  order  U312622727  P0006944501        1  1069
2 2015-07-01 00:00:03  order  U239012343  P0006018073        1  1680
3 2015-07-01 00:00:19  order U10007697373  P0002267974        1   285
4 2015-07-01 00:01:10  order  U296328517  P0016144236        1   550
```

```
5 2015-07-01 00:01:36   order   U300884570 P0014516980122        1   249
6 2015-07-01 00:01:48   order   U451050374     P0004134266        1  1780

> class(order.readr)
[1] "tbl_df"     "tbl"        "data.frame"
```

Managing data with a data.table

The two major advantages of a `data.table` as compared to a `data.frame` are the speed and clearer syntax of the former. Similar to a `data.frame`, we can perform operations to slice and subset a `data.table`. Here, we introduce some operations that you can perform on `data.table`.

Getting ready

Ensure that you completed the *Enhancing a data.frame with a data.table* recipe to load `purchase_view.tab` and `purchase_order.tab` as both `data.frame` and `data.table` into your R environment.

How to do it...

Perform the following steps to perform data manipulation on `data.table`:

1. First, use the `head` function to view the first three rows:

```
> head(purchase.dt[1:3])
                  Time Action        User     Product
1: 2015-07-01 00:00:01   view   U129297265 P0023468384
2: 2015-07-01 00:00:03   view   U321001337 P0018926456
3: 2015-07-01 00:00:05   view U10070718237 P0000063593

> head(purchase[1:3])
                  Time Action        User
1 2015-07-01 00:00:01   view   U129297265
2 2015-07-01 00:00:03   view   U321001337
3 2015-07-01 00:00:05   view U10070718237
4 2015-07-01 00:00:05   view   U393805241
5 2015-07-01 00:00:10   view U10004621134
6 2015-07-01 00:00:11   view   U370110321
```

2. Next, select the data by row number and column name:

```
> purchase.dt[1:3, User]
[1] "U129297265"   "U321001337"   "U10070718237"

> purchase[1:3,"User"]
[1] U129297265   U321001337   U10070718237
```

3. We can select multiple columns from `order.dt` with the `list` function:

```
> user.price <- order[1:3, c("User", "Price") ]
> head(user.price)
          User Price
1   U312622727  1069
2   U239012343  1680
3 U10007697373   285

> dt.user.price <- order.dt[1:3, list(User, Price) ]
> dt.user.price
           User Price
1:   U312622727  1069
2:   U239012343  1680
3: U10007697373   285
```

4. Alternatively, we can select multiple columns from the dataset using the . operation:

```
> dt.user.price2 <- order.dt[1:3, .(User, Price) ]
> head(dt.user.price2)
           User Price
1:   U312622727  1069
2:   U239012343  1680
3: U10007697373   285
```

5. Moving on, we can filter order data by its quantity and select the `Price` column from the dataset:

```
> dt.price <- order.dt[Quantity > 3, Price]
> head(dt.price)
[1]  19 159 529  49 536 550

> price <- order[order$Quantity > 3, "Price"]
> head(price)
[1]  19 159 529  49 536 550
```

6. We can remove NA prices from the `Price` attribute:

```
> dt.omit.price <- order.dt[, na.omit(Price)]
> head(dt.omit.price)
[1] 1069 1680  285  550  249 1780

> omit.price <- order[!is.na(order$Price), "Price"]
> head(omit.price)
[1] 1069 1680  285  550  249 1780
```

7. Next, we can remove some rows that contain NA prices from the dataset:

```
> dt.omit.price2 <- order.dt[na.omit(Price)]
> head(dt.omit.price2, 3)
                  Time Action       User       Product Quantity
Price
1: 2015-07-01 16:57:15  order U442229329   P0005772981        1
699
2: 2015-07-01 22:02:12  order U424420542   P0004281340        1
1399
3: 2015-07-01 08:36:34  order  U37738470   P0005756026        1
719
```

8. At this point, we can use `:=` to update data by reference:

```
> dt.omit.price2[,Avg_Price := Price/Quantity]
> head(dt.omit.price2[Quantity >=2], 3)
                 Time           Action           User     Product
Quantity Price Avg_Price
1: 2015-07-02 11:30:03  order U447614356 P0007510112     2   105
52.50000
2: 2015-07-01 18:12:51  order U367988353 P0005486876     3   179
59.66667
3: 2015-07-01 11:01:28  order U152432028 P0006173230     4   159
39.75000
```

9. We can also use `:=` to remove attribute by reference:

```
> dt.omit.price2[,avg_price:=NULL]
> head(dt.omit.price2[Quantity >=2], 3)
                  Time Action       User     Product Quantity
Price
1: 2015-07-02 11:30:03  order U447614356 P0007510112        2
105
```

```
2: 2015-07-01 18:12:51   order U367988353 P0005486876        3
179
3: 2015-07-01 11:01:28   order U152432028 P0006173230        4
159
```

10. To copy an entire table, we can use the `copy` function:

```
> dt.omit.price3 <- copy(dt.omit.price2)
```

11. Here, we illustrate how to extract the last user ID from the dataset:

```
> purchase.dt[.N, User]
[1] "U151479349"

> purchase[nrow(purchase), "User"]
[1] U151479349
```

How it works...

In this recipe, we demonstrated how you can slice and subset data in the form of `data.frame` and `data.table`. First, let's understand the basics of the `data.table` command. The `data.table` command is composed in the following form:

```
DT[i, j, by]
```

A brief description of the parameters is as follows:

- The `i` part is similar to the `WHERE` clause in SQL
- The `j` part is similar to `SELECT` in SQL
- The `by` part corresponds to the `GROUP BY` clause in SQL

In this section, we focused on how to slice and subset data using the `i` and `j` part of the `data.table`.

First, we can slice the data with a given index sequence. We find that we should put a series before the "`,`" comma to tell the `data.frame` to obtain the first three rows. On the other hand, `data.table` knows the given sequence is used to slice the first three rows from the dataset, and it does not require a comma when slicing rows from the dataset.

Besides slicing data with given row indices, we can select individual columns from the data. Similar to how a `data.frame` selects a column from the original data frame, we only need to place the column name after the comma. We can select individual columns from the original dataset. However, when we subset multiple columns, we need to wrap the column names into a list, or you can just use "`.`" to wrap up selected column names.

Moving on, we can perform data filtering by setting up a filtering condition at the location of the row. Here, we can either filter records with a quantity over three or omit records that do not contain NA values. When removing rows containing NA values, the placement of the filtering criteria is important. If the filter is applied to the `j` part, we can only obtain the price without NA values. Alternatively, if the filter is used in the `i` part, we can retrieve the rows without NA prices.

We also introduce some operations to update data efficiently within a `data.table`. In general, R copies object when the variable is modified. Therefore, if we simply need to add a column or change data of a `data.frame`, we will invoke a full object copy operation. The full object copy will pose a great performance issue when dealing with a table of a Gigabyte or larger in size. However, as `data.table` provides functions that enable users to update data by reference, we can avoid the performance impact when dealing with large datasets. Here, we first illustrated how to use the `:=` notation to update and drop attribute by reference. We then demonstrated how to use the `copy` function to copy a large object efficiently.

Finally, we introduced how to use `.N` to select the last row from the data table, which is more convenient as we do not need to calculate the number of rows beforehand.

There's more...

In this recipe, we showed you how to make a copy of the original `data.table` using the `copy` function. In `data.frame`, we can copy a new `data.frame` using the pure `<-` notation. However, if you use the `<-` notation to copy a `data.table`, this only copies the location reference. To make a full copy of a `data.table` object, you need to use the `copy` function. In the following example, we explain how the `<-` notation and the `copy` function work differently when copying an object:

1. First, let's create a new `data.table`:

   ```
   > dt <- data.table(product=c("p1", "p2", "p3"),
   price=c(100,200,300), category="beverage" )
   ```

2. We can copy the original `data.table` by reference, or make an explicit copy:

   ```
   > dt2 <- dt
   > dt.copy = copy(dt)
   ```

3. At this point, we can see how each `data.table` is affected if we update the data of the original `data.table`:

   ```
   > dt
      product price category
   1:      p1     1 beverage
   2:      p2     1 beverage
   3:      p3     1 beverage
   ```

```
> dt2
   product price category
1:      p1     1 beverage
2:      p2     1 beverage
3:      p3     1 beverage

> dt.copy
   product price category
1:      p1   100 beverage
2:      p2   200 beverage
3:      p3   300 beverage
```

4. Finally, we can use the `identical` function to compare the two `data.table`:

```
> identical(dt, dt2)
[1] TRUE

> identical(dt, dt.copy)
[1] FALSE
```

Performing fast aggregation with a data.table

Another advantage of a `data.table` is that we can easily aggregate data without the help of additional packages. In this recipe, we will illustrate how to perform data aggregation using `data.table`.

Getting ready

Ensure that you completed the *Enhancing a data.frame with a data.table* recipe to load `purchase_view.tab` and `purchase_order.tab` as both a `data.frame` and a `data.table` into your R environment.

How to do it...

Perform the following steps to perform data aggregation over `data.table`:

1. First, we can average the price of orders in total:

```
> order.dt[,mean(na.omit(Price))]
[1] 2012.119
```

2. Next, we can average the amount of orders per user:

```
> mean.price.by.user <- order.dt[,mean(na.omit(Price)), User]
> head(mean.price.by.user)
          User      V1
1:   U312622727  934.00
2:   U239012343 1298.75
3: U10007697373  247.00
4:   U296328517  745.00
5:   U300884570  249.00
6:   U451050374 1780.00
```

3. In order to name the aggregated data, we can surround the `mean` function with `.()`:

```
> mean.price.by.user2 <- order.dt[,.(Price=mean(na.omit(Price))),
User]
> head(mean.price.by.user2)
          User   Price
1:   U312622727  934.00
2:   U239012343 1298.75
3: U10007697373  247.00
4:   U296328517  745.00
5:   U300884570  249.00
6:   U451050374 1780.00
```

4. We can then calculate the average purchased amount by date:

```
> mean.price.by.date <- order.dt[,.(Price=mean(na.omit(Price))),
by=as.Date(Time)]
> head(mean.price.by.date)
      as.Date    Price
1: 2015-07-01 2380.242
2: 2015-07-02 2124.970
3: 2015-07-03 2316.778
4: 2015-07-04 1933.913
5: 2015-07-05 1792.385
6: 2015-07-06 1828.492
```

5. We can also calculate both the sum amount and number of users by date:

```
> price_sum_n_users.by.date <- order.dt[,.(Price_Sum = sum(na.
omit(Price)), N_Users= length(unique(User))) , by=as.Date(order.
dt$Time)]
> head(price_sum_n_users.by.date)
      as.Date Price_Sum N_Users
1: 2015-07-01   5524541    1773
2: 2015-07-02   4349813    1812
3: 2015-07-03   4114597    1524
4: 2015-07-04   4184987    1818
5: 2015-07-05   2530847    1191
6: 2015-07-06   3475964    1545

> price_sum_n_users.by.date2 <- order.dt[,.(Price_Sum = sum(na.
omit(Price)), N_Users=uniqueN(User)) , by=as.Date(order.dt$Time)]
> head(price_sum_n_users.by.date2)
      as.Date Price_Sum N_Users
1: 2015-07-01   5524541    1773
2: 2015-07-02   4349813    1812
3: 2015-07-03   4114597    1524
4: 2015-07-04   4184987    1818
5: 2015-07-05   2530847    1191
6: 2015-07-06   3475964    1545
```

6. Additionally, we can calculate the average cost per person to date:

```
> price_avg.by.date <- order.dt[,.(Price_Avg = sum(na.omit(Price))
/ uniqueN(User)) , by=.(Date=as.Date(order.dt$Time))]
> head(price_avg.by.date)
         Date Price_Avg
1: 2015-07-01  3115.928
2: 2015-07-02  2400.559
3: 2015-07-03  2699.867
4: 2015-07-04  2301.973
5: 2015-07-05  2124.976
6: 2015-07-06  2249.815
```

7. Next, we can retrieve the total purchased amount of each product by date:

```
> Price_Sum.by.Date_N_Product <- order.dt[,.(Price_Sum =
na.omit(sum(Price* Quantity))) , by=.(Date = as.Date(Time),
Product)]
```

8. Furthermore, we can sort the data by the sum of price in ascending order:

```
> sorted.price.asc <- Price_Sum.by.Date_N_Product[order(Price_
Sum)]
> head(sorted.price.asc)
         Date        Product Price_Sum
1: 2015-07-03    P0009305785        15
2: 2015-07-07    P0023454992        15
3: 2015-07-27 P0024828260122        15
4: 2015-07-30    P0002976595        15
5: 2015-07-04    P0022533744        19
6: 2015-07-01    P0015225991        25
```

9. Alternatively, we can sort the data by the sum of price in descending order:

```
> sorted.price.desc <- Price_Sum.by.Date_N_Product[order(-Price_
Sum)]
> head(sorted.price.desc)
         Date     Product Price_Sum
1: 2015-07-23 P0006584093   3520000
2: 2015-07-09 P0006993663    449400
3: 2015-07-10 P0007082051    446400
4: 2015-07-03 P0000143500    438888
5: 2015-07-16 P0000143511    438888
6: 2015-07-17 P0000143511    438888
```

How it works...

In this recipe, we showed you how to aggregate data with `data.table`. When using `data.frame`, we can only slice and subset data by row and by column. If we want to average the data of individual columns, we need to obtain the column from the `data.frame` first, and we then need to apply a summary statistic function over that single column. However, in `data.table`, we can perform summary statistics of a column just by placing the summary statistic function of this column in the `j` part. By doing this, we can easily calculate the average price from `order.dt`.

Now, let's illustrate how to use the by part in data.table. The by part is similar to the GROUP BY function in SQL expression. In other words, we can compute the j part by grouping values by the variable specified in the by part. Here, we can calculate the average amount spent for each user. Furthermore, if we want to specify the name of aggregated columns, we can place the column to be aggregated in the .() notation. We can also apply the transformation function onto a variable of the by part. At this point, we can calculate average shopping amount per day by transforming the time variable into the date.

Similar to SQL operation, we can obtain the summary statistic of multiple variables by grouping them by the same variable, or furthermore, grouping them by multiple variables. In the next step, we showed you how to retrieve the summation of shopping amount and the number of unique user grouping by shopping date. At this point, we can use both the length and unique function to obtain the number of unique users. On the other hand, you can use a more advanced function, uniqueN, to get the number of unique users. We can then obtain the total expanding per day by each person. The only thing that we need to do is to group the column by both date and product variables.

Finally, we can arrange the data.table in order. To achieve this, we wrap the sorting key column with the order function, and retrieve the result sorted by the sum of the price in ascending order. However, in order to sort the data by the sum of the price in descending order, we can put a minus in front of the column name.

There's more...

With the use of the := notation on the j part, we can create a new column using the aggregated results:

```
> order.dt[,':='(Avg_P_By_U= mean(na.omit(Price)) ), User]
> head(order.dt,3)
                   Time Action         User     Product Quantity Price
Avg_P_By_U
1: 2015-07-01 00:00:01 order   U312622727 P0006944501    1  1069     934.00
2: 2015-07-01 00:00:03 order   U239012343 P0006018073    1  1680   1298.75
3: 2015-07-01 00:00:19 order U10007697373 P0002267974 1    285    247.00
```

Merging large datasets with a data.table

In previous recipes, we demonstrated how to manipulate and aggregate data with a data.table. In addition to performing data manipulation on a single table, we often need to import additional features or correlate data from other data sources. Therefore, we can join two or more tables into one. In this recipe, we introduce some methods that we can use to merge two data.table.

Getting ready

Ensure that you completed the *Enhancing a data.frame with a data.table* recipe to load `purchase_view.tab` and `purchase_order.tab` as both `data.frame` and `data.table` into your R environment.

How to do it...

Perform the following steps to merge two `data.table`:

1. First, we generate a `product.dt` data table by calculating the number of purchased items:

```
> product.dt <- order.dt[,.(Buy = length(Action)),by=Product]
> head(product.dt[order(-Buy)])
         Product Buy
1:     P0005772981 821
2:     P0024239865 729
3:     P0004607050 584
4:     P0003425855 552
5:     P0014252066 438
6: P0006587250014 357
```

2. Next, we can calculate the number of views by product, and place the calculation result into `view.dt`:

```
> view.dt <- purchase.dt[, .(n_views = length(User)), by=Product]
> head(view.dt[order(-n_views)])
       Product n_views
1: P0004880654    8336
2: P0004405251    7506
3: P0006437900    6298
4: P0006437863    5917
5: P0024239865    4799
6: P0004906226    4742
```

3. Now, we can use the `merge` function to join `product.dt` and `view.dt` by `Product` column:

```
> merged.dt <- merge(view.dt, product.dt, by="Product")
> head(merged.dt)
       Product n_views Buy
```

```
1: P0000005913      14   3
2: P0000006020      13   1
3: P0000006591       3   1
4: P0000007744      13   2
5: P0000008352       4   1
6: P0000010194       2   1
```

4. Alternatively, we can use `setkey` to create a key on `data.table` first, and then perform an inner join of `view.dt` and `product.dt`:

```
> setkey(view.dt,Product)
> setkey(product.dt,Product)

> inner_join.dt <- view.dt[product.dt, nomatch=0]
> head(inner_join.dt)
       Product n_views Buy
1: P0000005913      14   3
2: P0000006020      13   1
3: P0000006591       3   1
4: P0000007744      13   2
5: P0000008352       4   1
6: P0000010194       2   1
```

5. As for left join, we can also perform the left join with either the `merge` function or in `data.table` style:

```
> left_join <- merge(view.dt, product.dt, by="Product", all.
x=TRUE)
> left_join.dt <- product.dt[view.dt]
> head(left_join.dt)
       Product Buy n_views
1: P0000005681  NA       1
2: P0000005740  NA       1
3: P0000005751  NA       1
4: P0000005762  NA       1
5: P0000005854  NA       1
6: P0000005865  NA       2
```

6. Likewise, we can perform right join with either the `merge` function or in the `data.table` style:

```
> right_join <- merge(view.dt, product.dt, by="Product", all.
y=TRUE)
> right_join.dt <- view.dt[product.dt]
> head(right_join.dt)
         Product n_views Buy
1:      P0000005913      14   3
2:      P0000006020      13   1
3:      P0000006591       3   1
4:      P0000007744      13   2
5: P0000008190011      NA   1
6:      P0000008352       4   1
```

7. Finally, we demonstrate how full outer join works with the `merge` function:

```
> full_outer_join <- merge(view.dt, product.dt, by="Product",
all=TRUE)
> head(full_outer_join)
       Product n_views Buy
1: P0000005681       1  NA
2: P0000005740       1  NA
3: P0000005751       1  NA
4: P0000005762       1  NA
5: P0000005854       1  NA
6: P0000005865       2  NA
```

How it works...

In a relational database, it is quite common to use the join operation to combine information from different sources. In this recipe, we illustrated how join works in `data.table`. Here, we first create two `data.table` objects. The first `product.dt` data table holds the information of how many items are bought per product. The second `view.dt` data table holds the information of how many times each product was viewed. At this point, we can perform a `merge` operation to join these two `data.table` into a single relation that includes purchase and view number information.

Here, we first used a `merge` function to perform the inner join of `product.dt` and `view.dt`. In `data.frame`, we can also use the `merge` function to combine two `data.frames`, but the `merge` function here belongs to `data.table`. We can use the `merge` function to perform an inner join, a left join, a right join, or a full outer join on any of the two `data.table`. What makes the type of join different is that in the left join, we have to make `all.x=TRUE`, in the right join, we have to make `all.y=TRUE`, and for the outer join, we have to set `all=TRUE`.

Besides using the `merge` function, we can consider that different types of join are similar to taking the interaction of two `data.table`. Therefore, if both `data.table` are sorted using the same column name as key first, then the `join` action would take binary search practice to merge the `data.table` of two sources. Otherwise, the default operation takes a vector search to merge the two `data.table`. Here, we can use `setkey` to sort two `data.table` first, and then use `data.table` syntax to merge two different `data.table`.

There's more...

Once we have used the `setkey` function to arrange data in order by key columns, we can extract and search the data of a value group:

1. First, we can create a `data.table` named `dt`:

```
> dt <- data.table(product=c("A", "B", "B", "A", "A"),
price=c(100,200,300,100,200), quantity=c(2,2,1,3,2) )
> dt
   product price quantity
1:       A   100        2
2:       B   200        2
3:       B   300        1
4:       A   100        3
5:       A   200        2
```

2. Next, we use `setkey` to sort `data.table` by `product` and `price`:

```
> setkey(dt, product, price)
> dt
   product price quantity
1:       A   100        2
2:       A   100        3
3:       A   200        2
4:       B   200        2
5:       B   300        1
```

3. We use the `mult` argument to extract the record with the lowest price of product `A` and `B`:

```
> dt[c("A","B"),mult="first"]
   product price quantity
1:       A   100        2
2:       B   200        2
```

4. Alternatively, we can use the `mult` argument to extract the record with the highest price of product `A` and `B`:

```
> dt[c("A","B"),mult="last"]
   product price quantity
1:       A   200        2
2:       B   300        1
```

5. Or, we can summarize the price of product `B` with the `sum` function:

```
> dt["B",.(sum(price), sum(quantity))]
    V1 V2
1: 500  3
```

Subsetting and slicing data with dplyr

In this recipe, we will introduce how to use `dplyr` to manipulate data. We first cover the topic of how to use the `filter` and `slice` functions to subset and slice data.

Getting ready

Ensure that you completed the *Enhancing a data.frame with a data.table* recipe to load `purchase_view.tab` and `purchase_order.tab` as both `data.frame` and `data.table` into your R environment.

You also need to make sure that you have a version of R higher than 3.1.2 installed on your operating system.

How to do it...

Perform the following steps to subset and slice data with `dplyr`:

1. Let's first install and load the `dplyr` package:

```
> install.packages("dplyr")
> library(dplyr)
```

2. Next, we can filter data by quantity number with the `filter` function:

```
> quantity.over.3 <- filter(order.dt, Quantity >= 3)
> head(quantity.over.3, 3)
                   Time Action       User     Product Quantity
Price
1: 2015-07-01 00:39:22  order U465146448 P0006173160        3
1076
2: 2015-07-01 00:59:16  order U465150411 P0007371652        3
119
3: 2015-07-01 02:26:31  order U333330653 P0010814576        3
289
```

3. Furthermore, we can filter data with multiple conditions:

```
> quantity.price.filter <-  filter(order.dt, Quantity >= 3, Price
> 1000)
> head(quantity.price.filter, 3)
                   Time Action       User     Product Quantity
Price
1: 2015-07-01 00:39:22  order U465146448 P0006173160        3
1076
2: 2015-07-01 13:29:58  order U148075646 P0023947464        3
2980
3: 2015-07-01 13:41:43  order U373160329 P0006663650        3
3791
```

4. On the other hand, we can filter data with the | or operator:

```
> quantity.price.or.filter <-  filter(order.dt, Quantity >= 3 |
Price > 1000)
> head(quantity.price.or.filter, 3)
                   Time Action       User     Product Quantity
Price
1: 2015-07-01 00:00:01  order U312622727 P0006944501        1
1069
2: 2015-07-01 00:00:03  order U239012343 P0006018073        1
1680
3: 2015-07-01 00:01:48  order U451050374 P0004134266        1
1780
```

5. We can also filter multiple categories from the dataset with the `%in%` operator:

```
> quantity.price.in.filter <-  filter(order.dt, Product %in%
c('P0006944501', 'P0006018073'))
> head(quantity.price.in.filter, 3)
                  Time Action       User     Product Quantity
Price
1: 2015-07-01 00:00:01  order U312622727 P0006944501        1
1069
2: 2015-07-01 00:00:03  order U239012343 P0006018073        1
1680
3: 2015-07-01 00:03:45  order U311808547 P0006944501        1
1069
```

6. Besides using the `head` function, we can use a `slice` function to obtain the first three rows from the `data.table`:

```
> slice(order.dt,1:3)
                  Time Action        User     Product Quantity
Price
1: 2015-07-01 00:00:01  order  U312622727 P0006944501        1
1069
2: 2015-07-01 00:00:03  order  U239012343 P0006018073        1
1680
3: 2015-07-01 00:00:19  order U10007697373 P0002267974        1
285
```

7. Finally, we demonstrate how we can use the `slice` function and the `n()` notation to obtain the last two rows of `order.dt`:

```
> slice(order.dt,(n()-1):n())
                  Time Action      User        Product Quantity
Price
1: 2015-07-30 23:59:37  order U14085643 P0009890670031        1
245
2: 2015-07-30 23:59:56  order U33015290    P0014252055        1
1493
```

How it works...

To perform more advanced descriptive analysis, we must know how to use the `plyr` package to reshape data and obtain summary statistics. However, the insufficient performance and ambiguous syntax of `plyr` has many disadvantages. Here, we introduced how to use a faster and simpler package, `dplyr`, to make the descriptive analysis easier and more approachable. In this recipe, we introduced how to apply the `filter` and `slice` functions on a `data.table`. However, besides `data.table`, we can also apply the functions contained in `dplyr` on a `data.frame`.

We begin by introducing how to use the `filter` function to subset data. Like all `data.frame` and `data.table` operations, we can filter data with a single condition. Or, we can filter data with multiple conditions. It is also possible to use the `%in%` operator to filter data by multiple categories.

Finally, we introduced the `slice` function, which can slice data by row index. To obtain the last few records, we can use `n()` notation in the `slice` function.

There's more...

Besides conducting data analysis on `data.frame` and `data.table`, we can connect `dplyr` to SQLite and perform SQL operations:

1. First, install and load the `RSQLite` package:

   ```
   > install.packages("RSQLite")
   > library(RSQLite)
   ```

2. Next, copy the `orderdt` data table into `order.sqlite`:

   ```
   > orderdt <- fread("purchase_order.tab", header=TRUE, sep='\t')
   > my_db <- src_sqlite("order.sqlite", create = T)
   > order_sqlite <-copy_to(my_db, orderdt, temporary = FALSE)
   ```

3. At this point, we can perform SQL operations on `orderdt` in `order.sqlite`:

   ```
   > tbl(my_db, sql("SELECT Product FROM orderdt"))
   ```

Sampling data with dplyr

As a single machine cannot efficiently process big data problems, a practical approach is to take samples that we can effectively use to draw conclusions. Here, we will show you how to use `dplyr` to sample from data.

Getting ready

Ensure that you installed and loaded `data.table` in your R session. You also need to complete the *Enhancing a data.frame with a data.table* recipe to load `purchase_view.tab` and `purchase_order.tab` as both `data.frame` and `data.table` into your R environment.

How to do it...

Perform the following steps to sample data with dplyr:

1. First, we can sample six rows from the data:

```
> set.seed(123)
> sample_n(order.dt, 6, replace = TRUE)
                  Time Action       User     Product Quantity
Price
1: 2015-07-10 09:22:37  order  U46651253 P0004306934        1
750
2: 2015-07-25 21:42:34  order U232322558 P0014273055        1
3688
3: 2015-07-13 22:55:33  order  U14804834 P0013147260        1
32900
4: 2015-07-29 08:48:18  order U364096419 P0003425855        1
999
5: 2015-07-30 12:53:09  order U480262356 P0024027371        1
149
6: 2015-07-02 06:28:22  order U465433065 P0006173543        1
830
```

2. Then, we can obtain only 10% of the data with the sample_frac function:

```
> sample.dt <- sample_frac(order.dt, 0.1, replace = TRUE)
> nrow(sample.dt)
[1] 5477
```

How it works...

Data sampling can randomly subset rows from original data, and we can use the sampled data to perform statistical inference. There are two methods that we can use to perform data sampling with dplyr: we can use sample_n to randomly select n rows from order.dt. If we want to sample data with replacement, we can specify TRUE at replace argument; and we can use sample_frac to sample data of a particular percentage from the original dataset. In this case, we sample 10% of the data with replacement from the original dataset. Then, we use nrow to obtain the number of rows of the sample. The sample contains 5,477 rows, which is around 10% of the original dataset.

There's more...

We can also specify the sample weighting in `sample` argument of either the `sample_n` or `sampel_frac` function. However, the length of weight parameter input should match the row number of the input dataset:

```
> df <- data.frame(a=seq(1,10,1), b = c(rep(1,8),rep(2,2)))
> set.seed(123)
> sample_n(df, 5, weight=df$b)
  a b
9 9 2
4 4 1
5 5 1
1 1 1
2 2 1
```

Selecting columns with dplyr

In the last recipe, we introduced how to use the `filter` and `slice` functions to subset and slice data by rows. In this recipe, we will present how to select particular columns from the dataset using the `select` function.

Getting ready

Ensure that you completed the *Enhancing a data.frame with a data.table* recipe to load `purchase_view.tab` and `purchase_order.tab` as both `data.frame` and `data.table` into your R environment.

You also need to make sure that you have a version of R higher than 3.1.2 installed on your operating system.

How to do it...

Perform the following steps to select columns from the dataset:

1. First, let's select the `Quantity` and `Price` columns from the dataset:

   ```
   > select.quantity.price <-  select(order.dt, Quantity, Price)
   > head(select.quantity.price, 3)
      Quantity Price
   ```

```
1:          1  1069
2:          1  1680
3:          1   285
```

2. Alternatively, we can rule out the `Price` column by placing a minus sign in front:

```
> select.not.price <-  select(order.dt, -Price)
> head(select.not.price, 3)
                  Time Action         User
1: 2015-07-01 00:00:01  order   U312622727
2: 2015-07-01 00:00:03  order   U239012343
3: 2015-07-01 00:00:19  order U10007697373
       Product Quantity
1: P0006944501        1
2: P0006018073        1
3: P0002267974        1
```

3. However, if we want to select all columns, we can place `everything()` within the `select` function:

```
> select.everything <- select(order.dt, everything())
> head(select.everything,3)
                  Time Action         User     Product Quantity
Price
1: 2015-07-01 00:00:01  order   U312622727 P0006944501        1
1069
2: 2015-07-01 00:00:03  order   U239012343 P0006018073        1
1680
3: 2015-07-01 00:00:19  order U10007697373 P0002267974        1
285
```

4. We can also select multiple columns in sequence:

```
> select.from.user.to.quantity <-  select(order.dt, User:Quantity)
> head(select.from.user.to.quantity, 3)
           User     Product Quantity
1:   U312622727 P0006944501        1
2:   U239012343 P0006018073        1
3: U10007697373 P0002267974        1
```

5. Moving on, we can select columns containing certain words or phrases:

```
> select.contains.p <- select(order.dt, contains('P') )
> head(select.contains.p,3)
       Product Price
1: P0006944501  1069
2: P0006018073  1680
3: P0002267974   285
```

6. Finally, we demonstrate how to use the `select` and `filter` functions together:

```
> select.p.price.over.1000 <- select(filter(order.dt, Price >=
1000 ), contains('P') )
> head(select.p.price.over.1000, 3)
       Product Price
1: P0006944501  1069
2: P0006018073  1680
3: P0004134266  1780
```

How it works...

Besides selecting individual rows from the dataset, we can use the `select` function in `dplyr` to select a single or multiple columns from the dataset. In this recipe, we begin by selecting both the `Quantity` and `Price` columns from `order.dt`. Alternatively, we can obtain data without a particular column by placing a minus sign ahead of the column. To select every column from the dataset, simply place `everything()` within the `select` function.

Moving on, to select columns from `User` to `Quantity`, we can put `User` and `Quantity` in a sequence form. As a result, we can obtain a subset with `User`, `Product`, and `Quantity` in return.

It is also possible to select columns containing specific words using the `contains` function. Here, we can obtain a subset with column names containing `P` in return.

Besides using `select` or `filter` solely, we can wrap them up in a function call. Therefore, we can filter `order.dt` by `quantity` first, and then select columns with a column name containing the `P` character.

There's more...

To select multiple columns with the column name starting with `a` and in a range sequence, we can use `num_range` inside the `select` function:

Here, let's use the `num_range` function to select columns with a name equal to `a1` and `a2`:

```
> set.seed(123)
> df <- data.frame(a1=rnorm(3), a2=rnorm(3), b1=1, b2=NA, b3="str")
> select(df, num_range("a", 1:2))
          a1         a2
1 -0.5604756 0.07050839
2 -0.2301775 0.12928774
3  1.5587083 1.71506499
```

Chaining operations in dplyr

To perform multiple operations on data using `dplyr`, we can wrap up the function calls into a larger function call. Or, we can use the `%>%` chaining operator to chain operations instead. In this recipe, we introduced how to chain operations when using `dplyr`.

Getting ready

Ensure that you completed the *Enhancing a data.frame with a data.table* recipe to load `purchase_view.tab` and `purchase_order.tab` as both `data.frame` and `data.table` into your R environment.

How to do it...

Perform the following steps to subset and slice data with `dplyr`:

1. In R, to sum up a sequence from `1` to `10`, we can wrap the series of `1` to `10` with the `sum` function:

   ```
   > sum(1:10)
   [1] 55
   ```

2. Alternatively, we can use a chaining operator to chain operations:

   ```
   > 1:10 %>% sum()
   [1] 55
   ```

3. To select and filter data with `dplyr`, we can wrap the filtered data in a `select` function:

   ```
   > select.p.price.over.1000 <- select(filter(order.dt, Price >=
   1000 ), contains('P') )
   > head(select.p.price.over.1000, 3)
          Product Price
   ```

```
1: P0006944501  1069
2: P0006018073  1680
3: P0004134266  1780
```

4. Or, we can use the `%>%` operator to chain the operation:

```
> chain.operations <- select(order.dt, contains('P') ) %>%
filter(Price >= 1000 )
> head(chain.operations,3)
        Product Price
1: P0006944501  1069
2: P0006018073  1680
3: P0004134266  1780
```

5. Furthermore, it is possible to continue chaining operations with the %>% operator:

```
> select(order.dt, contains('P') ) %>% filter(Price >= 1000 ) %>%
select(Price) %>% sum()
[1] 89959970
```

How it works...

In the last step of the previous recipe, we demonstrated how to select and filter a dataset in a function call. However, when we need to perform multiple operations, the tangled function calls become complex. To create a more readable syntax, we can use the `%>%` chaining operator from `magrittr` to chain all `dplyr` operations.

First, we illustrated how to apply the `sum` function onto a sequence from `1` to `10`. Similar to all other function calls, we just need to surround the series with the `sum` function. Besides wrapping the series, we can chain the operation with the `%>%` chaining operator. At this point, we can now use the chaining operator to chain multiple function calls made by `dplyr`. Finally, we demonstrated how to chain a data filtering and column selection operation with the chaining operator. It is possible to continue chaining operations if required.

There's more...

Instead of using the `%>%` operator, we can use `%.%` as a chaining operator. However, as the `%.%` operator is marked as deprecated, we should switch the use of `%.%` to `%>%` for future package compatibility. The following command can be used to view the chain operator document:

```
> ?"%.%"
```

Arranging rows with dplyr

Arranging rows in order may help us rank data by value or gain a more structured view of data in the same category. In this recipe, we will introduce how to arrange rows with `dplyr`.

Getting ready

Ensure that you completed the *Enhancing a data.frame with a data.table* recipe to load `purchase_view.tab` and `purchase_order.tab` as both `data.frame` and `data.table` into your R environment.

How to do it...

Perform the following steps to arrange data with `dplyr`:

1. To arrange data by price, pass `Price` to the `arrange` function:

```
> order.dt %>% arrange(Price) %>% head(3)
                  Time Action        User     Product Quantity
Price
1: 2015-07-30 23:54:01  order  U166076125 P0003659246        3
10
2: 2015-07-03 09:19:24  order U1012162712 P0009305785        1
15
3: 2015-07-07 23:45:09  order  U423898356 P0023454992        1
15
```

2. We can also arrange rows by `price` in descending order:

```
> order.dt %>% arrange(desc(Price)) %>% head(3)
                  Time Action         User     Product Quantity
Price
1: 2015-07-03 16:01:57  order U10120098943 P0000143500        1
438888
2: 2015-07-16 17:05:30  order U10062834851 P0000143511        1
438888
3: 2015-07-17 17:57:13  order U10120098943 P0000143511        1
438888
```

3. It is possible to sort data by two column variables:

```
> order.dt %>% arrange(Price, desc(Quantity)) %>% head(3)
                  Time Action       User     Product Quantity
Price
```

```
1: 2015-07-30 23:54:01  order U166076125 P0003659246       3
10
2: 2015-07-12 01:32:29  order U465000350 P0002976595      24
15
3: 2015-07-12 01:50:35  order U464992246 P0002976595      24
15
```

How it works...

To sort data by column in `dplyr`, we just needs to pass the sorted key name into the `arrange` function. Here, we demonstrated that we can sort data by price in ascending order. By applying the `desc` function on price, we can arrange data by price in descending order. Moreover, we can sort data by multiple keys in an `arrange` function.

There's more...

In the `arrange` function, we can use the `desc` function to sort data in descending order. However, we can use the `desc` function to transform a vector into reverse order by placing a minus sign in front:

```
> s <- c(1,3,2,4,6)
> desc(s)
[1] -1 -3 -2 -4 -6
```

Eliminating duplicated rows with dplyr

To avoid counting duplicate rows, we can use the `distinct` operation in SQL. In `dplyr`, we can also eliminate duplicated rows from a given dataset.

Getting ready

Ensure that you completed the *Enhancing a data.frame with a data.table* recipe to load `purchase_view.tab` and `purchase_order.tab` as both `data.frame` and `data.table` into your R environment.

How to do it...

Perform the following steps to distinct duplicate rows with `dplyr`:

1. First, we illustrate how to obtain unique products from the dataset:

```
> order.dt %>% select(Product) %>% distinct() %>% head(3)
       Product
```

```
1: P0006944501
2: P0006018073
3: P0002267974
```

2. We can also `distinct` duplicated rows containing multiple columns:

```
> distinct.product.user.dt <- order.dt %>% select(Product, User)
%>% distinct()
> head(distinct.product.user.dt, 3)
       Product          User
1: P0006944501   U312622727
2: P0006018073   U239012343
3: P0002267974 U10007697373
```

3. At this point, let's compare the number of rows before and after performing data distinction:

```
> nrow(order.dt)
[1] 54772

> nrow(distinct.product.user.dt)
[1] 50381
```

How it works...

To count unique items or users from our data, we need to remove duplicated rows. In `dplyr`, we can use the `distinct` function to eliminate duplicated rows. First, we select the `product` column from the dataset, and then exclude duplicated products. This should return a collection of unique products. We then demonstrate that we removed duplicated rows of multiple columns. We first select both `Product` and `User` columns from the dataset and then perform `distinct` on the returned collection.

Finally, we can use `nrows` to count the rows of data before and after making data distinction. Here, we find that the number of rows of unique combinations of product and user is less than the original number of rows of the dataset.

There's more...

We can also use the `unique` function in the base package to obtain distinct values from the dataset:

```
> data.frame(a=c(1,2,1,1,2)) %>% select(a) %>% unique()
  a
1 1
2 2
```

Adding new columns with dplyr

Besides performing data manipulation on existing columns, there are situations where a user may need to create a new column for more advanced analysis. In this recipe, we will introduce how to add a new column using `dplyr`.

Getting ready

Ensure that you completed the *Enhancing a data.frame with a data.table* recipe to load `purchase_view.tab` and `purchase_order.tab` as both `data.frame` and `data.table` into your R environment.

How to do it...

Perform the following steps to add a new column to an existing dataset:

1. First, we calculate the average price of each purchase and add the created result as a column back to the original dataset:

```
> order.dt %>%  select(Quantity, Price) %>% mutate(avg_price=
Price/Quantity) %>% head()
   Quantity Price avg_price
1:        1  1069      1069
2:        1  1680      1680
3:        1   285       285
4:        1   550       550
5:        1   249       249
6:        1  1780      1780
```

2. Alternatively, we can use the `transmute` function to create a new column on `order.dt`, and drop the rest of variables:

```
> transmute(order.dt, Avg_Price= Price/Quantity) %>% head()
   Avg_Price
1:      1069
2:      1680
3:       285
4:       550
5:       249
6:      1780
```

How it works...

This recipe shows you how to add a new column to an existing dataset. First, we demonstrated that we can create a new column named `avg_price` by dividing the price by the quantity per shopping transaction. Here, we use the `mutate` function in `dplyr` to create the new column based on the calculation. Besides keeping original variables, we can choose to drop existing variables using the `transmute` function.

There's more...

We can also make use of the `transform` function in the base package to add a new column from the dataset:

```
> order.dt %>% select(Quantity, Price) %>% transform(Avg_Price= Price/
Quantity) %>% head()
   Quantity Price Avg_Price
1:        1  1069      1069
2:        1  1680      1680
3:        1   285       285
4:        1   550       550
5:        1   249       249
6:        1  1780      1780
```

Summarizing data with dplyr

Besides manipulating a dataset, the most important part of `dplyr` is that we can easily obtain summary statistics from the data. In SQL operation, we can use the `GROUP BY` function for this purpose, and it is possible to perform a similar operation in `dplyr`. In this recipe, we will show you how to summarize data with `dplyr`.

Getting ready

Ensure that you completed *Enhancing a data.frame with a data.table* recipe to load `purchase_view.tab` and `purchase_order.tab` as both `data.frame` and `data.table` into your R environment.

How to do it...

Perform the following steps to summarize data with `dplyr`:

1. First, use the `summarize` and `group_by` functions to obtain the total purchase amount of each product:

```
> order.dt %>%
+    select(User, Price) %>%
+    group_by(User) %>%
+    summarise(sum(Price)) %>%
+    head()
          User sum(Price)
1   U312622727       1868
2   U239012343       5195
3 U10007697373        494
4   U296328517       1490
5   U300884570        249
6   U451050374       1780
```

2. To obtain summary statistics of multiple columns, we can use the `summarize_each` function:

```
> order.dt %>%
+    select(User, Price, Quantity) %>%
+    filter(! is.na(Price)) %>%
+    group_by(User) %>%
+    summarise_each(funs(sum), Price, Quantity) %>%
+    head()
          User Price Quantity
1   U312622727  1868        2
2   U239012343  5195        4
3 U10007697373   494        3
4   U296328517  1490        2
5   U300884570   249        1
6   U451050374  1780        1
```

3. Furthermore, we can use `summarize_each` to obtain the minimum and maximum price spent by each user:

```
> order.dt %>%
+   select(User, Price) %>%
+   filter(! is.na(Price)) %>%
+   group_by(User) %>%
+   summarise_each(funs(max(., na.rm=TRUE), min(., na.rm=TRUE)), Price) %>%
+   head()
         User  max  min
1   U312622727 1069  799
2   U239012343 2999  200
3 U10007697373  285  209
4   U296328517  940  550
5   U300884570  249  249
6   U451050374 1780 1780
```

4. Additionally, we can obtain the amount of each product bought using the `n()` notation:

```
> purchase.dt %>%
+   select(User, Product) %>%
+   group_by(Product) %>%
+   summarise_each(funs(n())) %>%
+   head()
     Product User
1 P0023468384    1
2 P0018926456  198
3 P0000063593   84
4 P0005541535    5
5 P0022135540    9
6 P0001249080  173
```

5. We can use the `n_distinct` function to count the amount of products bought by unique users:

```
> purchase.dt %>%
+   select(User, Product) %>%
+   group_by(Product) %>%
+   summarise_each(funs(n_distinct(User))) %>%
+   head()
```

```
      Product User
1 P0023468384    1
2 P0018926456  104
3 P0000063593   51
4 P0005541535    1
5 P0022135540    7
6 P0001249080  155
```

How it works...

A great advantage of `dplyr` is that it provides common data manipulation functions, such as `select` and `filter`. Using the chaining operator, we can perform a SQL-like operation to summarize data. In this recipe, we introduced how to obtain summary statistics with some `dplyr` functions.

In the first step of this recipe, we presented how to obtain the sum of purchase amount per user using the `summarize` and `group_by` functions. Furthermore, to obtain the summary statistics of multiple columns, we can use `summarize_each` as the summary function by passing `funs(sum)` as the first argument to the `summarize_each` function. We can use `summarize_each` to calculate both the total purchase price and purchased quantity per user. To perform different summarization operations on the same column, we just need to pass all different summarization functions into the `funs` function. In step 3 of this recipe, we retrieved both the minimum and maximum paid price by each user with the `summarize_each` function.

Finally, we can use the `n()` function to count the amount of each product purchased. If we want to obtain products bought by the number of unique users, we can pass `n_distinct` instead.

There's more...

If we group data by a particular variable using the `group_by` function, we may not sort data freely by any given column due to the existing grouping. Therefore, we can use the `ungroup` function to remove existing grouping, and we can then sort summarized data by any variable:

1. First, generate a sample `data.frame` with user, product, and price information:

```
> sample.df <-data.frame(user    =c("U1", "U1", "U1", "U3"),
+                        product=c("A" , "B" , "A" , "B" ),
+                        price  =c(200 ,  100,  300 , 300 ))
```

2. Next, obtain the sum of purchase amount by user and product, and then sort the data by sum of the purchase amount:

```
> sample.df %>%
+   group_by(user,product) %>%
+   summarise(price_sum = sum(price)) %>%
+   arrange(price_sum)
Source: local data frame [3 x 3]
Groups: user [2]

    user product price_sum
  (fctr)  (fctr)     (dbl)
1     U1       B       100
2     U1       A       500
3     U3       B       300
```

3. We can check what happens if we use the `ungroup` function to remove existing grouping:

```
> sample.df %>%
+   group_by(user,product) %>%
+   summarise(price_sum = sum(price)) %>%
+   ungroup() %>%
+   arrange(price_sum)
Source: local data frame [3 x 3]

    user product price_sum
  (fctr)  (fctr)     (dbl)
1     U1       B       100
2     U3       B       300
3     U1       A       500
```

Merging data with dplyr

In a SQL operation, we can perform a join operation to combine two different datasets. In `dplyr`, we have the same join operation that enables us to merge data easily. In this recipe, we explain how join works in `dplyr`.

Getting ready

Ensure that you completed the *Enhancing a data.frame with a data.table* recipe to load `purchase_view.tab` and `purchase_order.tab` as both `data.frame` and `data.table` into your R environment.

How to do it...

Perform the following steps to merge data with `dplyr`:

1. First, we generate a `product.dt` data table by calculating the amount of purchased items:

```
> product.dt <- order.dt[,.(Buy = length(Action)),by=Product]
> head(product.dt[order(-Buy)])
          Product Buy
1:     P0005772981 821
2:     P0024239865 729
3:     P0004607050 584
4:     P0003425855 552
5:     P0014252066 438
6: P0006587250014 357
```

2. Next, we can calculate the number of views by product, and put the calculation result into `view.dt`:

```
> view.dt <- purchase.dt[, .(n_views = length(User)), by=Product]
> head(view.dt[order(-n_views)])
       Product n_views
1: P0004880654    8336
2: P0004405251    7506
3: P0006437900    6298
4: P0006437863    5917
5: P0024239865    4799
6: P0004906226    4742
```

3. We can use the `inner_join` function to join `product.dt` and `view.dt` by the `Product` column:

```
> merged.dt <- inner_join(view.dt, product.dt, by="Product")
> head(merged.dt)
       Product n_views Buy
```

```
1 P0000005913      14   3
2 P0000006020      13   1
3 P0000006591       3   1
4 P0000007744      13   2
5 P0000008352       4   1
6 P0000010194       2   1
```

4. Alternatively, we can also perform left join with the `left_join` function:

```
> left_join.dt <- left_join(view.dt, product.dt, by="Product")
> head(left_join.dt)
      Product n_views Buy
1 P0000005681       1  NA
2 P0000005740       1  NA
3 P0000005751       1  NA
4 P0000005762       1  NA
5 P0000005854       1  NA
6 P0000005865       2  NA
```

5. Likewise, we can perform right join with the `right_join` function:

```
> right_join.dt <- right_join(view.dt, product.dt, by="Product")
> head(right_join.dt)
         Product n_views Buy
1    P0006944501      19   8
2    P0006018073     101   4
3    P0002267974      65  10
4    P0016144236      NA   1
5 P0014516980122      NA   1
6    P0004134266    1605 313
```

6. Finally, we demonstrate how full join works using the `full_join` function:

```
> full_join.dt <- full_join(view.dt, product.dt, by="Product")
> head(full_join.dt)
      Product n_views Buy
1 P0023468384       1  NA
2 P0018926456     198  NA
3 P0000063593      84  NA
4 P0005541535       5  NA
5 P0022135540       9  NA
6 P0001249080     173   1
```

How it works...

In this recipe, we illustrated how join works in `dplyr`. Here, we first created two `data.table` objects. The first `product.dt` data table holds the information of how many items are bought per product. The second `view.dt` data table contains information of how many times each product was viewed. At this point, we can perform a merge operation to join these two `data.tables` into a single relation with both purchase amount and view number.

We can now use the `inner_join` function to perform inner join of `product.dt` and `view.dt`. Next, we can perform a left join, right join, or full join on any of two `data.table` or `data.frame`.

There's more...

Besides joining tables on matched value, we can perform an `anti_join` to retrieve all rows from the first table that do not have any matching values in the second table:

```
> first.df  <- data.frame(a=c("A","B"), b=c(10,20))
> second.df <- data.frame(a=c("B","C"), c=c("P","A"))
> anti_join(second.df,first.df, by="a")
  a c
1 C A
> anti_join(first.df,second.df, by="a")
  a  b
1 A 10
```

5
Visualizing Data with ggplot2

This chapter covers the following topics:

- Creating basic plots with `ggplot2`
- Changing aesthetics mapping
- Introducing geometric objects
- Performing transformations
- Adjusting plot scales
- Faceting
- Adjusting themes
- Combining plots
- Creating maps

Introduction

When analyzing data, our primary goal is to efficiently and precisely deliver the findings to our audience. An easy way to present data is to display it in a table format. However, for larger datasets, it becomes challenging to visualize data in this format.

For example, the following table contains regional sales data:

Region	**Jul-12**	**Aug-12**	**Sep-12**	**Oct-12**	**Nov-12**	**Dec-12**
Alberta	22484.08	65244.19	15946.36	38593.39	34123.56	34753.98
British Columbia	23785.05	51533.77	44508.33	57687.6	19308.37	43234.77

In table format, it is hard to see which region's sales performed best. Thus, to make the data easier to read, it may be preferable to present the data in a chart or other graphical format. The following figure is a graph of the data from the table, which makes it much easier to determine which region performed best each month in terms of sales:

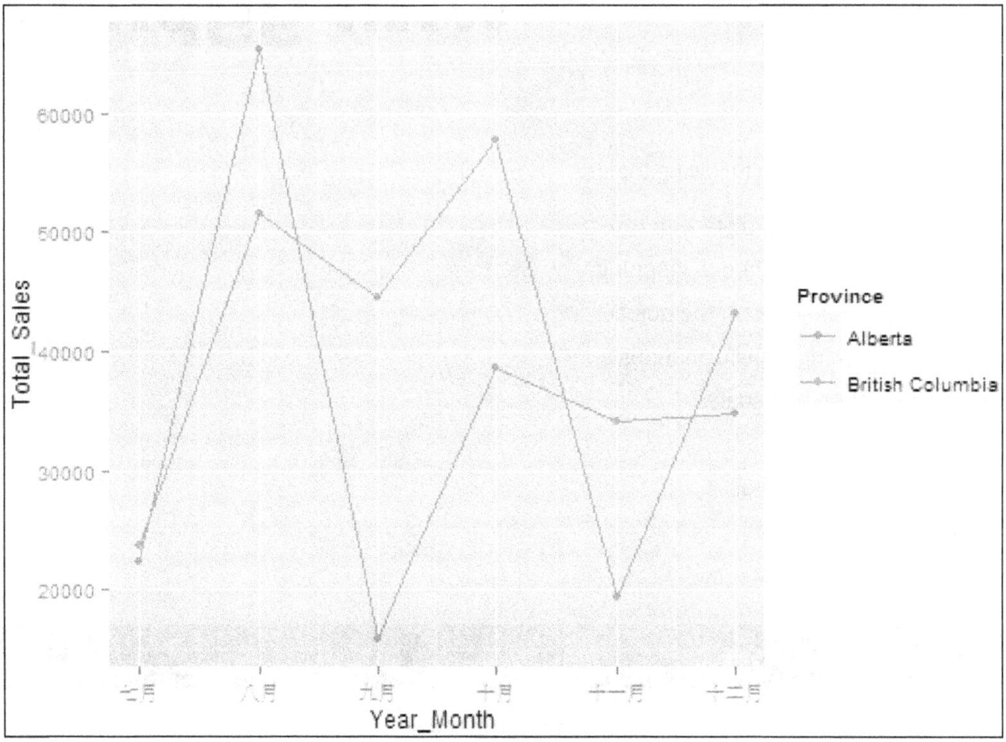

Figure 1: Sales amount by region

One of the most attractive features of R is that it already has many visualization packages. With its `graphics` package, one can easily make a line chart, bar graph, and more, from a given dataset with the generic `plot` function. The `plot` function can quickly present the fundamental characteristics of the dataset. However, it is quite inflexible when more sophisticated plots are required. Another great graphics package is `lattice`, which enables the user to chart more advanced charts than the `graphics` package. There are also other functions that permit the user to make more sophisticated plots. For example, if you would like to plot *Figure 1* with the lattice package, you would need to use the `xyplot` function:

```
> library(lattice)
> xyplot( Alberta + British.Columbia ~ Date, type='b',
+         data= sum_price_by_province,
+         xlab = "Year_Month",
+         ylab = "Total_Sales",
+         key=list(space="right",
+                  lines=list(col=c("red", "blue"), lwd=3),
+                  text=list(c("Alberta", "British Columbia") )
+         ),
+         col= c("red", 'blue')
+         )
```

To chart a plot with more flexible grammar syntax, one can use the `ggplot2` package. Here, the prefix `gg` of `ggplot2` originates from *The Grammar of Graphics* (Wilkinson, 2005). The main feature of this package is that it does not provide a particular plotting method for different types of plots. However, one can use fundamental elements (such as lines and polygons) to compose a complicated chart with simple grammar.

In this chapter, we introduce how to plot various charts with `ggplot2`. We start by creating a basic plot with `ggplot2`. We then cover the fundamental topics of using geometric objects and setting aesthetic mapping. Next, we begin to create a more sophisticated graph by performing data transformations, configuring scales, making subplots with a particular variable, adding a theme, and integrating different subplots into one. Finally, we will teach you how to create a map with assigned geographic features.

Creating basic plots with ggplot2

In this recipe, we demonstrate how to use *The Grammar of Graphics* to construct our very first ggplot2 chart with the superstore sales dataset.

Getting ready

First, download the superstore_sales.csv dataset from the https://github.com/ywchiu/rcookbook/raw/master/chapter7/superstore_sales.csv GitHub link.

Next, you can use the following code to download the CSV file to your working directory:

```
> download.file('https://github.com/ywchiu/rcookbook/raw/
master/chapter7/superstore_sales.csv', 'superstore_sales.csv')
```

You will also need to load the dplyr package to manipulate the superstore_sales dataset.

How to do it...

Please perform the following steps to create a basic chart with ggplot2:

1. First, install and load the ggplot2 package:

   ```
   > install.packages("ggplot2")
   > library(ggplot2)
   ```

2. Import superstore_sales.csv into an R session:

   ```
   > superstore <-read.csv('superstore_sales.csv', header=TRUE)
   > superstore$Order.Date <- as.Date(superstore$Order.Date)
   > str(superstore)
   'data.frame':8399 obs. of  9 variables:
    $ Order.ID        : int  3 293 293 483 515 515 613 613 643 678
   ...
    $ Order.Date      : Date, format: "2010-10-13" "2012-10-01"
   "2012-10-01" ...
    $ Order.Quantity  : int  6 49 27 30 19 21 12 22 21 44 ...
    $ Sales           : num  262 10123 245 4966 394 ...
    $ Profit          : num  -213.2 457.8 46.7 1199 30.9 ...
    $ Unit.Price      : num  38.94 208.16 8.69 195.99 21.78 ...
   ```

```
 $ Province         : Factor w/ 13 levels "Alberta","British
Columbia",..: 8 8 8 8 8 8 8 8 8 8 ...
 $ Customer.Segment: Factor w/ 4 levels "Consumer","Corporate",..:
4 1 1 2 1 1 2 2 2 3 ...
 $ Product.Category: Factor w/ 3 levels "Furniture","Office
Supplies",..: 2 2 2 3 2 1 2 2 2 2 ...
```

3. Next, we can summarize the sales amount by year, month, and province:

```
> sum_price_by_province <- superstore %>%
+   filter(Order.Date > '2012-01-01') %>%
+   select(Sales, Province, Order.Date) %>%
+   group_by(Year_Month = as.Date(strftime(Order.
Date,"%Y/%m/01")), Province) %>%
+   summarise(Total_Sales = sum(Sales))

> head(sum_price_by_province)
Source: local data frame [6 x 3]
Groups: Year_Month [1]

  Year_Month              Province Total_Sales
      (date)                (fctr)       (dbl)
1 2012-01-01               Alberta   45517.604
2 2012-01-01      British Columbia   17429.967
3 2012-01-01              Manitoba   51071.193
4 2012-01-01         New Brunswick    8085.410
5 2012-01-01          Newfoundland    1666.000
6 2012-01-01 Northwest Territories    1845.185
```

4. Subset the sales data in `British Columbia` and `Alberta` from 2012-07-01:

```
> sample_sum <- sum_price_by_province %>% filter(Year_Month >
'2012-07-01', Province %in% c('Alberta', 'British Columbia' ) )
```

5. Moving on, we can start creating a canvas by mapping `Year_Month` to the *x* axis, `Total_Sales` to the *y* axis, and `Province` to `colour`:

```
> g <- ggplot(data=sample_sum, mapping=aes(x=Year_Month, y=Total_
Sales, colour=Province)) + ggtitle("Pure Canvas")
> g
```

6. Add point geometry onto the canvas:

```
> g <- g + geom_point()+ ggtitle("With Point Geometry")
> g
```

7. Add line geometry onto the canvas:

```
> g <- g + geom_line()+ ggtitle("With Line Geometry")
> g
```

8. Add a label and title to the canvas:

```
> g <- g + xlab("Year Month") + ylab("Sale Amount") +
ggtitle("Sale Amount By Region")
> g
```

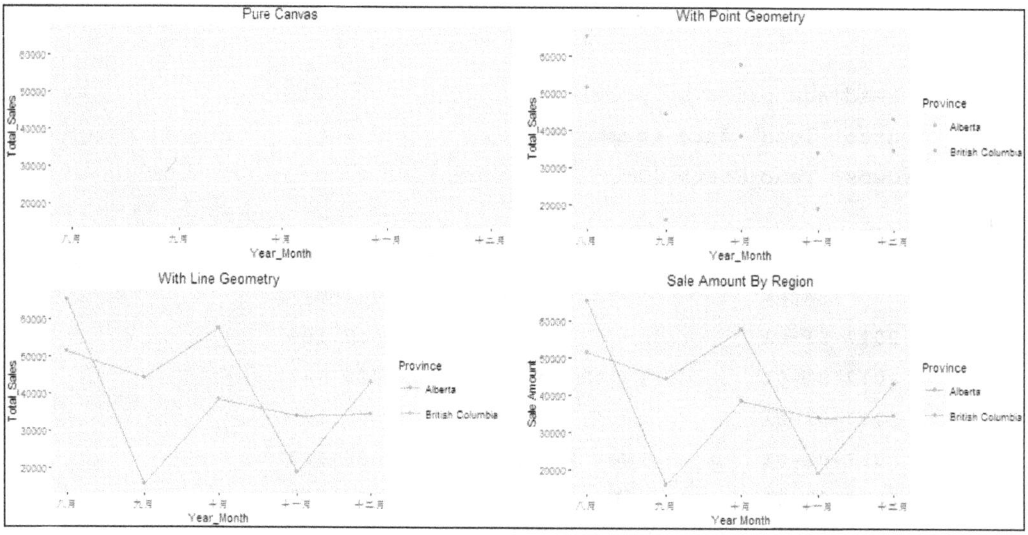

Figure 2: Plotting sales amount by region with ggplot2

How it works...

In `ggplot2`, the data is charted by mapping the element from mathematical space to physical space. Therefore, we can use simple elements to build a complicated figure. To put it simply, plotting in `ggplot2` is similar to using a painting tool. First, we need to create a canvas. We can then add a layer with geometric objects and an aesthetic mapping layer onto the canvas with + notation.

In this recipe, we first install and load the `ggplot2` package, and then import `superstore_sales.csv` into an R session. Next, we use `dplyr` to summarize superstore sales data by year, month, and province. We can then obtain the sales data in `British Columbia` and `Alberta` from 2012-07-01 with the `filter` function.

As the data is now ready, we can create our first canvas with the `ggplot2` function. Here, to assign an *x* axis and *y* axis and separate the data by province, we configure aesthetic mapping by setting `x=Year_Month`, `y=Total_Sales`, and `colour=Province`. Then, when we call the first plot, we find a canvas marked with an *x* axis and *y* axis.

Moving on, we can add layers with a geometric object onto the canvas with the + notation. In this example, we first add point geometry onto the canvas. We can now format the plot as a scatterplot, with the red point indicating Alberta's sales figures and the green point representing British Columbia's sales results. Furthermore, we can continue adding line geometry onto the canvas. This allows us to determine the sales trend of each region from the generated line chart. Last, we can also add an *x* label, a *y* label, and a title to the canvas with the + notation.

There's more...

It is important to understand the grammar of graphics before using the `ggplot` function to create graphs. As an alternative, one can use `qplot` (quick plot) to make a `ggplot2` chart. The `qplot` function is similar to the `plot` function, which hides many details, and one can use the consistent calling scheme, `qplot`, to create various types of graphs:

```
> qplot(Year_Month, Total_Sales, data=sample_sum, colour=Province,
geom=c("line", "point"), main="qplot example")
```

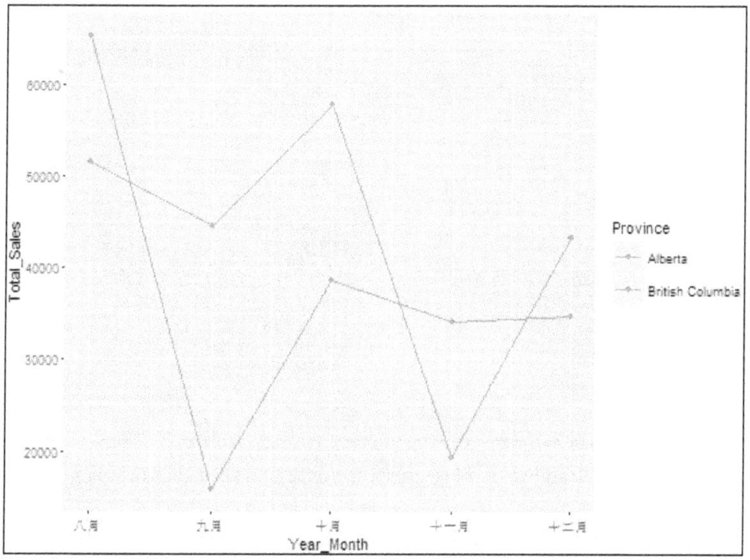

Figure 3: A qplot example

Changing aesthetics mapping

Aesthetics mapping describes how data variables are mapped to the visual property of a plot. In this recipe, we discuss how to modify aesthetics mapping on geometric objects.

Getting ready

Ensure you have installed and loaded `ggplot2` into your R session. Also, you need to complete the previous steps by storing `sample_sum` in your R environment.

How to do it...

Please perform the following steps to add aesthetics to the plot:

1. First, create a scatterplot by mapping `Year_Month` to the *x* axis, `Total_Sales` to the *y* axis, and `Province` to color:

```
> g <- ggplot(data=sample_sum, mapping=aes(x=Year_Month, y=Total_
Sales, colour=Province)) + ggtitle('With geom_point')
> g + geom_point()
```

2. Set the aesthetics mapping on the geometric object:

```
> g2 <- ggplot(data=sample_sum) + geom_point(mapping=aes(x=Year_
Month, y=Total_Sales, colour=Province)) + ggtitle('With Aesthetics
Mapping')
> g2
```

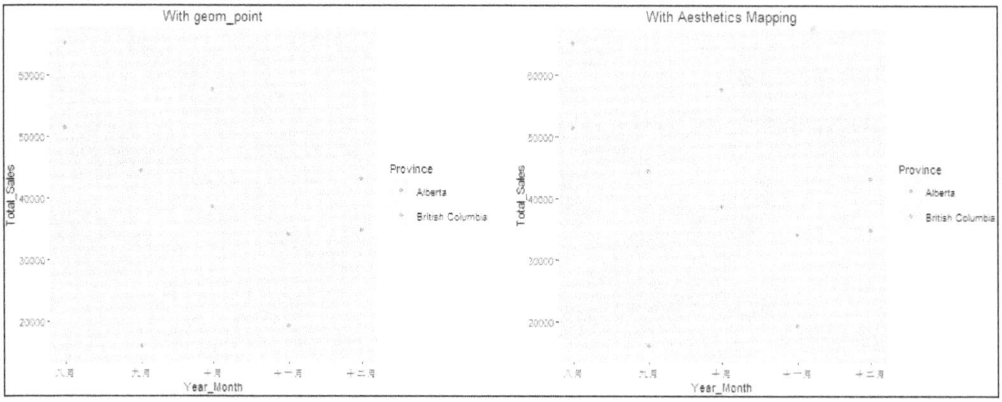

Figure 4: Scatterplots using `geom_point` and aesthetics mapping

3. Adjust the point size of the scatterplot by setting the size property in the geometric object:

```
> g + geom_point(aes(size=5)) + ggtitle('Adjust Point Size')
```

4. Moving on, we can adjust the aesthetics property by geometric object:

```
> g + geom_point(size=5, colour="blue") + geom_line()
```

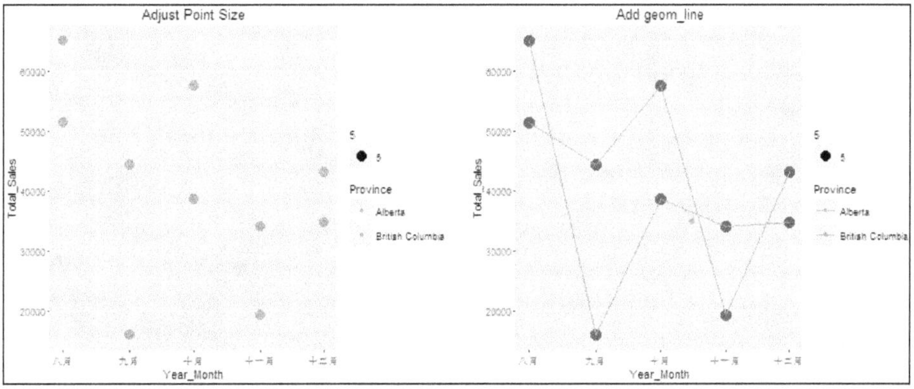

Figure 5: Adjust the point size and add an additional geom_line

5. Also, we can override the position of the *y* axes:

```
> g + geom_point(aes(y=Total_Sales/10000)) + ggtitle('Override
y-axes')
```

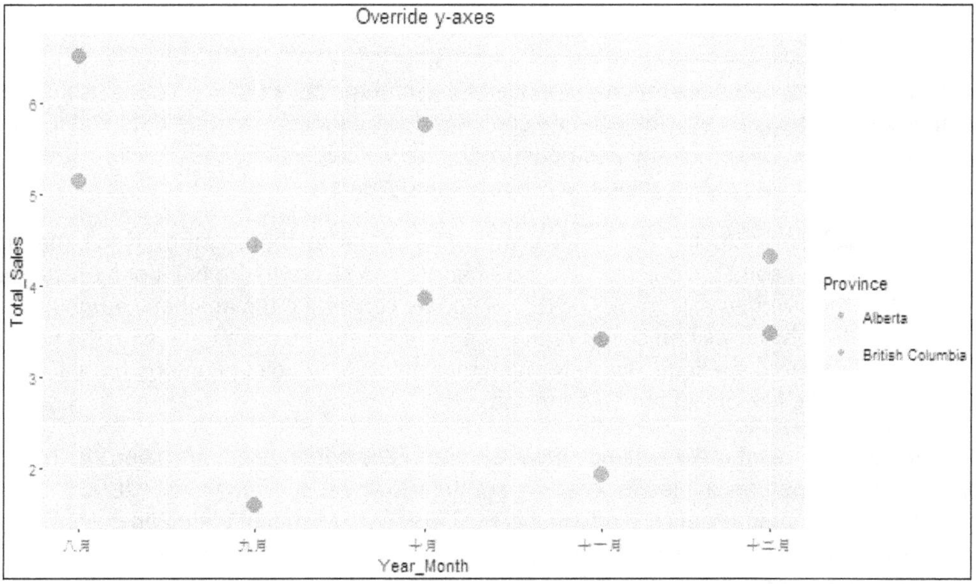

Figure 6: Override y axes

6. Moreover, we can remove the aesthetics property by setting the property value to `NULL`:

```
> g + geom_point(aes(colour=NULL))  + ggtitle('Remove Aesthetics
Property')
```

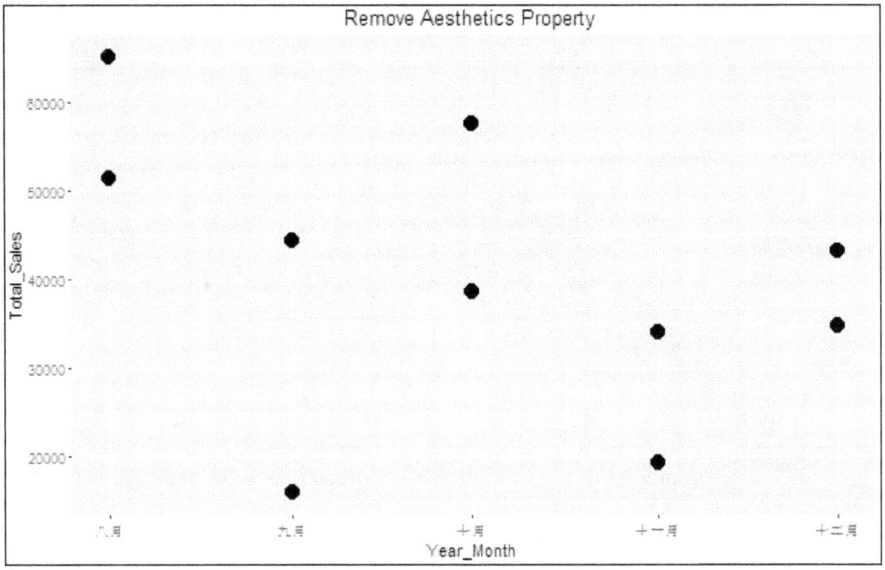

Figure 7: Remove the aesthetics property

How it works...

Aesthetics mapping describes the appearance of a geometric object in `ggplot2`. To change the appearance of a geometric object, one can use the `aes` function to map data variables to visual properties. By setting the aesthetics mapping, we can change the position, size of elements, shape of items, and color of a given geometric object.

In this recipe, we first demonstrate that we can configure the aesthetics property in either the plot side or on the geometric object. Next, we illustrate how to resize the point of a scatterplot by setting up the size option in `geom_point`. Since we can set up the aesthetic mapping by the geometric object, we can check what happens if we set the color to blue on the point geometry object. From the figure, we determine that the color of point geometry turns blue, but the color of the line object remains the same.

Moreover, we can transform aesthetic mappings of the geometric object and override the original visual properties. In step 5, when we set the *y* location to *Total_Sales / 10000*, the scale of the *y* axis changed accordingly in *Figure 6*. Besides overriding the original aesthetics properties, we can remove the visual properties by setting the property to `NULL`. As a result, all colored points turn black in *Figure 7*.

There's more...

You can map an aesthetic property to a variable, or set the value to the layer parameter. In the following figures, by setting the color to `blue` in the `aes` function, the plot creates a new variable containing `blue` values and all points on the plot now become the same color, but the color of the points may not be blue. On the other hand, if we set the color to `blue` in the layer parameter, one can see all points on the canvas turn blue:

```
> g  + geom_point(aes(colour="blue")) + ggtitle('Set Color to Blue in Aes Function')
> g  + geom_point(colour="blue") + ggtitle('Set Color to Blue in Layer')
```

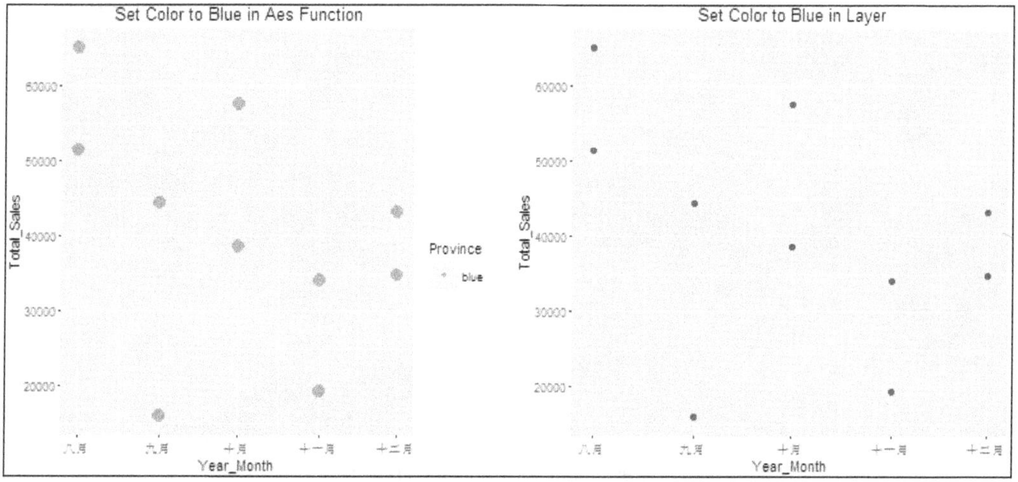

Figure 8: Comparing the effect of setting the color in the `aes` function and the layer parameter

Introducing geometric objects

Geometric objects are elements that we mark on the plot. One can use the geometric object in `ggplot2` to create either a line, bar, or box chart. Moreover, one can integrate these simple geometric objects and aesthetic mapping to create a more professional plot. In this recipe, we introduce how to use geometric objects to create various charts.

Getting ready

Ensure you have completed the previous steps by storing `sample_sum` in your R environment.

How to do it...

Perform the following steps to create a geometric object in `ggplot2`:

1. First, create a scatterplot with the `geom_point` function:

```
> g <- ggplot(data=sample_sum, mapping=aes(x=Year_Month, y=Total_
Sales, col=Province )) + ggtitle('Scatter Plot')
> g + geom_point()
```

2. Use the `geom_line` function to plot a line chart:

```
> g+ geom_line(linetype="dashed")
```

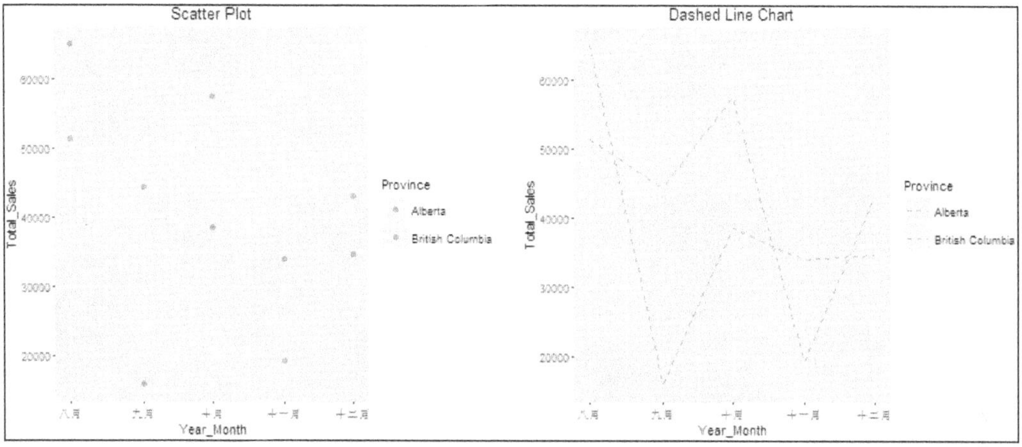

Figure 9: Scatterplot and dashed line chart

3. Use the `geom_bar` function to make a stack bar chart:

```
> g+geom_bar(stat = "identity", aes(fill=Province) , position =
"stack") + ggtitle('Stack Position')
```

4. When we set the `position` to `fill`, we can draw a stacked bar plot with normalized bar height:

```
> g+geom_bar(stat = "identity", aes(fill=Province), position =
"fill") + ggtitle('Fill Position')
```

5. By configuring `position` to `dodge`, we can arrange the bars side by side:

```
> g+geom_bar(stat = "identity", aes(fill=Province), position =
"dodge") + ggtitle('Dodge Position')
```

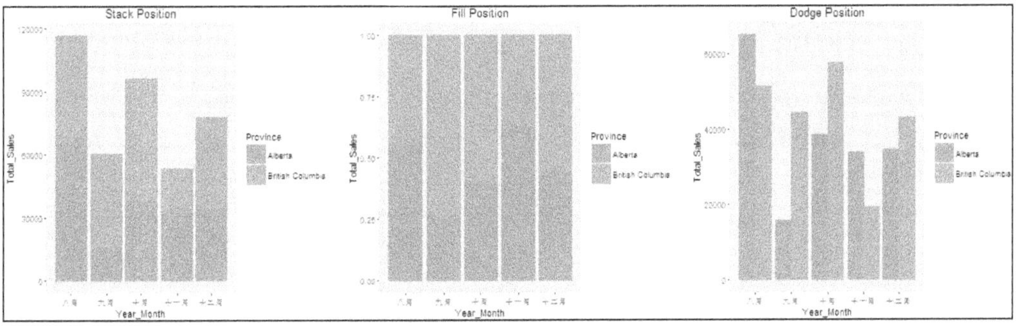

Figure 10: Bar plot in different positions

6. Also, one can use the `geom_boxplot` function to present a boxplot:

```
> g+geom_boxplot(aes(x=Province)) + xlab("Province") +
ggtitle('Boxplot')
```

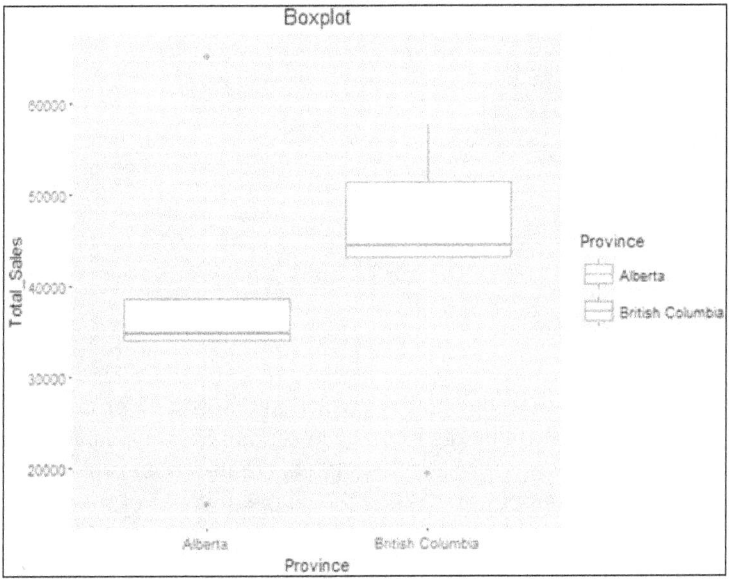

Figure 11: A Boxplot example

7. Moreover, one can use the `geom_histogram` function to create a histogram:

```
> set.seed(123)
> norm.sample = data.frame(val=rnorm(1000))
> ggplot(norm.sample, aes(val)) + geom_histogram(binwidth = 0.1) +
ggtitle('Histogram')
```

8. Finally, we can make a density plot with the `geom_density` function:

```
> ggplot(norm.sample, aes(val)) + geom_density()+ ggtitle('Density
Plot')
```

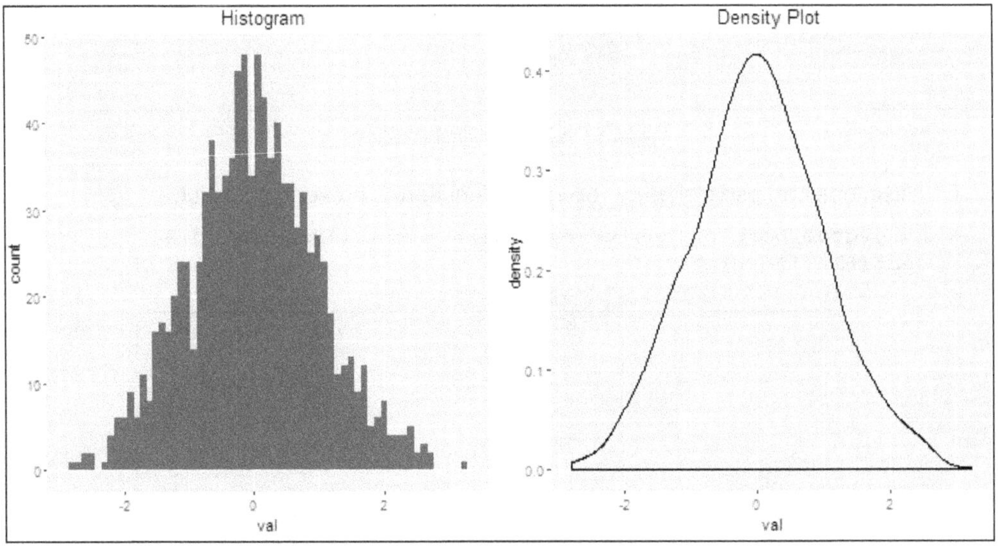

Figure 12: Histogram versus density plot

How it works...

In `ggplot2`, the marks shown on the chart are called **geometric objects**; one can use geometric objects such as lines and dots to produce various chart types. In this recipe, we first introduce how to use the `geom_point` function to create a scatterplot. Here, we can set shape and size in aesthetic mapping to change the shape and size of the point. We can then use a `geom_line` function to create a line chart. One can specify the `linetype` to `blank`, `solid`, `dashed`, `dotted`, `dotdash`, `lingdash`, and `twodash`.

The `geom_bar` function can also be used to create a bar chart. To change the bar chart layout, configure the `position` option within `geom_bar`. Setting the `position` to `stack` will stack elements on top of one another. Also, if the `position` option is set to `fill`, the bar height would be normalized. Alternatively, setting the `position` to `dodge` will arrange the bars side by side.

Moreover, we can draw a boxplot with the `geom_box` function. To present the box by province, we configure *x="Province"*.

Last, we can also make a histogram and density plot with `geom_histogram` and `geom_density`. This allows us to plot the probability distribution of a dataset.

There's more...

The `ggplot2` package does not provide any specific geometric objects that can be used to make a pie chart. However, one can create a pie chart by using the `coord_polar` function to make a stacked bar chart in polar coordinates:

1. First, generate the total sales amount of each province by using `dplyr`:

```
> sample_stat <- sum_price_by_province  %>% select(Province,
Total_Sales) %>% group_by(Province) %>% summarise(sales_
stat=sum(Total_Sales))

> head(sample_stat)

Source: local data frame [6 x 2]


                 Province sales_stat
                   (fctr)      (dbl)
1                 Alberta  416216.34
2        British Columbia  440368.13
3                Manitoba  368265.55
4           New Brunswick  161298.07
5            Newfoundland   29392.34
6 Northwest Territories  169592.82
```

2. Create a bar plot with the `geom_bar` function:

```
> g <- ggplot(sample_stat, aes(x = "", y=sales_stat,
fill=Province)) + geom_bar(stat = "identity") + ggtitle('Bar Chart
With geom_bar')

> g
```

3. Last, use the `coord_polar` function to create a pie chart from the stacked bar plot:

```
> g +  coord_polar("y", start=0) + ggtitle('Pie Chart')
```

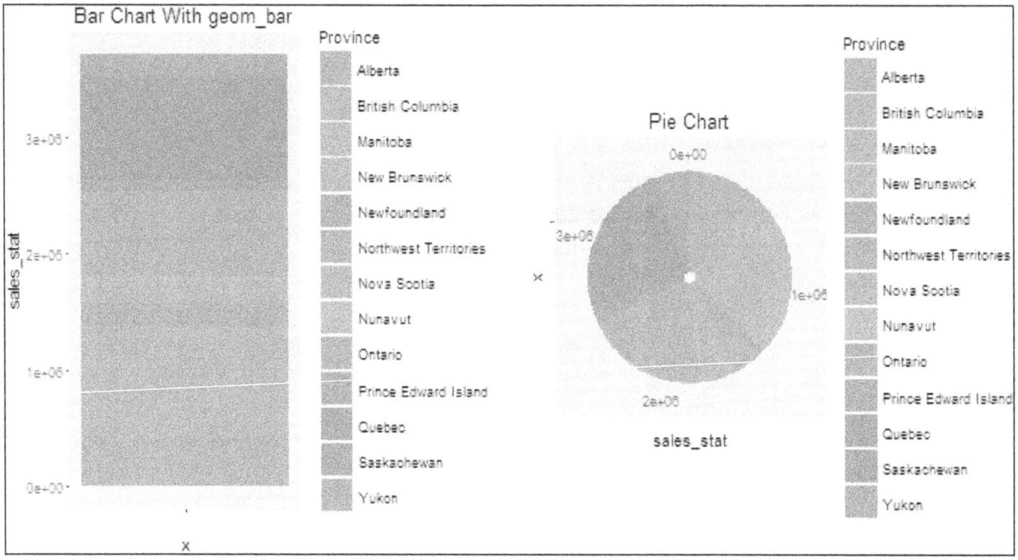

Figure 13: Creating a pie chart with `ggplot2`

Performing transformations

Besides mapping particular variables to either the *x* or *y* axis, one can first perform statistical transformations on variables, and then remap the transformed variable to a specific position. In this recipe, we introduce how to perform variable transformations with `ggplot2`.

Getting ready

Ensure you have completed the previous steps by storing `sum_price_by_province`, and `sample_sum` in your R environment.

How to do it...

Perform the following steps to perform statistical transformation in ggplot2:

1. First, create a dataset named sample_sum2 by filtering sales data from Alberta and British Columbia:

```
> sample_sum2 <- sum_price_by_province %>% filter(Province %in%
c('Alberta', 'British Columbia' ) )
```

2. Create a line plot with a regression line, using the geom_point and geom_smooth functions:

```
> g <- ggplot(data=sample_sum2, mapping=aes(x=Year_Month, y=Total_
Sales, col=Province ))
> g + geom_point(size=5) + geom_smooth() + ggtitle('Adding
Smoother')
```

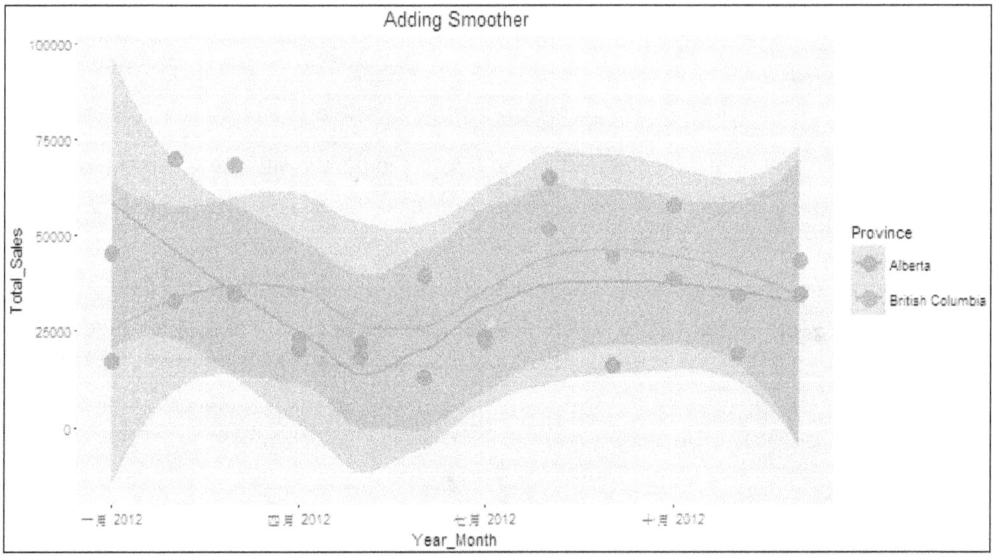

Figure 14: Adding Smoother

3. Alternatively, you can use the stat_smooth function to add a regression line onto the plot:

```
> g + geom_point() + stat_smooth()
```

4. We can specify the regression method in the `method` argument:

```
> library(MASS)
> g + geom_point() + geom_smooth(method=rlm)
```

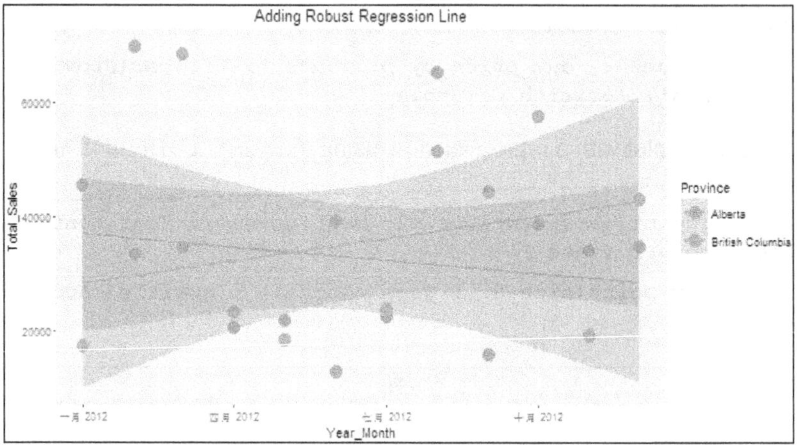

Figure 15: Adding a robust regression line

5. Also, we can add average sales by month with red points with the `stat` argument:

```
> g + geom_point() + geom_point(stat = "summary", fun.y = "mean",
colour = "red", size = 4)
```

6. We can use the `stat_summary` function to obtain a summary aesthetic:

```
> g + geom_point(size = 3) + geom_point(stat = "summary", fun.y =
"mean", colour = "red", size = 4) + ggtitle('Adding Mean Points')
```

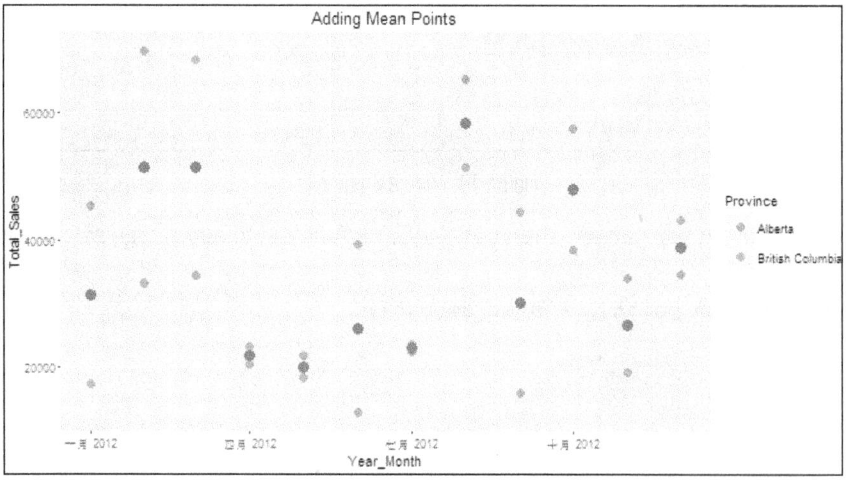

Figure 16: Adding mean points

How it works...

In addition to creating a chart from a dataset, one can use `ggplot2` to add statistical analysis onto the plot. To do so, one can either specify the `stat` argument within the `geom` function, or the `stat` function can be used as an alternative.

In this recipe, we first demonstrate how to add a regression line onto the data layer. To add a regression line, one can use either the `geom_smooth` or `stat_smooth` functions. By doing so, `ggplot2` will perform regression analysis on a given data point, and add a **LOWESS** (**locally weighted scatterplot smoothing**) line with a semi-transparent ribbon onto the plot. One can alternatively perform different regression methods by specifying the `method` argument within the `geom_smooth` function. Here, we load the `MASS` library first and then add a robust regression line onto the plot.

Last, we illustrate how to calculate the mean of *y* with condition on *x*. We need to pass the `stat` argument to the `geom` function, and then specify *stat= "summary"* and *fun.y="mean"*. As a result, we can see the average value of *y* is plotted as red points on the plot. As an alternative to passing the `stat` argument to the `geom` function, one can use `stat_summary` to generate the mean points.

There's more...

To plot confidence intervals, one can follow the summary functions from the `Hmisc` package within `stat_summary`:

- `mean_cl_normal()`: This returns the sample mean and confidence interval based on the t-distribution
- `mean_sdl()`: This returns the mean and confidence interval of plus or minus a constant multiplied by the standard deviation
- `mean_cl_boot()`: This returns the confidence interval for the population means without assuming normality
- `median_hilow()`: This returns the sample median, upper, and lower quantiles with equal tail areas

Adjusting scales

Besides setting aesthetic mapping for each plot or geometric object, one can use scale to control how variables are mapped to the visual property. In this recipe, we introduce how to adjust the scale of aesthetics in `ggplot2`.

Getting ready

Ensure you have completed the previous steps by storing `sample_mean` in your R environment.

How to do it...

Perform the following steps to adjust the scale of aesthetic magnitude in `ggplot2`:

1. First, make a scatterplot by setting `size=Total_Sales`, `colour=Province`, `y=Province`, and conditional on `Year_Month`. Resize the point with the `scale_size_continuous` function:

```
> g <- ggplot(data=sample_sum, mapping=aes(x=Year_Month,
y=Province, size=Total_Sales, colour = Province ))
```

```
> g + geom_point(aes(size=Total_Sales)) + scale_size_
continuous(range=c(1,10)) + ggtitle('Resize The Point')
```

2. Repaint the point in gradient color with the `scale_color_gradient` function:

```
> g + geom_point(aes(colour=Total_Sales)) + scale_color_
gradient()+ ggtitle('Repaint The Point in Gradient Color')
```

3. We can adjust the shape of the point by using `Province`:

```
> g+geom_point(aes(shape=Province) )  + scale_shape_
manual(values=c(5,10)) + ggtitle('Adjust The Shape of The Point')
```

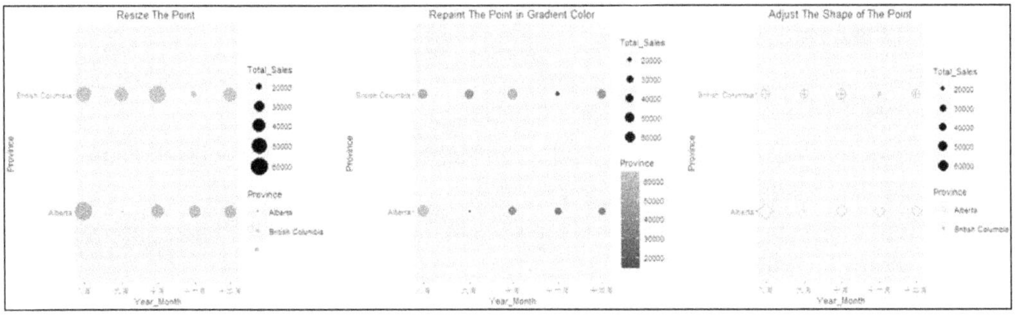

Figure 17: Adjusting scale of aesthetic magnitude for a scatterplot

4. For a bar plot, one can refill the color of the bar by using the `scale_fill_brewer` function:

```
> g2 <- ggplot(data=sample_sum, mapping=aes(x=Year_Month, y=Total_
Sales, colour = Province ))
```

```
> g2+geom_bar(stat = "identity", aes(fill=Province), position =
"dodge") + scale_fill_brewer(palette=2) + ggtitle('Refill Bar
Colour')
```

5. Moreover, we can rescale the *y* axis with the `scale_y_continuous` function:

```
> g2+geom_bar(stat = "identity", aes(fill=Province), position
= "dodge")  + scale_y_continuous(limits = c(1,100000),
trans="log10") + ggtitle('Rescale y Axes')
```

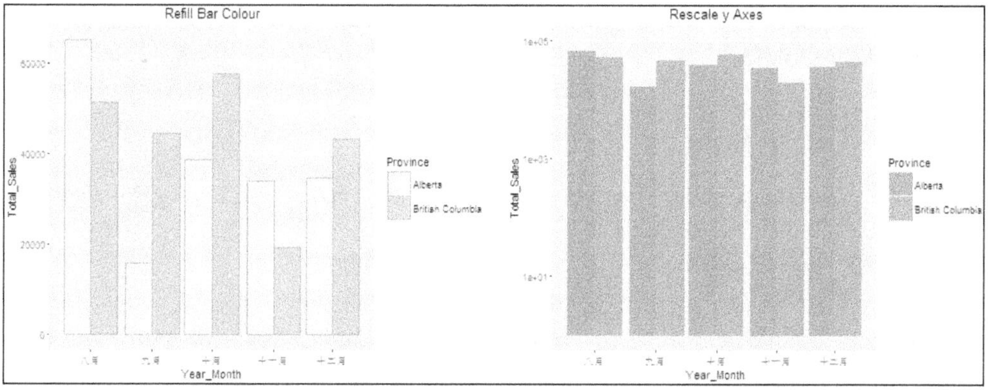

Figure 18: Adjusting scale of aesthetic magnitude on a bar plot

How it works...

In this recipe, we introduced how to map aesthetics to geometric objects in `ggplot2`. As we can express the aesthetic of data object in magnitude, we can rescale the *x* axis, *y* axis, color, fill, and even shape with the `scale_*` function.

This recipe starts by creating a scatterplot with point geometry from the superstore dataset. We can rescale the size of the point with the `scale_size_*` function. Here, we can configure the range of the size with the `range` parameter. Furthermore, we can repaint the point with color in gradient with the `scale_color_*` function. We can also reshape the point with the `scale_shape_*` function.

For a bar plot, we can change the bar color of the bar plot with the `scale_fill_*` function. To change the color scheme, specify the palette with the `palette` argument. Moreover, we can rescale the *y* axis with the `scale_y_*` function. In addition to limiting the *y* axis scale with the `limits` argument, one can perform magnitude transformation by passing the `trans` argument to `scale_y_continus` function.

See also

To style the visual properties of geometric objects with the color palette, examine the color palette options in the `RColorBrewer` document:

```
> ?RColorBrewer
```

Faceting

When performing data exploration, it is essential to compare data across different groups. Faceting is a technique that enables the user to create graphs for subsets of data. In this recipe, we demonstrate how to use the `facet` function to create a chart for multiple subsets of data.

Getting ready

Ensure you have completed the previous steps by storing `sample_sum` in your R environment.

How to do it...

Please perform the following steps to create a chart for multiple subsets of data:

1. First, use the `facet_wrap` function to create multiple subplots by using the `Province` variable as the condition:

```
> g <- ggplot(data=sample_sum, mapping=aes(x=Year_Month, y=Total_
Sales, colour = Province ))
```

```
>  g+geom_point(size = 5) +  facet_wrap(~Province) +
ggtitle('Create Multiple Subplots by Province')
```

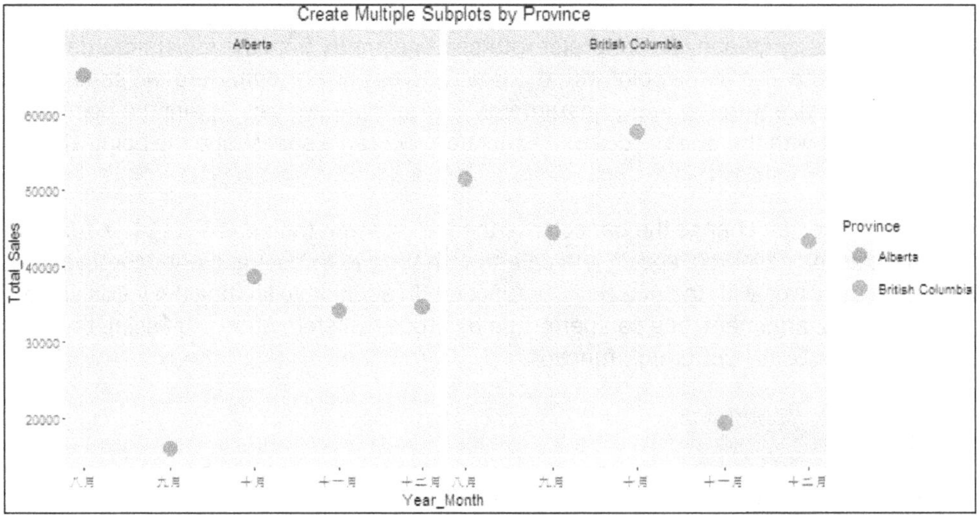

Figure 19: Create multiple subplots by province

2. On the other hand, we can change the layout of the plot in a vertical direction if we set the number of columns to `1`:

```
> g+geom_point() +  facet_wrap(~Province, ncol=1) +
ggtitle('Multiple Subplots in Vertical Direction')
```

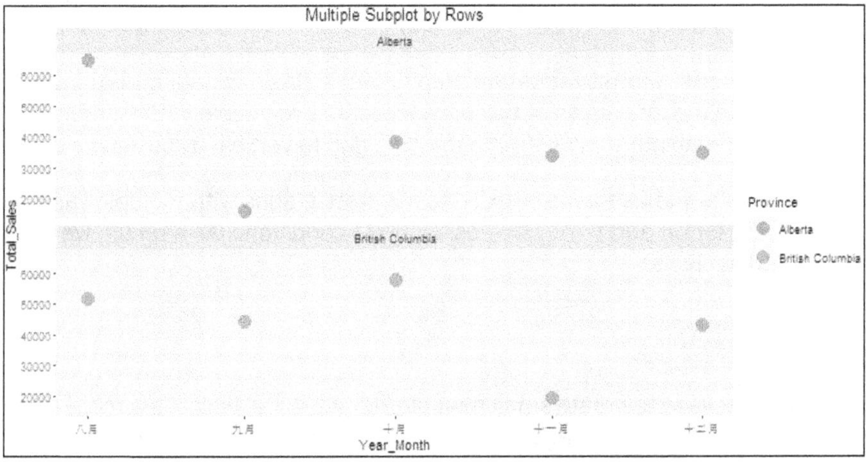

Figure 20: Create multiple subplots in a vertical direction

3. To facet by more variables, one can use the `facet_grid` function:

```
> g <- ggplot(data=sample_sum, mapping=aes(x="Total Sales",
y=Total_Sales, col=Province ))

> g+geom_bar(stat = "identity", aes(fill=Province)) + facet_
grid(Year_Month ~ Province) + ggtitle('facet_grid Example')
```

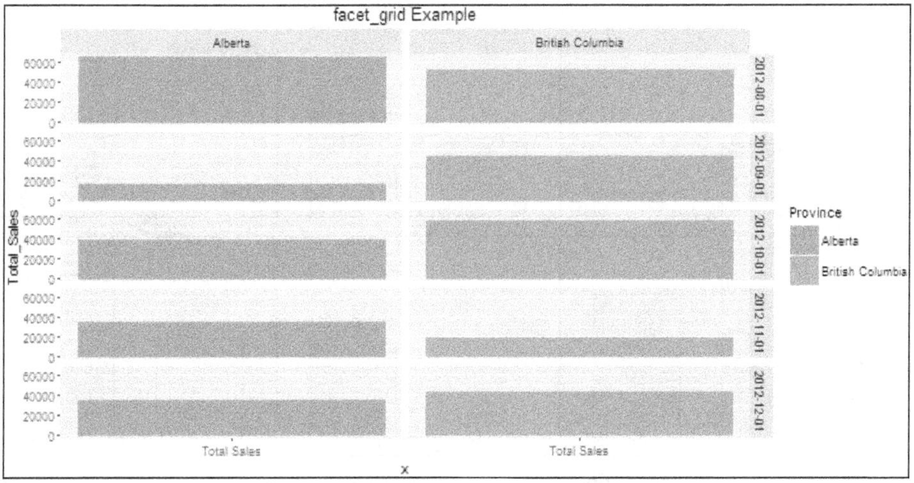

Figure 21: facet_grid example

How it works...

Faceting is a technique that enables the user to create small multiples. In other words, faceting separates the original plot into multiple subplots based on single or multiple variables. This helps facilitate comparisons of the same plot under different conditions. In this recipe, we introduce how to use `facet_wrap` and `facet_grid` to separate the plot.

We start by using `facet_wrap` to separate graphs with a single variable, `Province`. This allows us to separate the total sales of two provinces into two subplots. By default, the layout of a subplot is horizontal, but we can switch the direction to vertical by assigning `ncol` to `1`.

Last, we introduced how to use `facet_grid` to separate graphs with multiple variables. Here, we use both `Year_Month` and `Province` as separate conditions. As a result, we obtain a plot containing 10 subplots.

There's more...

Another powerful package to create small multiples is `lattice`, which is a more advanced package than the `graphics` package. The `lattice` has the ability to display the relationship of multiple variables.

For example, to create small multiples with `Province` as `x` value, `Total_Sales` as `y` value, and `Year_Month` as the separate condition, the following command can be used:

```
> library(lattice)
> barchart(Total_Sales ~ Province| Year_Month , data=sample_sum)
```

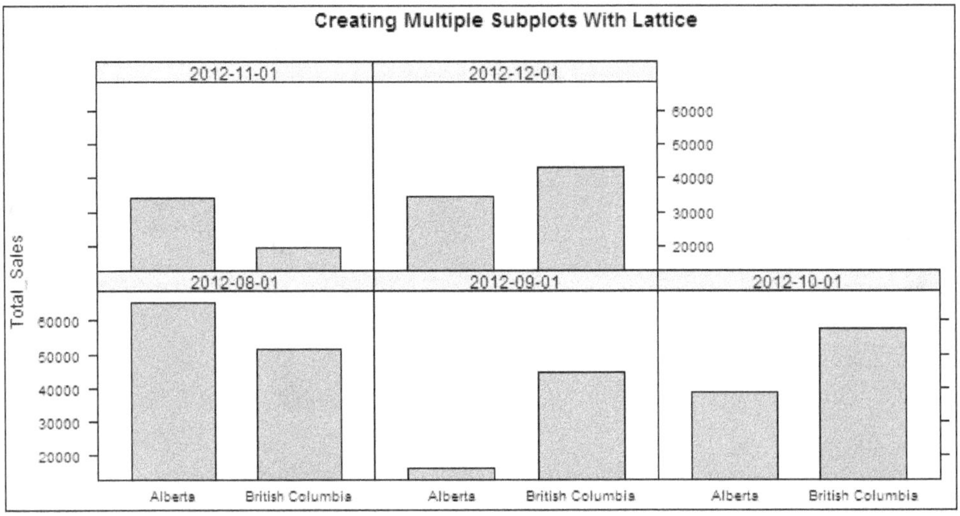

Figure 22: Creating multiple subplots with Lattice

Adjusting themes

Besides deciding the visual property of a geometric object with aesthetic mapping, one can adjust the background color, grid lines, and other non-data properties with the theme. We introduce how to change the theme in this recipe.

Getting ready

Ensure you have completed the previous steps by storing `sample_sum` in your R environment.

How to do it...

Please perform the following steps to adjust the theme in `ggplot2`:

1. We can use different `theme` functions to adjust the theme of the plot:

```
> g <- ggplot(data=sample_sum, mapping=aes(x=Year_Month, y=Total_
Sales, colour = Province ))
> g+geom_point(size=5) + theme_bw()+ ggtitle('theme_bw Example')
> g+geom_point(size=5) + theme_dark()+ ggtitle('theme_dark
Example')
```

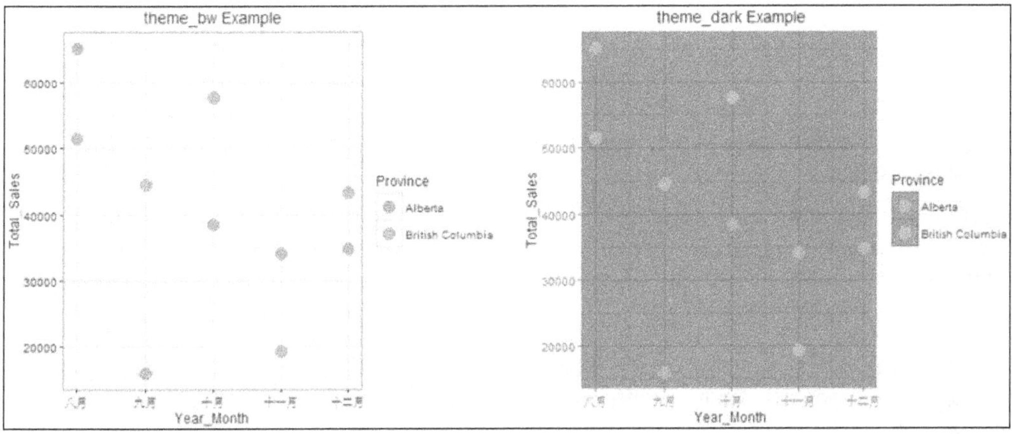

Figure 23: Scatterplot with different themes

2. We can set the theme freely with the `theme` function:

```
> g +geom_point(size=5) +
+ theme(
+    axis.text = element_text(size = 12),
```

```
+    legend.background = element_rect(fill = "white"),
+    panel.grid.major = element_line(colour = "yellow"),
+    panel.grid.minor = element_blank(),
+    panel.background = element_rect(fill = "blue")
+ ) + ggtitle('Customized Theme')
```

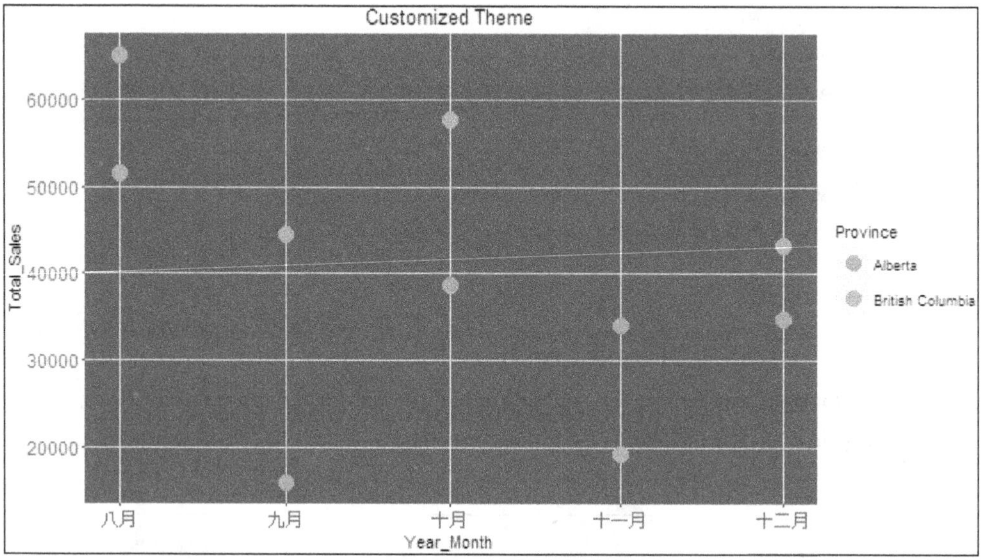

Figure 24: Customized theme

How it works...

A user can adjust the layout, color, text font, and other attributes of a non-data object using the theme system in `ggplot2`. By default, `ggplot2` provides many themes, and one can adjust the current theme with the `theme_*` function. Alternatively, to customize a theme, one can use the `theme` function.

In this recipe, we first make a `ggplot2` plot and then set the theme with the `theme_bw` and `theme_dark` functions. Applying different themes will change the background color, grid line, and legend.

Last, we illustrate how we can customize our theme with the `theme` function. In this sample, we revise the background color of legend, grid line, axis text, and panel background with the `element_*` function.

There's more...

To know more about the different options in terms of setting theme elements, one can read the `theme` documentation using the `help` function or `?` notation:

```
> ?theme
```

Within the theme function, one can choose to use the following four `element_*` functions to modify theme elements:

```
element_text()
element_line()
element_rect()
element_blank()
```

Combining plots

To create an overview of a dataset, we may need to combine individual plots into one. In this recipe, we introduce how to combine individual subplots into one plot.

Getting ready

Ensure you have installed and loaded `ggplot2` into your R session. Also, you need to complete the previous steps by storing `sample_sum` in your R environment.

How to do it...

Please perform the following steps to combine plots in `ggplot2`:

1. First, we need to load the `grid` library into an R session:

   ```
   > library(grid)
   ```

2. We can now create a new page:

   ```
   > grid.newpage()
   ```

3. Moving on, we can create two `ggplot2` plots:

   ```
   > g <- ggplot(data=sample_sum, mapping=aes(x=Year_Month, y=Total_
   Sales, colour = Province ))
   >   plot1 <- g + geom_point(size=5) + ggtitle('Scatter Plot')
   >   plot2 <- g + geom_line(size=3) + ggtitle('Line Chart')
   ```

4. Next, we can push the visible area with a layout of two columns in one row, using the `pushViewport` function:

```
> pushViewport(viewport(layout = grid.layout(1, 2)))
```

5. Last, we can put the chart onto the visible area by row and column position:

```
> print(plot1, vp =viewport(layout.pos.row = 1, layout.pos.col =
1))
```

```
> print(plot2, vp =viewport(layout.pos.row = 1, layout.pos.col =
2))
```

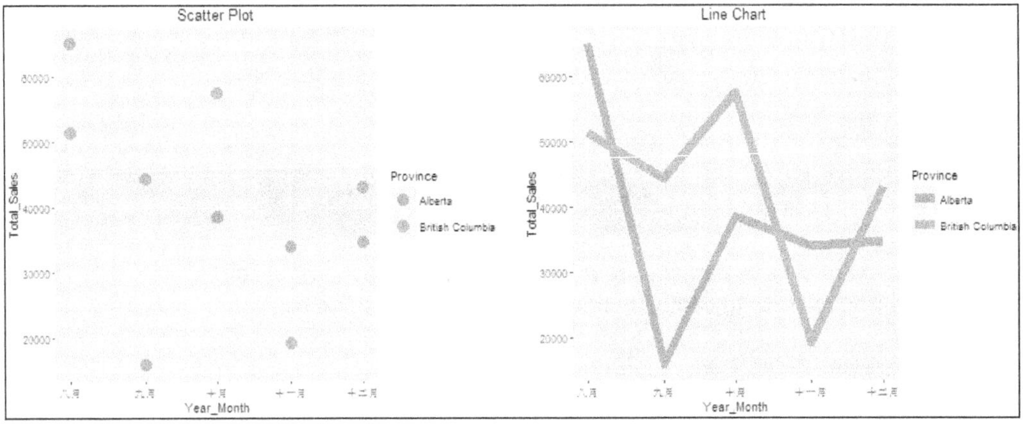

Figure 25: Combining plots

How it works...

Unlike the base R graphics system, one can easily use the `par` or `layout` functions to combine multiple plots. Using `ggplot2` requires combining multiple plots into an overall graph with the help of the `grid` package.

First, we load the `grid` package into an R session. We then use the `grid.newpage` function to create a canvas. Next, we create a scatterplot and line chart, and store them separately in `plot1` and `plot2`. We can now `push` the visible area with layout information, using the `pushViewport` function. Last, we can overlay the two plots onto the visible area by column with the `print` function. Within the `print` function, we can configure the position of the visible area (viewport) with the `viewport` function.

There's more...

Instead of using the `grid` package, if you prefer using simple syntax such as `par` to combine multiple plots into one graph, you can use the `grid.arrange` function from the `gridExtra` package:

1. First, load the `gridExtra` package into R:

```
> install.packages("gridExtra")
> library("gridExtra")
```

2. You can use `grid.arrange` to arrange two plots column by column in one graph:

```
> grid.arrange(plot1,plot2, ncol=2)
```

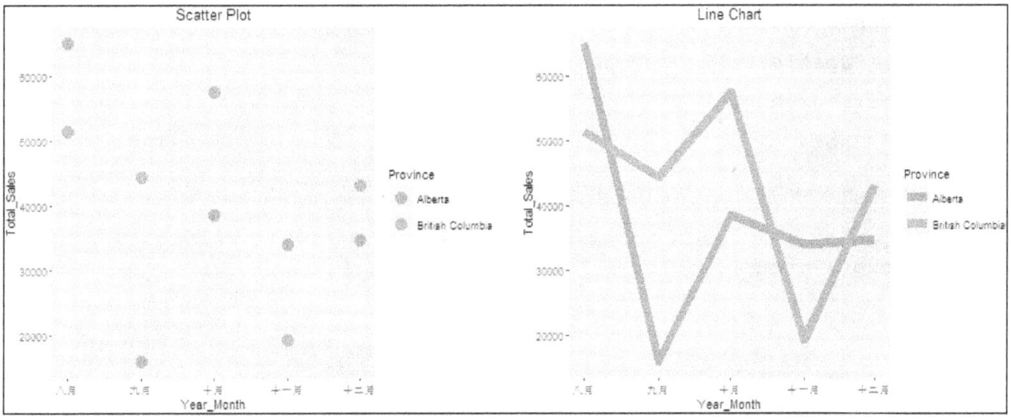

Figure 26: Combining plots with the `gridExtra` package

Creating maps

One can use a map to visualize the geographical relationship of spatial data. Here, we introduce how to create a map from a shapefile with `ggplot2`. Moreover, we introduce how to use `ggmap` to download map data from an online mapping service.

Getting ready

Ensure you have installed and loaded `ggplot2` into your R session. Please download all files from the following GitHub link folder:

`https://github.com/ywchiu/rcookbook/tree/master/chapter7`

How to do it...

Perform the following steps to create a map with ggmap:

1. First, load the ggmap and maptools libraries into an R session:

```
> install.packages("ggmap")
> install.packages("maptools")
> library(ggmap)
> library(maptools)
```

2. We can now read the .shp file with the readShapeSpatial function:

```
> nyc.shp <- readShapeSpatial("nycc.shp")
> class(nyc.shp)
[1] "SpatialPolygonsDataFrame"
attr(,"package")
[1] "sp"
```

3. At this point, we can plot the map with the geom_polygon function:

```
> ggplot() +  geom_polygon(data = nyc.shp, aes(x = long, y = lat,
group = group), color = "yellow", size = 0.25)
```

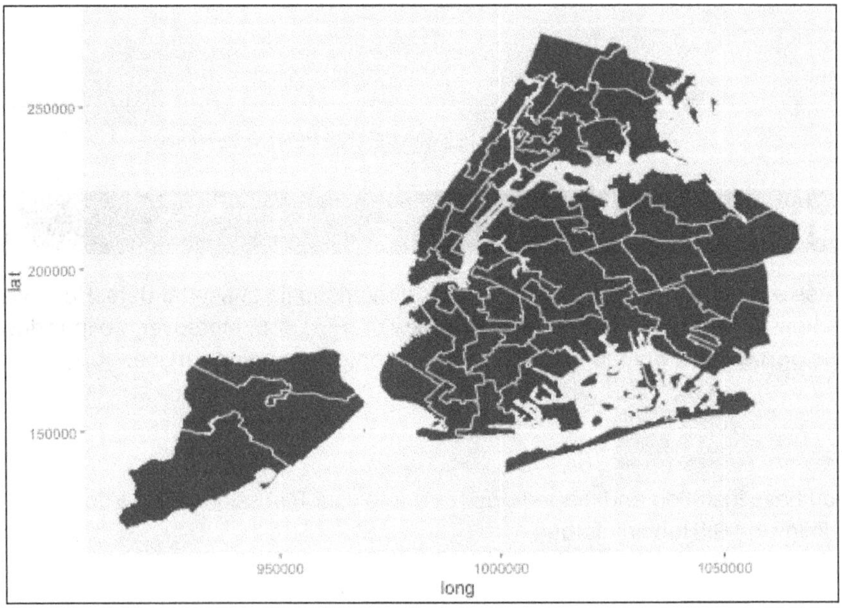

Figure 27: Plotting a map with the geom_polygon function

4. Moving on, we can read Wi-Fi hotspot data from the `wifi_hotspot.shp` file:

```
> wifi.hotspot <- readShapeSpatial("wifi_hostspot.shp")
> class(wifi.hotspot)
[1] "SpatialPointsDataFrame"
attr(,"package")
[1] "sp"
```

5. Create a scatterplot with the `geom_point` function:

```
> ggplot() + geom_point(data=as.data.frame(wifi.hotspot),
aes(x,y), color="red")
```

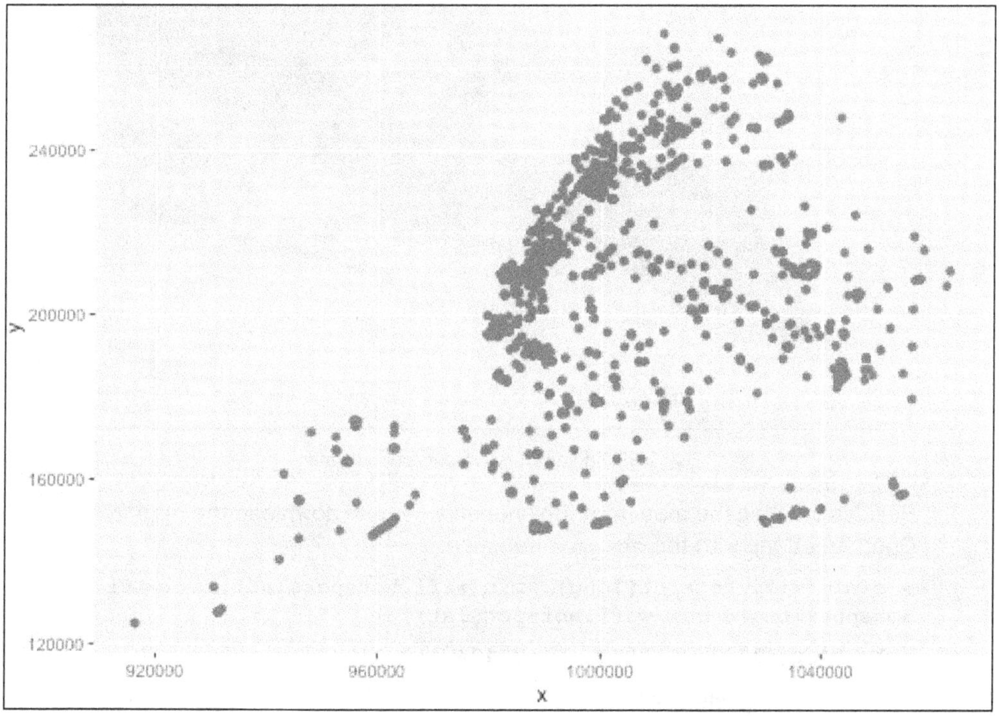

Figure 28: Marking Wi-Fi hotspots with the geom_point function

6. We can also add the Wi-Fi hotspot points onto the map:

```
> ggplot() +  geom_polygon(data = nyc.shp, aes(x = long, y =
lat, group = group), color = "yellow", size = 0.25)  + geom_
point(data=as.data.frame(wifi.hotspot), aes(x,y), color="red")
```

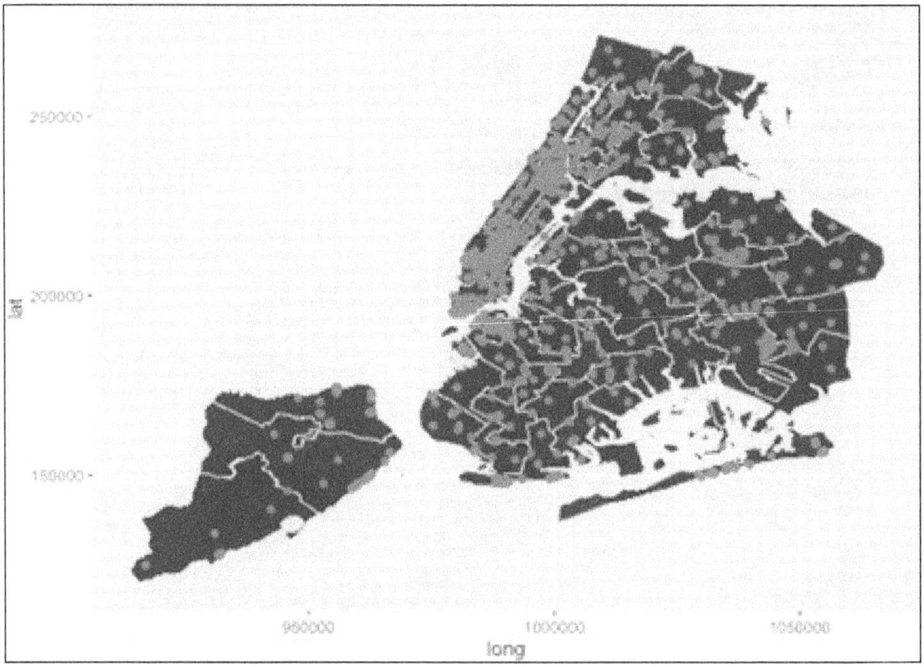

Figure 29: Adding Wi-Fi hotspots onto the map

7. Besides creating the map from the shapefile, we can download the map from OpenStreetMap with the `get_map` function:

```
> c(min(wifi.hotspot$long), min(wifi.hotspot$lat),max(wifi.
hotspot$long), max(wifi.hotspot$lat))

[1] -74.24411  40.50953 -73.71484  40.90372

> map <- get_map(location = c(min(wifi.hotspot$long), min(wifi.
hotspot$lat),max(wifi.hotspot$long), max(wifi.hotspot$lat)),
source="osm")

> ggmap(map)
```

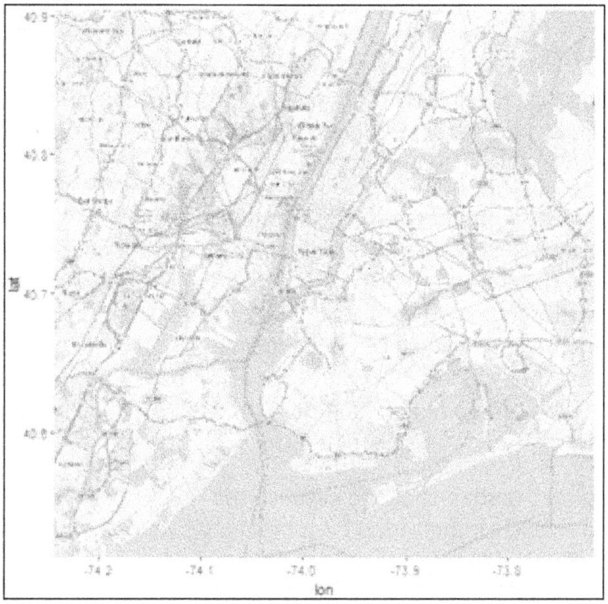

Figure 30: Obtaining an OpenStreetMap

8. Moreover, we can add Wi-Fi hotspots in red points onto the downloaded map:

```
> ggmap(map) + geom_point(data=as.data.frame(wifi.hotspot),
aes(x=long,y=lat), color="red")
```

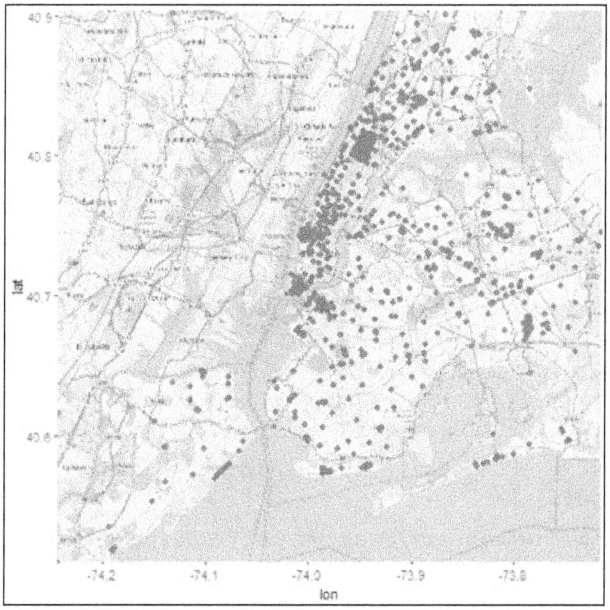

Figure 31: Adding Wi-Fi hotspots onto OpenStreetMap

How it works...

Maps are the most intuitive tools for presenting geographical information, and are particularly useful for understanding the spatial relationship between data. In this recipe, we first illustrate how to use `ggplot2` to create a map from a shapefile.

At the start of the recipe, we import a New York shapefile, which is a geospatial vector data format for GIS software. As a map is made of polygons with coordinate information, one can use the `geom_polygon` function in `ggplot2` to create a simple map with some aesthetic properties. Next, we read `wifi_hostspot.shp` into an R session. By using a `class` function, one can locate the imported `wifi_hotspot.shp` file in `SpatialPointsDataFrame` format. Therefore, one can use the `geom_point` function to mark the Wi-Fi hotspots in red points on an empty canvas. Moreover, one can add the red points onto the New York map by using some of the `ggplot2` functions.

Besides reading data from a shapefile, one can use the `get_map` function from the `ggmap` package to download a raster format map from mapping services such as Google Maps, OpenStreetMap, and Stamen Maps. Here, we extract the minimum latitude and longitude, and maximum latitude and longitude, from the geographical information of Wi-Fi hotspots. Next, we use the `get_map` function from `ggamp` to download the New York map with the coordinates from OpenStreetMap. As we have downloaded the map from OpenStreetMap, we can use `ggmap` to plot the map. Moreover, we can add the Wi-Fi hotspot points onto the map with the `geom_point` function.

There's more...

To make data from different geographic coordinate systems consistent, it is necessary to perform geographical transformation. In R, `rgdal` can be used for this purpose, which is an interface that binds to the **Geospatial Data Abstraction Library** (**GDAL**). Before installing `rgdal`, one needs to download the GDAL library from `http://trac.osgeo.org/gdal/wiki/DownloadSource` and PROJ.4 from `http://trac.osgeo.org/proj/`. For Mac users, the `.dmg` file is available from `http://www.kyngchaos.com/`. Once this is done, `rgdal` can be downloaded from CRAN.

If you are unable to install `rgdal` from CRAN on a Mac, you can download the `tar.gz` from `https://cran.r-project.org/web/packages/rgdal/index.html`, and then use the following commands to install `rgdal`:

```
R CMD INSTALL /rgdal_1.1-3.tar.gz

--configure-args='--with-gdal-config=/Library/Frameworks/GDAL.framework/
Programs/gdal-config

  --with-proj-include=/Library/Frameworks/PROJ.framework/Headers

  --with-proj-lib=/Library/Frameworks/PROJ.framework/unix/lib'
```

6
Making Interactive Reports

This chapter covers the following topics:

- Creating R Markdown reports
- Learning the markdown syntax
- Embedding R code chunks
- Creating interactive graphics with `ggvis`
- Understanding basic syntax and grammar
- Controlling axes and legends
- Using scales
- Adding interactivity to a `ggvis` plot
- Creating an R Shiny document
- Publishing an R Shiny report

Introduction

After completing data analysis, it is important to document the research results and share the findings with others. The most common methods involve documenting results through text, slides, or web pages. However, these formats are generally limited to only sharing the results and don't often include the research process. As a result, other researchers cannot reproduce the research, making it hard to fully understand how the author conducted the analysis. Without knowing how to replicate the research, the authenticity of the study may be questioned.

R scripts can be used to perform data analysis and generate figures, and it is possible to create a reproducible report by copying all the code and images into a document. However, as this is quite labor intensive, there is a risk of making errors. A better solution is to automate the documentation process. This allows the user to dynamically generate a report, in any format, which records both scripts and the analysis results.

There are a variety of methods that can be used to create a reproducible report with R packages. However, the user-friendly GUI of RStudio is one of the best choices, and allows the user to create a repeatable and dynamic report with R Markdown. In this chapter, we introduce how to create, record, and compile scripts with an R Markdown workflow. In some cases, exploring data from a static figure may be difficult. To simplify this process, making the chart interactive allows the user to quickly adjust the visual properties or data mapping. We therefore introduce how to create an interactive chart with the `ggvis` package. Lastly, we explain how to embed the interactive figure into a Shiny document and publish the Shiny document to `http://www.shinyapps.io/`.

Creating R Markdown reports

RStudio has an R Markdown workflow built in; we can use its GUI to create markdown reports in HTML, PDF, slide, or Microsoft Word format. In this recipe, we will introduce how to build an R Markdown report with RStudio.

Getting ready

Ensure you have installed the latest versions of R and RStudio on your operating system. If you have not yet installed RStudio, please visit the following URL to download the most recent version:

`https://www.rstudio.com/products/rstudio/download/`

How to do it...

Please perform the following steps to create an R Markdown report:

1. First, click on **File** | **New File** | **R Markdown** on the menu bar:

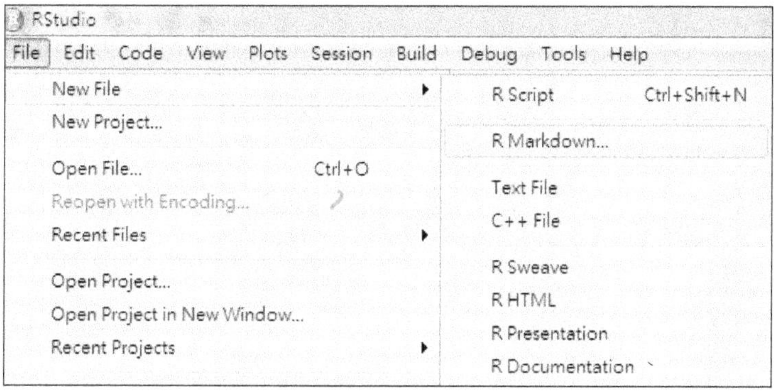

Figure 1: Create an R Markdown report

2. Next, select the document type in the left-hand side menu, then fill in the **Title** and **Author**, and choose the output format in the right-hand side menu:

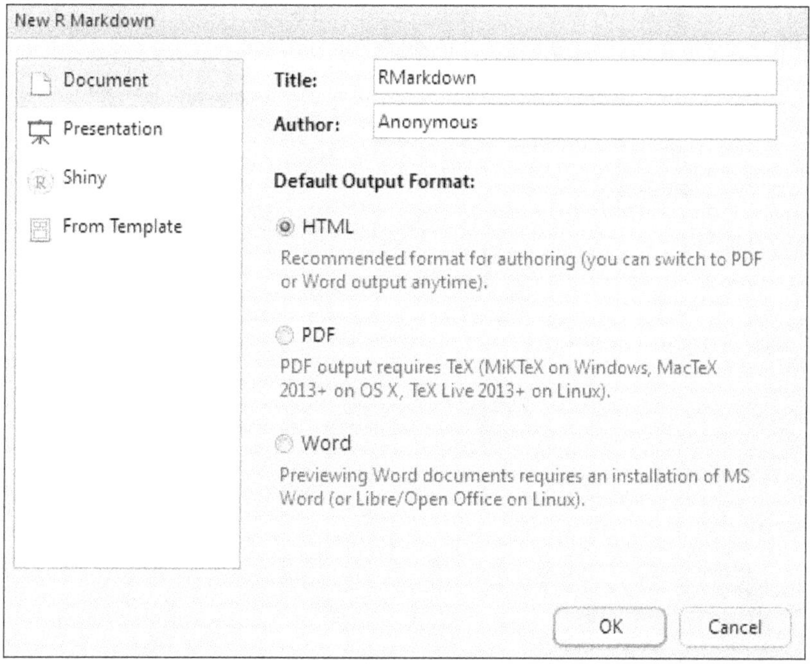

Figure 2: Fill in the meta information of an R Markdown report

3. Moving on, we can now see a markdown document with a YAML header and content in markdown syntax. You can now click on **Knit HTML** to compile and render a report:

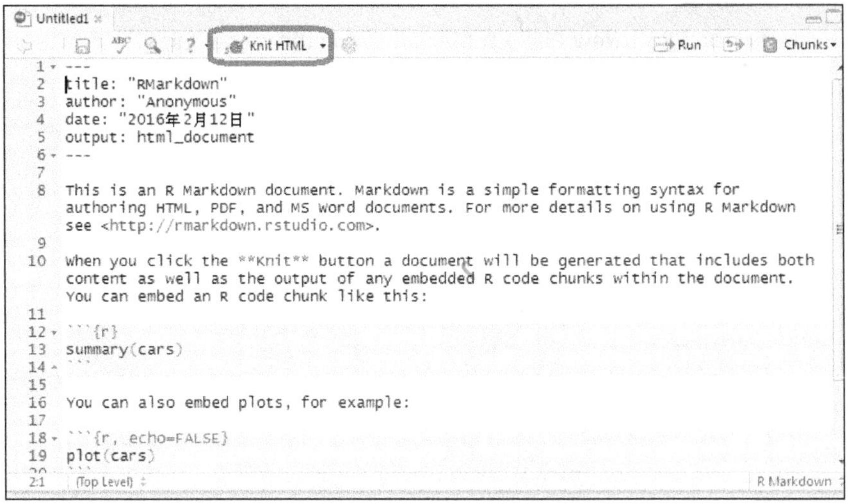

Figure 3: Compile an R Markdown report

4. Lastly, the system will request a file name to be entered. Once the file name is given, RStudio will generate a `<filename>.rmd` and a `<filename>.html` file. The formatted report will be displayed in a pop-up window:

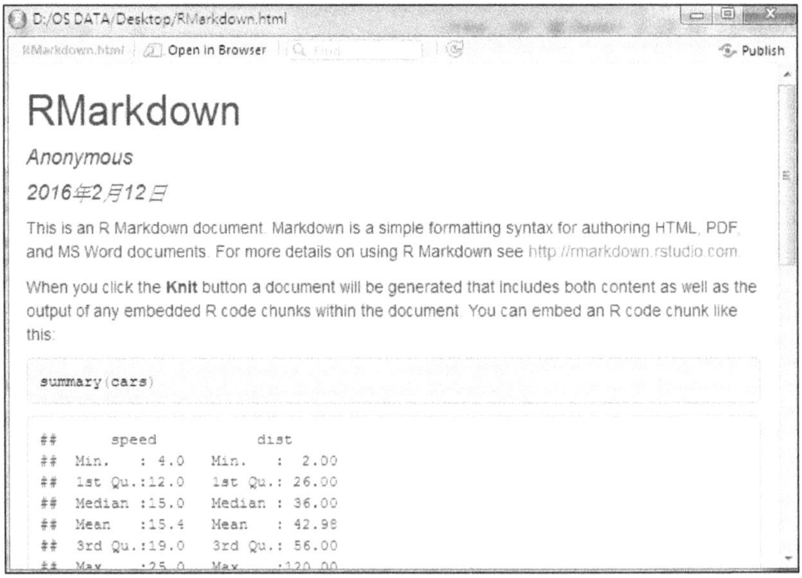

Figure 4: A compiled R Markdown report

How it works...

Creating an R Markdown report with RStudio is a straightforward process. First, make sure that RStudio is installed on your operating system. Next, click on **R Markdown** in the menu bar to create a new document. We can choose to make a document, presentation file, Shiny application, or a document from the default template. Here, we decide to make a markdown document. Next, we need to fill in the **Title** and **Author**, and choose the output format for our new document. At this point, you will see a new file with R Markdown syntax and simple instructions on how to generate an R Markdown report.

From the new R Markdown file, one can find a YAML header in the following format:

```
---
title: "RMarkdown"
author: "Anonymous"
date: "2016/02/12"
output: html_document
---
```

The title, author, and output format is what we have typed and selected when we create our markdown report. You can change the title, author, and output format by revising the YAML header. Therefore, if you would like to generate a PDF document, just alter the `output` from `html_document` to `pdf_document`. Below the YAML headers, one can find markdown syntax with some embedded R code. We can type our report in markdown syntax and embed R code in between the code chunks.

Finally, one can click on **Knit HTML** (or any icon initial with **Knit**) to generate an HTML report. To change the output format, click the upside down caret besides **Knit HTML** to choose another output format:

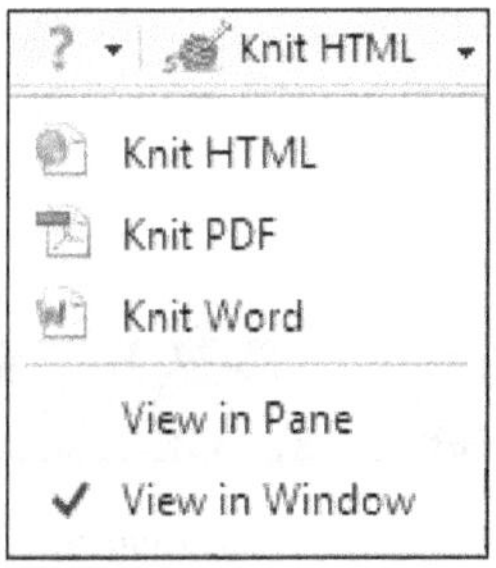

Figure 5: Choosing output format

There's more...

To edit the document options, click the gear icon besides **Knit HTML**, shown in the following image. This will open the R Markdown document options panel:

Figure 6: To open the R Markdown document option panel

One can edit the output format, figures, and some advanced features in the control panel shown:

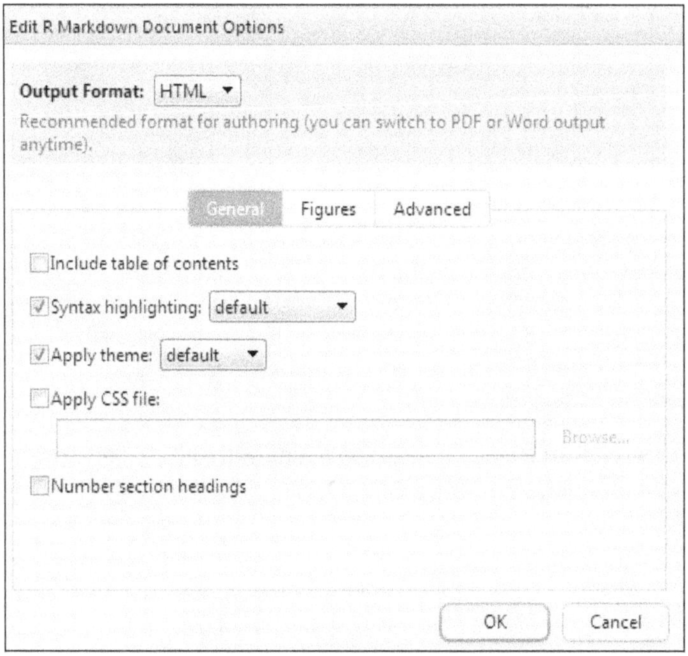

Figure 7: Configuration of advanced features

Learning the markdown syntax

Markdown is a lightweight markup language; it allows anyone to quickly create a formatted document in plain text instead of using complicated HTML. In this recipe, we introduce how to write a report with markdown syntax.

Getting ready

Ensure you have installed the latest version of R and RStudio on your operating system. Also, you need to have created and opened a new R Markdown (`.rmd`) file in RStudio.

How to do it...

Please perform the following steps to write simple markdown syntax:

1. First, create headings by placing a hashtag # in front of the words:

Markdown	Preview

`# H1`
`## H2`
`### H3`
	H1
	H2
	H3

2. Next, emphasize words by style with bold, italics, and strikethrough:

Markdown	Preview

This is plan text:
`*italics1*`
`_italics2_`
`**bold1**`
`__bold2__`
`~~Strikethrough~~`
	This is plan text
	italics1
	italics2
	bold1
	bold2
	~~Strikethrough~~

3. For list items, we can list items in an ordered item list or an unordered item list:

Markdown	Preview
1. First Ordered list item 2. Second Ordered List Item ❑ Unordered sub-list ▸ Unordered List Item 1. Ordered sub-list-1 2. Ordered sub-list-2	1. First Ordered list item 2. Second Ordered List Item ◦ Unordered sub-list • Unordered List Item 1. Ordered sub-list-1 2. Ordered sub-list-2

3. Moreover, we can also use markdown to create a table:

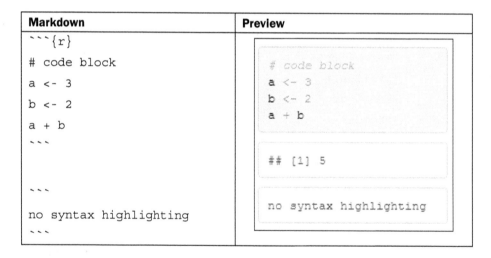

<table>
<tr><th>Markdown</th><th>Preview</th></tr>
<tr><td>| NAME	AGE	GENDER
Andy	23	M
Brian	33	M
Judy	18	F |</td><td>NAME AGE GENDER Andy 23 M Brian 33 M Judy 18 F</td></tr>
</table>

4. To embed R code, we can use three single quotes to wrap our R code. RStudio will compile the code and present the execution result below the script. If one does not specify the programming language in the code block, the code block will not highlight any of the code syntaxes:

<table>
<tr><th>Markdown</th><th>Preview</th></tr>
<tr><td>```{r}
# code block
a <- 3
b <- 2
a + b
```

```
no syntax highlighting
```</td><td># code block
a <- 3
b <- 2
a + b

[1] 5

no syntax highlighting</td></tr>
</table>

5. Moreover, one can add links and images in the R Markdown report:

Markdown	Preview

`This is [R](https://www.r-project.org/ "Title") download link.`
 	
`![alt text](https://www.r-project.org/Rlogo.png "Logo Title Text 1")`
	This is the R download link:

How it works...

The most attractive feature of a markdown report is that it enables the user to create a well-formatted document with plain text and simple markup syntax. Typically, a markdown file uses `.md` as its file extension. However, as RStudio needs to compile the R code and outputs the result right below the code block, it uses `.rmd` as its file extension. Therefore, when we click on **Knit HTML**, RStudio will automatically compile the R code and render R code, execution result, and markdown syntax into an HTML document.

In this recipe, we introduced some basic syntax that can be used to create a markdown report. First, we present how to create headings by placing a hashtag, `#`, in front of words. By using one hashtag, we can render a `H1` heading. If we place two hashtags in front of the word, we can make a `H2` heading.

Next, we covered how to emphasize words. If we use plain text in markdown, it will render the plain text. However, if we wrap the words with an asterisk or underscore, it formats the text in italics type. Alternatively, if we close the word with two asterisks or underscores, it will change the format to bold type. Furthermore, we can format the text with a strikethrough line by wrapping the word with tilde symbols.

Moving on, we can organize our points into a list item. If we place a number and a dot at the beginning of a paragraph, we can create an ordered list item. Alternatively, if we place a leading asterisk (`*`) in front, we can create an unordered list item. To create a sub-list, we need to place four leading spaces in the front of the list item.

We can also use markdown syntax to structure our data in a table. Creating a table is easy; we just need to separate each column with a pipe (`|`) symbol and append a new row from a new line. To differentiate column names and data values, simply place three dashes (`---`) between the column names and values.

Furthermore, we can embed R code in the R Markdown report. By using the `knitr` package, one can compile the R code and render the execution result in an HTML report. Here, we use three backticks (`) to create a code chunk. If we embed `{R}` in between the code chunks, when we compile the code, R will execute the R script and present the execution result below the code block. On the other hand, if we do not specify the programming language type in the code chunk, the `knitr` package will treat the code within the block as plain text.

Last, we demonstrate how to create an external link in a markdown report. To achieve this, we use brackets to wrap the link name and use parenthesis to wrap the link URL. To render an external image, we need to add a leading `! [alt text]` in front of the URL link, and pass the word to display as the second parameter in the parentheses.

There's more...

Besides using markdown syntax to create a report, one can use LaTeX syntax instead. To generate a LaTeX report with R, one can use the `Sweave` package. Similar to R Markdown, the `Sweave` package enables the user to combine R code with formatted text into one reproducible document.

For more information, please read the documentation for the `Sweave` package:

Sweave: Dynamic generation of statistical reports using literate data analysis by *Friedrich Leisch*. *Proceedings in Computational Statistics* by *Wolfgang Härdle* and *Bernd Rönz*, editors, Compstat 2002, pages 575-580. Physica Verlag, Heidelberg, 2002. ISBN 3-7908-1517-9.

Embedding R code chunks

In an R Markdown report, one can embed R code chunks into the report with the `knitr` syntax. In this recipe, we introduce how to create and control the output with different code chunk configurations.

Getting ready

Ensure you have installed the latest version of R and RStudio on your operating system. Also, you need to have created and opened a new R Markdown (`.rmd`) file in RStudio.

How to do it...

Please perform the following steps to create an R code chunk in the markdown report:

1. First, create a basic code chunk with the `knitr` syntax:

Markdown	Preview

``{r}
# code block 	
a <- 3
b <- 2
a + b
``
	# code block
	a <- 3
	b <- 2
	a + b
	
	## [1] 5

2. We can hide the script by setting `echo=FALSE`:

Markdown	Preview

``{r, echo=FALSE}
# code block 	
a <- 3
b <- 2
a + b
``
	## [1] 5

3. Alternatively, we can stop evaluating the code by setting `eval=FALSE`:

Markdown	Preview

``{r, eval=FALSE}
# code block 	
a <- 3
b <- 2
a + b
``
	# code block
	a <- 3
	b <- 2
	a + b

4. Moreover, we can choose not to render both evaluation result and script by setting `include=FALSE`:

Markdown	Preview
` ```{r, include=FALSE}` `# code block` `a <- 3` `b <- 2a + b` ` ``` `	

5. You can create inline code by closing the code with a single backtick:

Markdown	Preview
``Inline code demo: `r a <- 3;b <- 2;a + b` ``	`Inline code demo: 5`

6. Moving on, we can render a figure in the markdown report with a given figure width and height:

Markdown	Preview
` ```{r, fig.width=6, fig.height=4}` `data(iris)` `plot(iris)` ` ``` `	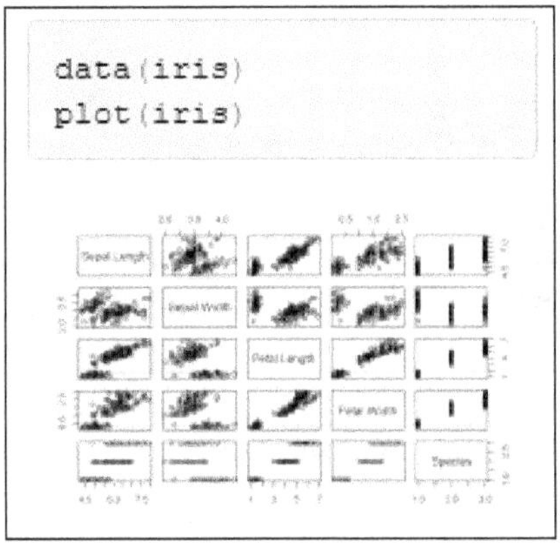

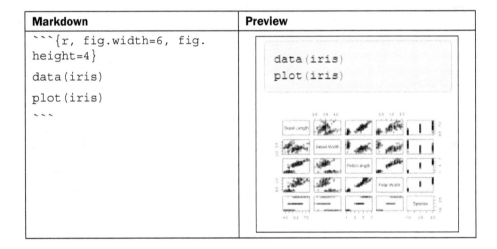

7. We can turn off the warning message by setting `warning=FALSE`:

<table>
<tr><th>Markdown</th><th>Preview</th></tr>
<tr><td>

```
```{r, fig.width=6, fig.
height=4,warning=FALSE}
library(ggplot2)
g <- ggplot(data=iris,
mapping=aes(x=Sepal.
Length, y=Sepal.Width,
colour=Species))
print(g + geom_point())
```
```

</td><td>
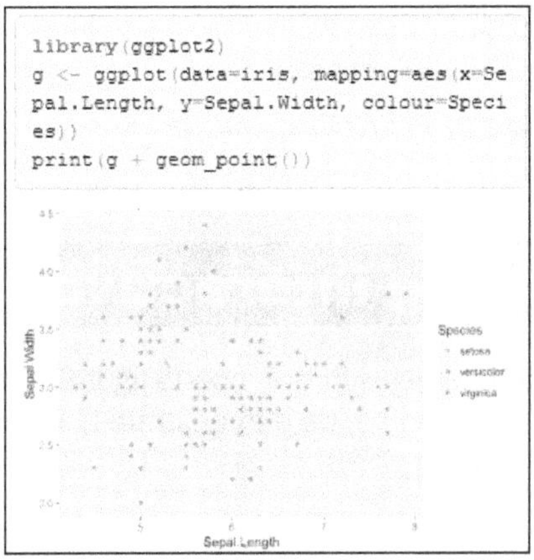

</td></tr>
</table>

How it works...

In an R Markdown report, we can embed R code chunks with the `knitr` package. To create an R code chunk, we begin by placing three backticks followed by an `r` in braces. We then insert our R code after the braces. Last, we close the code chunks with three backticks. After following these steps, we can click `knit *` to start compiling the RMD report. Besides creating code chunks, one can begin the inline code with one backtick followed by an `r`, and end the code line with one backtick.

When creating R code chunks, we can pass parameters into braces to control the rendering of the code. Here we list some parameters that can be passed:

- `eval`: Determines whether to execute the script
- `echo`: Determines whether to render the script
- `warning`: Decides whether to print out the warning message
- `include`: Determines whether to present both the code script and the execution result

We can control the figure width and height by passing `fig.width` and `fig.height` into braces. We can also plot the figure on a markdown report with either the base R `graphics` package or the `ggplot2` package.

There's more...

To control the global option of the code chunk, one can use `opts_chunk` from the `knitr` package:

```
```{r, include=FALSE}
knitr::opts_chunk$set(fig.width=6, fig.height=4, fig.path='images/')
```
```

Creating interactive graphics with ggvis

In order to interact with the reports figures, one can create an interactive graphic with `ggvis`. In this recipe, we demonstrate how to build our first interactive plot from the real estate dataset.

Getting ready

Before starting this recipe, you should download the `RealEstate.csv` dataset from the following GitHub link:

https://github.com/ywchiu/rcookbook/blob/master/chapter8/RealEstate.csv

How to do it...

Please perform the following steps to create an interactive plot with `ggvis`:

1. Install and load the `ggvis` package:

   ```
   > install.packages("ggvis")
   > library(ggvis)
   ```

2. Import `RealEstate.csv` into an R session:

   ```
   > house <- read.csv('RealEstate.csv', header=TRUE)
   > str(house)
   'data.frame': 781 obs. of  8 variables:
    $ MLS         : int  132842 134364 135141 135712 136282 136431
   137036 137090 137159 137570 ...
    $ Location    : Factor w/ 54 levels " Arroyo Grande",..: 21 44 44
   39 50 42 50 50 39 22 ...
    $ Price       : num  795000 399000 545000 909000 109900 ...
    $ Bedrooms    : int  3 4 4 4 3 3 4 3 4 3 ...
   ```

```
$ Bathrooms   : int  3 3 3 4 1 3 2 2 3 2 ...
$ Size        : int  2371 2818 3032 3540 1249 1800 1603 1450 3360
1323 ...
$ Price.SQ.Ft: num  335 142 180 257 88 ...
$ Status      : Factor w/ 3 levels "Foreclosure",..: 3 3 3 3 3 3 3
3 3 3 ...
```

3. Next, we can create a scatterplot with **Size** as the *x* axis and **Price** as the *y* axis:

```
> house %>% ggvis(~Size, ~Price) %>% layer_points()
```

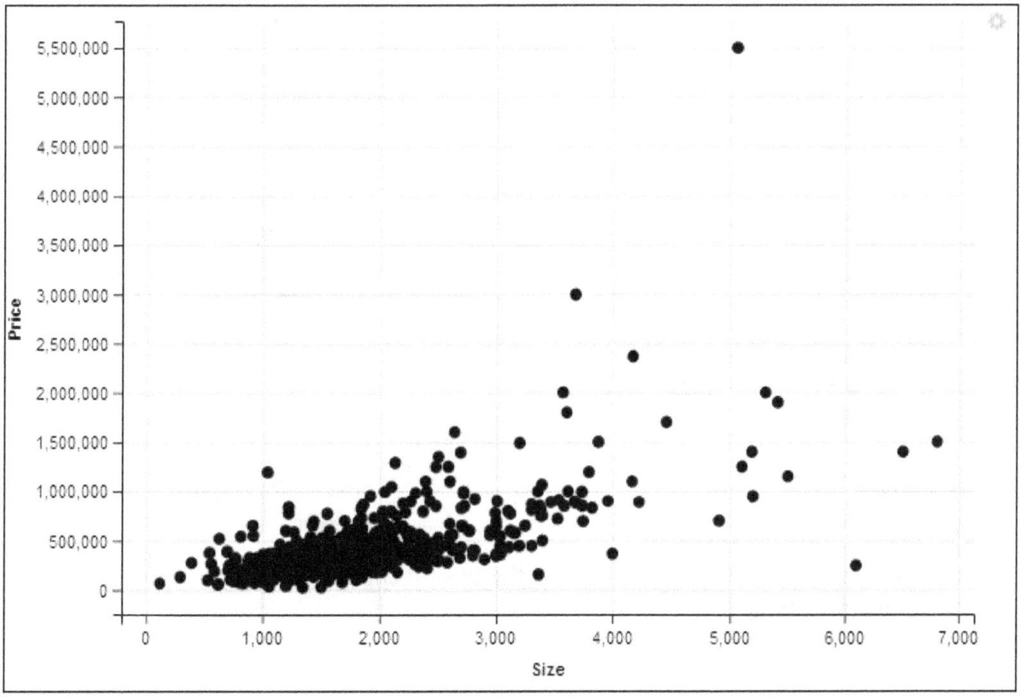

Figure 8: A scatterplot of house size against price

4. Moreover, we can use the `filter` function of `dplyr` to subset data first:

```
> house %>% filter(Price >= 500000) %>% ggvis(~Size, ~Price) %>%
layer_points()
```

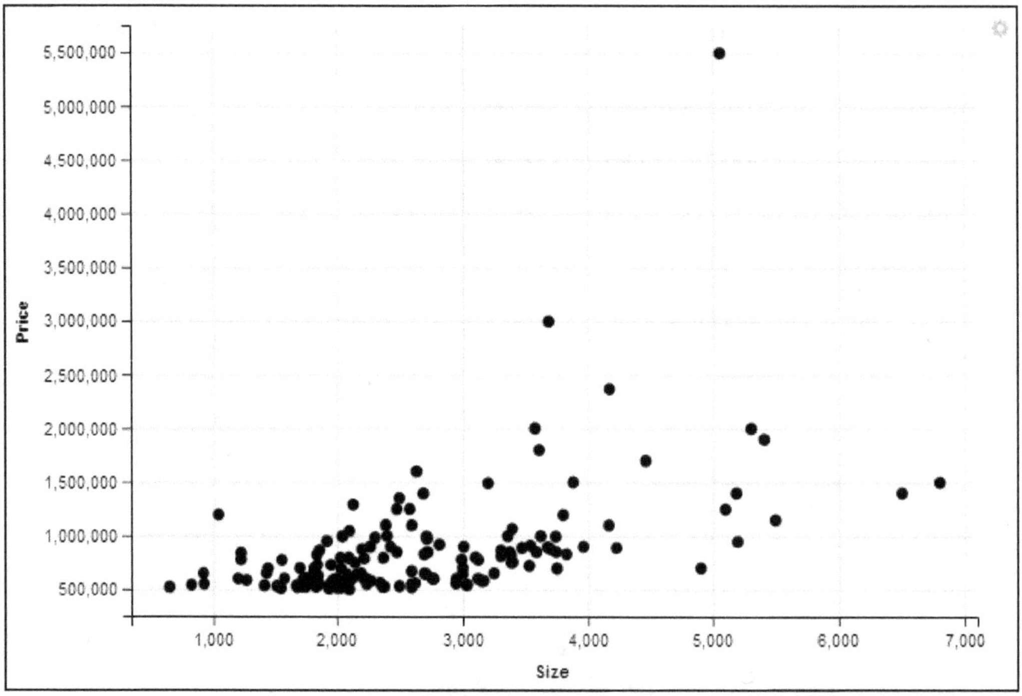

Figure 9: A scatterplot filtered with the dplyr function

5. Create an interactive plot that allows the user to choose the color from a drop-down menu:

```
> house %>% ggvis(~Size, ~Price, fill := input_select(choices =
c("red", "blue", "green"), selected = "blue", label = "Color"))
%>% layer_points()
```

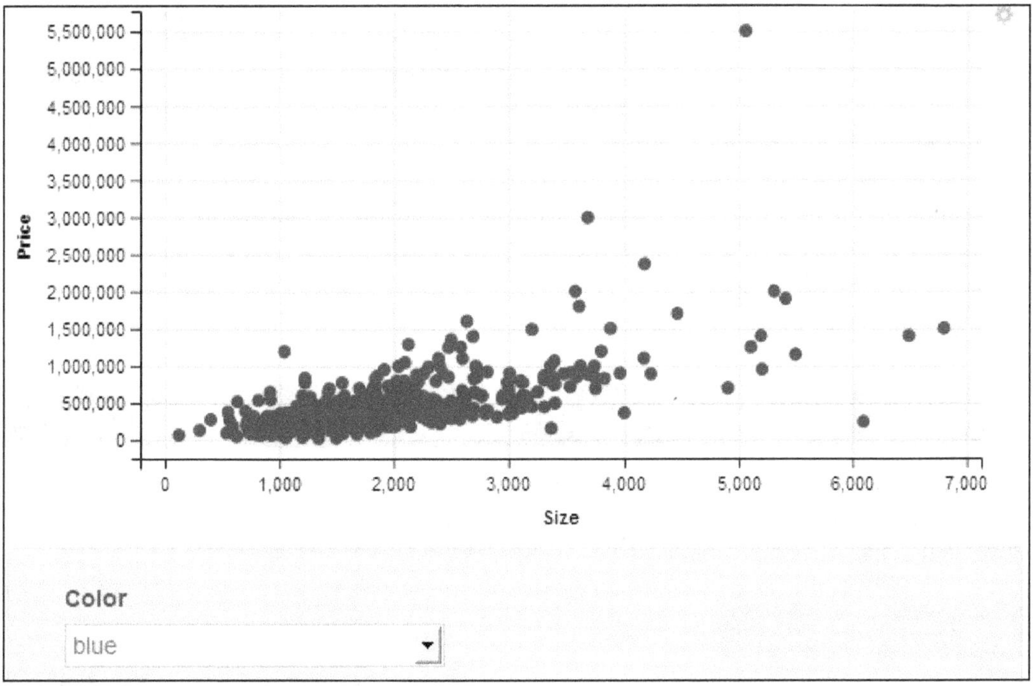

Figure 10: An interactive scatterplot that allows the user to change the color of the points

How it works...

The `ggvis` package creates HTML output with CSS and JavaScript. Thus, one can embed `ggvis` graphics into web applications or HTML reports. Unlike `ggplot2`, which only creates static image, `ggvis` enables the user to manipulate the data visualization through web form interaction.

To create a `ggvis` plot, we need to first install and load the `ggvis` library into an R session. We can then read the real estate dataset into an R data frame using the `read.csv` function. We can find some basic housing features in the real estate dataset with the `str` function.

As we wish to explore the relationship between house size and price, we need to make a scatterplot from these two features with `ggvis`. To create a basic scatterplot, we first need to load the dataset `house`, and then use the pipe function `%>%` from `magrittr` to pipe the data to the `ggvis` function. Here, the `ggvis` function is used to create a basic plot object, and we can then pass house size as *x* position and house price as *y* position into the `ggvis` function, by placing a tilde (~) in front of the variable name. Last, we can use the `layer_point` function to make the scatterplot.

Using the pipe operator in the `dplyr` package allows us to pipe the data to a `filter` function, and then pipe the filtered data to the `ggvis` function. We complete our figure by adding a drop-down menu of color choice in the **fill** option with the `input_select` function.

There's more...

If you would like to export a `ggvis` plot in SVG format, you need to first install the `vega` (a visualization grammar) runtime. For more information about `vega`, please visit the following GitHub page:

`https://github.com/vega/vega`

Understanding basic syntax and grammar

The `ggvis` uses similar grammar and syntax to `ggplot2`, and we can use this basic syntax to create figures. In this recipe, we cover how to use `ggvis` syntax and grammar to build advanced plots.

Getting ready

Ensure you have installed and loaded `ggvis` into your R session. Also, you need to complete the previous steps by storing `house` in your R environment.

How to do it...

Please perform the following steps to create plots with `ggvis`:

1. First, create a scatterplot by mapping **Size** to the *x* axis and **Price** to the *y* axis. Furthermore, we can assign different colors or shapes to points with different statuses:

```
> house %>% ggvis(~Size, ~Price, fill=~Status, size=10,
shape=~Status) %>% layer_points()
```

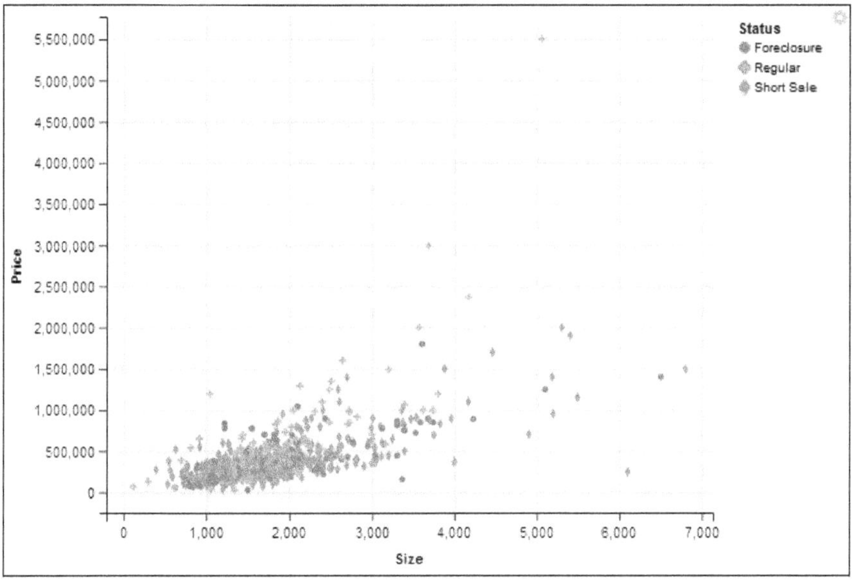

Figure 11: A scatterplot in different shapes and colors by housing status

2. Also, we can use the `add_props` function to change the fill color to red:

```
> house %>% ggvis(~Size, ~Price, fill=~Status, size=10,
shape=~Status) %>% layer_points() %>% add_props(fill:="red") %>%
layer_points()
```

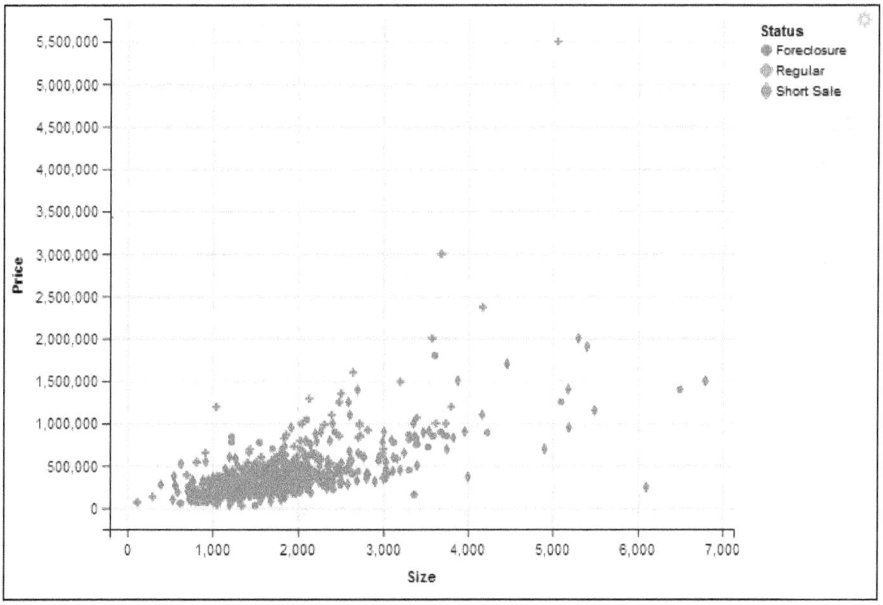

Figure 12: Turn the data points red

3. Next, we can render a smooth regression line with a standard error ribbon onto the current points:

```
> house %>% ggvis(~Size, ~Price) %>% layer_points(fill=~Status)
%>% layer_smooths(stroke:= "red", strokeWidth:=3, se=TRUE)
```

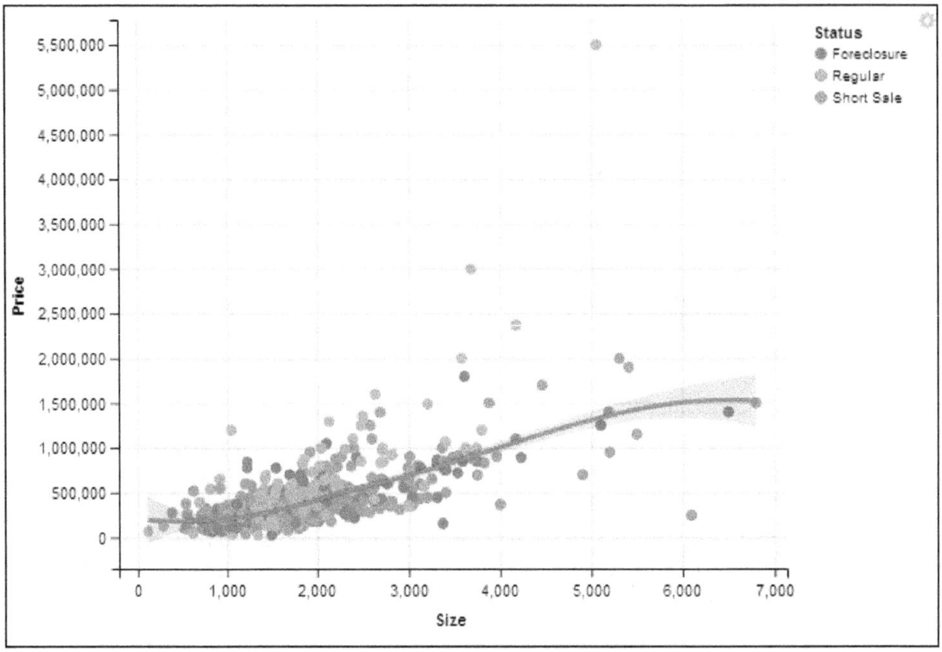

Figure 13: Render the regression line onto the scatterplot

4. Moreover, we can render a linear regression line with standard error ribbon on the data points:

```
> house %>% ggvis(~Size, ~Price) %>% layer_points(fill=~Status)
%>%   layer_model_predictions(formula= Price ~ Size, model = "lm",
se = TRUE)
```

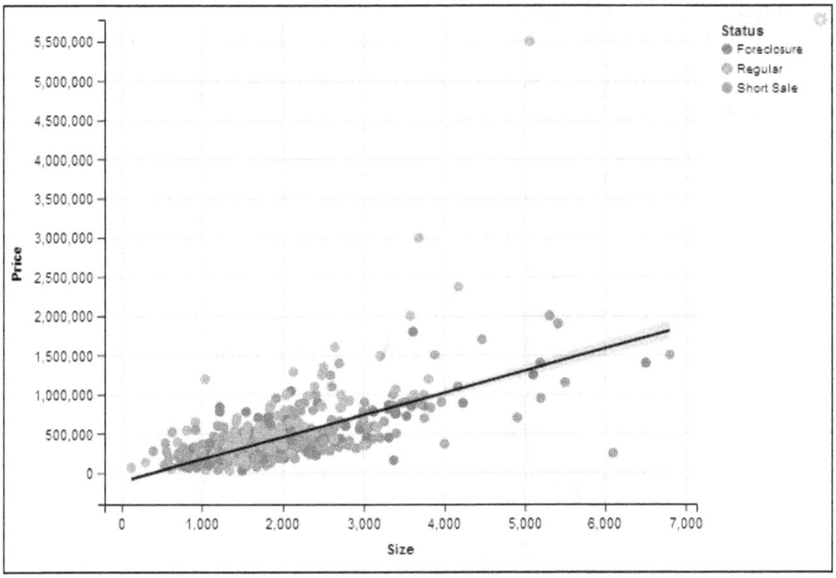

Figure 14: Adding a linear regression line onto the plot

5. We can group the data by status and draw separate linear regression lines by property status:

```
> house %>% ggvis(~Size, ~Price) %>% layer_points(fill=~Status)
%>% group_by(Status) %>% layer_model_predictions(model = "lm",
stroke=~Status, strokeWidth:=3, se=TRUE)
```

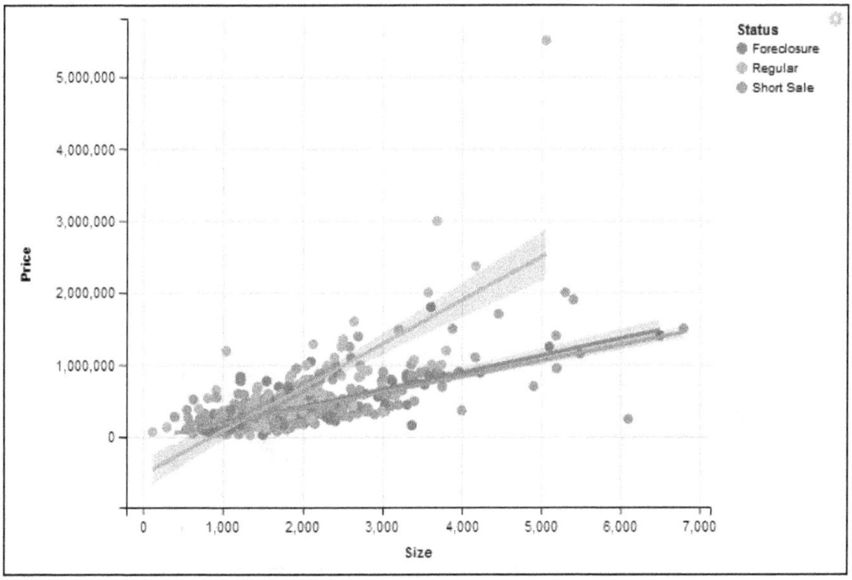

Figure 15: Adding multiple regression lines based on housing status

6. Moving on, we can make a blue histogram from the `Size` attribute of the house dataset:

```
> house %>% ggvis(~Size, fill:="blue") %>% layer_
histograms(width=300)
```

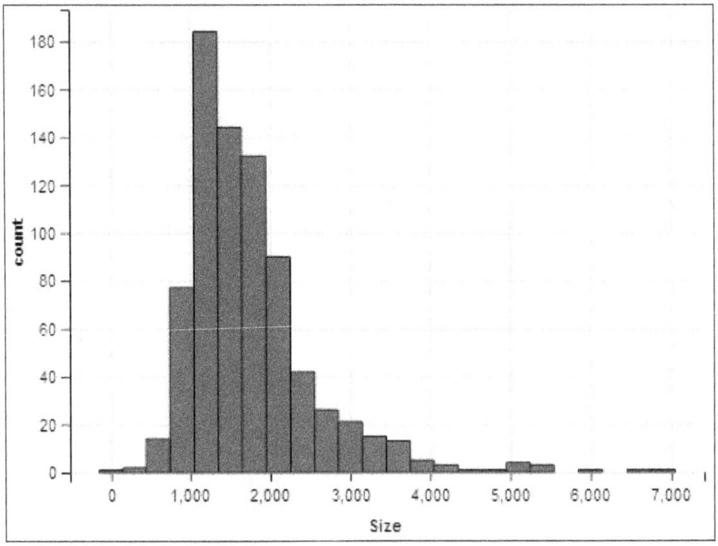

Figure 16: Making a histogram from the house dataset

7. We can then create a bar plot with **Status** as the *x* axis and **Price** as the *y* axis:

```
> house %>% ggvis(~Status, ~Price, fill=~Status) %>% layer_
bars(width=0.5)
```

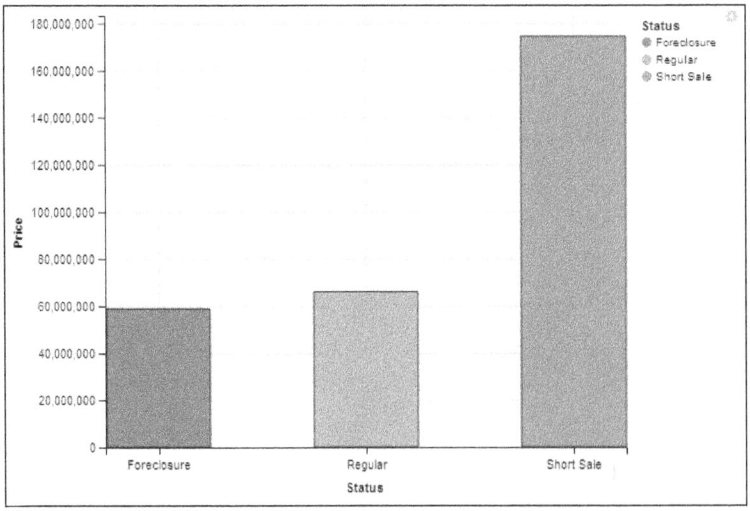

Figure 17: Creating a bar plot with housing status against price

8. Moreover, we can plot a boxplot **Status** as the *x* axis and **Price** as the *y* axis:

```
>house %>% ggvis(~Status, ~Price, fill=~Status)  %>% layer_
boxplots()
```

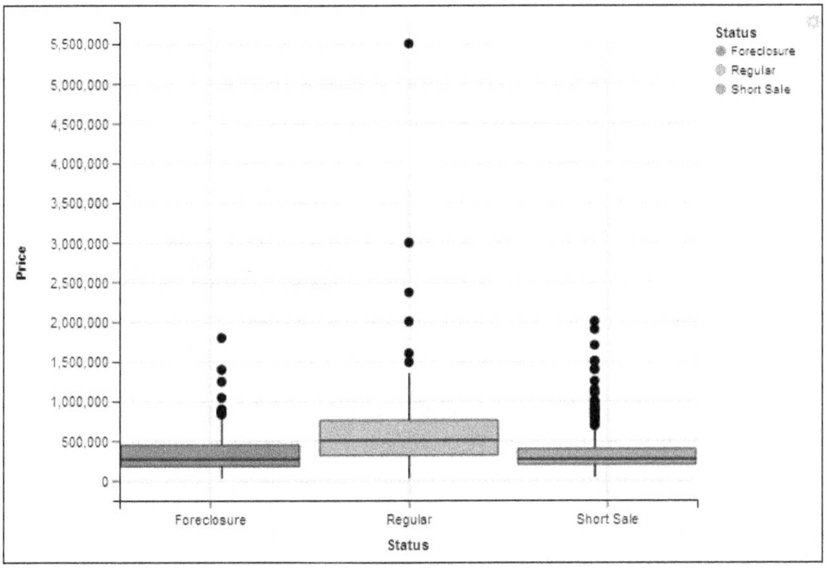

Figure 18: Creating a boxplot with status against price

9. Last, we can make a line plot with **Status** as the *x* axis and **Price** as the *y* axis:

```
> house %>% ggvis(~Size, ~Price) %>% layer_lines(strokeWidth:=5,
stroke:="blue")
```

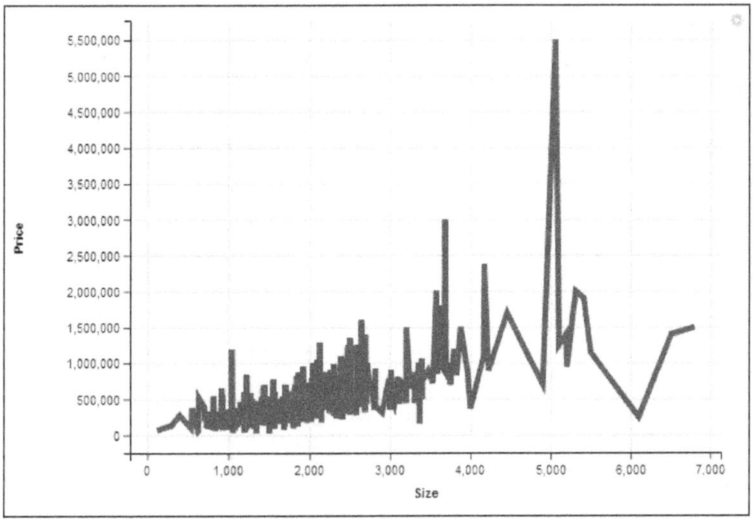

Figure 19: Creating a line chart with size against price

How it works...

In ggvis, one can use a simple layer to create lines, points, and other geometry objects in the plot. To construct a ggvis figure, use the following grammar format:

```
<data>  %>%
  ggvis( ~ <x property> , ~ <y property>, fill = ~ <fill property>,
...) %>%
  layer_<marks>()
```

To a dplyr or ggplot2 user, the previous grammar format may appear familiar. Similar to dplyr, one can use the pipe (%>%) function to construct a data pipeline. Also, similar to ggplot2, one can build a figure with aesthetic mapping and geometry objects.

In the first step, we passed house size as x property and house price as y property, with a point size of 10, into the ggvis function. We also set the color and style of the point in accordance with the status property variable. Last, we used the layout_points function to render the scatterplot. In this example, we find that it is possible to modify the visual property in the ggvis function. However, we can also alter the property with the add_prop function. In the next step, we used := notation to turn all data points red.

In ggplot2, one important feature is to add a regression line (we discuss regression lines further in *Chapter 11, Supervised Machine Learning*) onto the plot with the geom_smooth function. In ggvis, we can also add a smooth regression line with the layour_smooths function. Here, we set the stroke color to red, the stroke width to 3, and add the standard error ribbon by setting se equal to TRUE. Moreover, we can add a linear regression line with the layer_model_predictions function. However, as our dataset contains real estate data of different property statuses, we need to make separate prediction lines. Thus, we can use the group_by function to separate data into different groups in accordance with the **Status** variable. We can then create different prediction lines based on the status of the property.

Moving on, we can use different layer_<marks> to create various charts. In the samples, we created a histogram with layer_histograms. Next, we made a bar plot with layer_bars. We then plotted a boxplot with layer_boxplots. Last, we made a line graph with layer_lines.

There's more...

While ggvis provides interactive plots, one can also control the graph through web forms. However, there are some disadvantages of ggvis compared to ggplot. One obvious drawback of ggvis is that the available plot types are limited. Unlike ggplot2, ggvis does not offer coordinates, so it is not possible to make a pie chart. Also, one cannot use ggvis to create contour charts or dendrograms. Therefore, ggplot2 cannot be fully replaced by ggvis.

Controlling axes and legends

Besides making different plots with various layer types, one can control the axes and legends of a ggvis plot. In this recipe, we demonstrate how to set the appearance properties of both axes and legends.

Getting ready

Ensure you have installed and loaded ggvis into your R session. You also need to complete the previous steps by storing house in your R environment.

How to do it...

Please perform the following steps to control axes and legends in ggvis:

1. First, we use the add_axis function to control the axis orientation and label of a ggvis plot:

```
> house %>% ggvis(~Size, ~Price) %>% layer_points() %>%
+   add_axis("x", title = "Real Estate Square Feet", orient="top")
%>%
+   add_axis("y", title = "Real Estate Price", title_offset = 80)
```

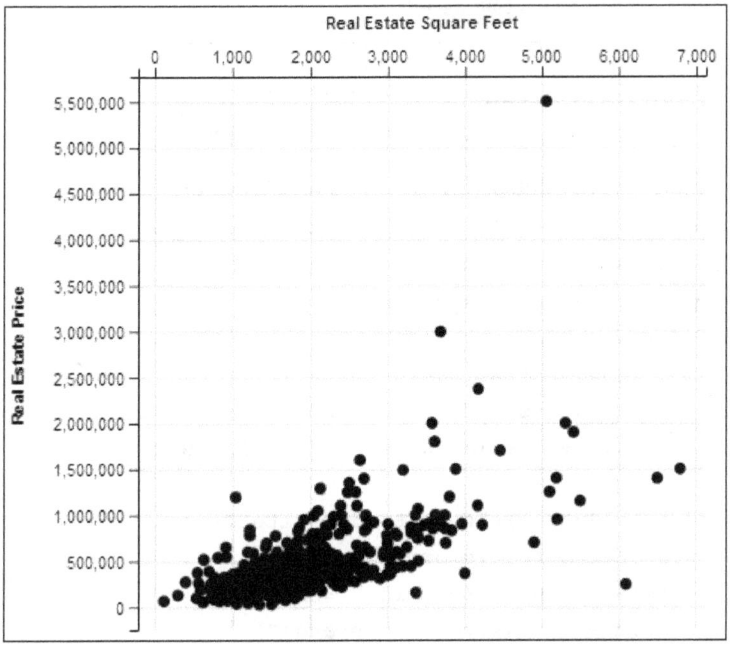

Figure 20: Creating an axis on the ggvis plot

2. We can control the scale of the axis:

```
> house %>% ggvis(~Size, ~Price) %>% layer_points() %>%
+   add_axis("x", title = "Real Estate Square Feet",
+            subdivide = 4, values = seq(0,8000, by=1000)) %>%
+   add_axis("y", title = "Real Estate Price",
+            subdivide = 5, values = seq(0,5500000, by=1000000),
title_offset = 80)
```

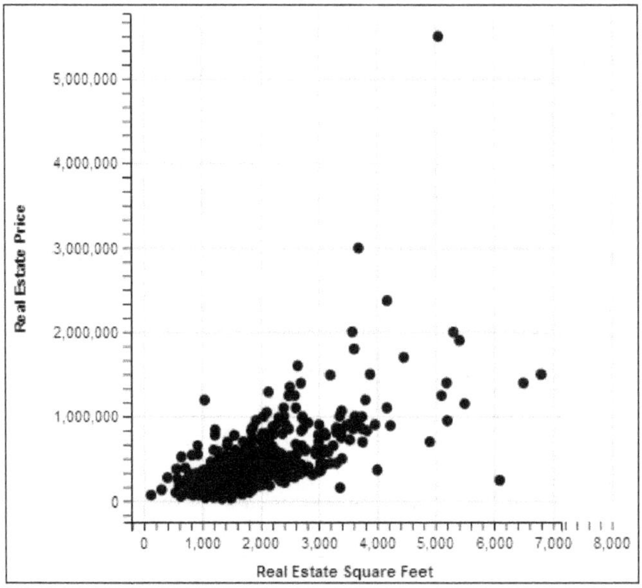

Figure 21: Controlling the axis of the plot

3. Furthermore, we can customize the visual properties of the axis:

```
> house %>% ggvis(~Size, ~Price) %>% layer_points() %>%
+   add_axis("x", title = "Real Estate Square Feet",
+     properties = axis_props(
+       axis = list(strokeWidth = 2),
+       grid = list(stroke = "blue"),
+       ticks = list(stroke = "red", strokeWidth = 2),
+       labels = list(align = "left", fontSize = 10)
+   )) %>%
+   add_axis("y", title = "Real Estate Price", title_offset = 80,
+     properties = axis_props(
+       axis = list(strokeWidth = 2),
+       grid = list(stroke = "orange"),
+       ))
```

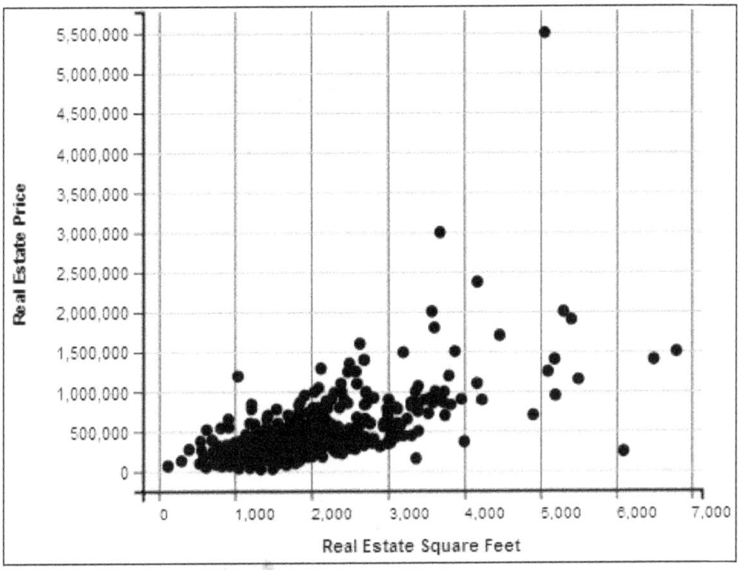

Figure 22: Adding grid lines onto the plot

4. Moving on, we can control the legend of the ggvis plot:

```
> house %>% ggvis(~Status, ~Price, fill=~Status) %>% layer_
bars(width=0.5)  %>%  add_legend("fill", title="Sale Status")
```

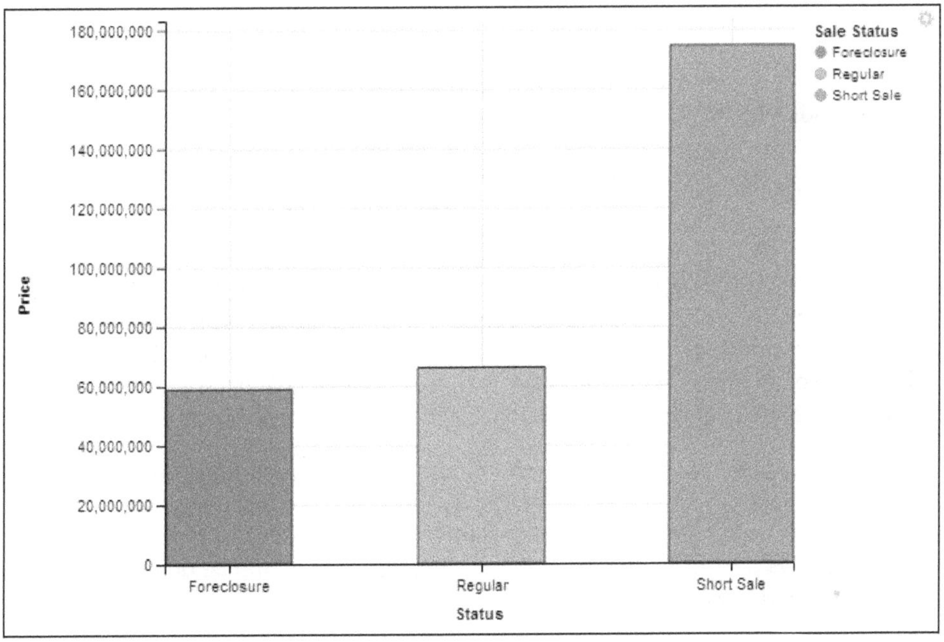

Figure 23: Adding the legend onto the chart

5. Also, we can hide the legend with the `hide_legend` function:

```
> house %>% ggvis(~Status, ~Price, fill=~Status) %>% layer_
bars(width=0.5)  %>%  hide_legend("fill")
```

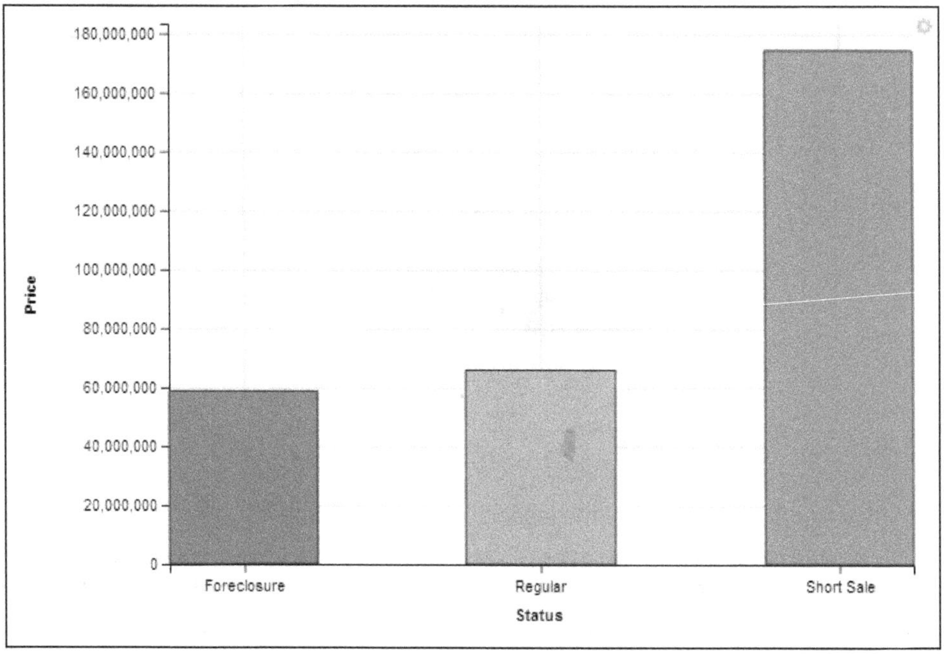

Figure 24: Hiding the legend of the chart

How it works...

In `ggvis`, we can control how axes and legends are displayed in a `ggvis` figure with the `*_axis and *_legend` functions. In this recipe, we start by making a scatterplot with the `layer_point` function. We then use the `add_axis` function to modify the label name, orientation, and label offset of the scatterplot. Next, we illustrate how to change the tick size and padding of the axis. Here, we configure a value option to set the axis limit, and we use `subdivide` to determine the number of minor ticks between each major tick. Besides changing the appearance of an axis, we can customize more axis properties with the `axis_props` function. Here, we can change properties such as the axis, grid, ticks, and label of the scatterplot.

We can modify the appearance of the legend with the `add_legend` function. In this step, we first make a bar plot from the price information. We then use the `add_legend` function to add a legend beneath the bar plot. This function allows us to set the `fill` points on the legend, with **Sale Status** as the legend title. Besides controlling the appearance of a legend, one can hide it with the `hide_legend` function.

There's more...

To modify the visual properties of a legend, one can set `properties` with `legend_props` to control the style:

```
> house %>%
+    ggvis(~Status, ~Price, fill=~Status) %>%
+    layer_bars(width=0.5)  %>%
+    add_legend("fill", title="Sale Status",
+               properties = legend_props(
+                 title = list(fontSize = 14),
+                 labels = list(fontSize = 12),
+                 symbol = list(shape = "square", size = 100)
+               ))
```

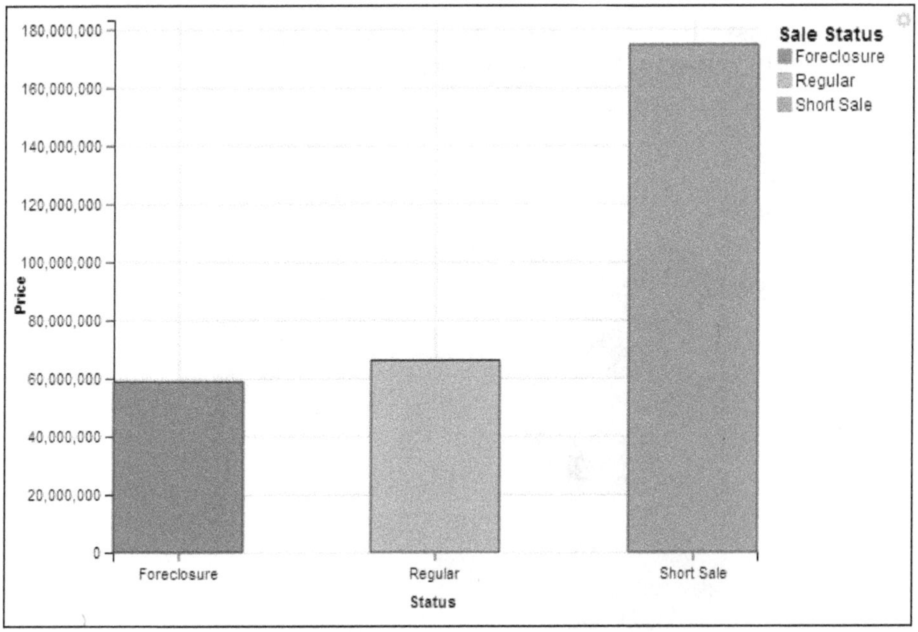

Figure 25: Configure the legend properties

Using scales

Besides changing the appearance of axes and legends, we can rescale the mapping of the data and how it should be displayed on the plot with the scale function. In this recipe, we introduce how to scale data in ggvis.

Getting ready

Ensure you have installed and loaded ggvis into your R session. Also, you need to complete the previous steps by storing house in your R environment.

How to do it ...

Please perform the following steps to rescale data in ggvis:

1. First, create a bar plot and then replace the linear scale with a power scale:

```
> house %>% ggvis(~Status, ~Price, fill=~Status) %>% layer_bars()
%>% scale_numeric("y", trans = "pow", exponent = 0.2)
```

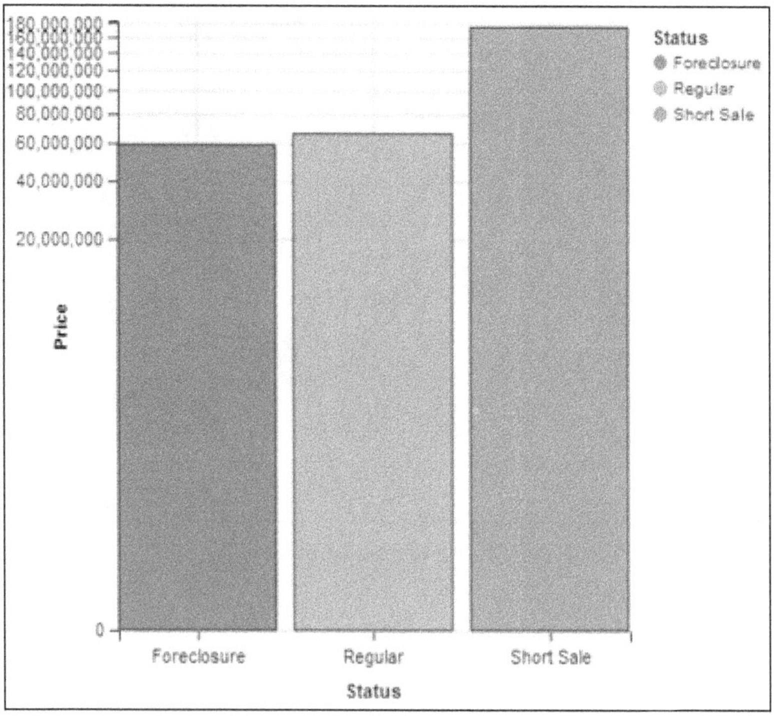

Figure 26. Adjusting the scale of the bar plot

2. We can change the fill color with the `scale_nominal` function:

```
> house %>% ggvis(~Status, ~Price, fill=~Status) %>% layer_bars()
%>% scale_nominal("fill", range = c("pink", "green", "lightblue"))
```

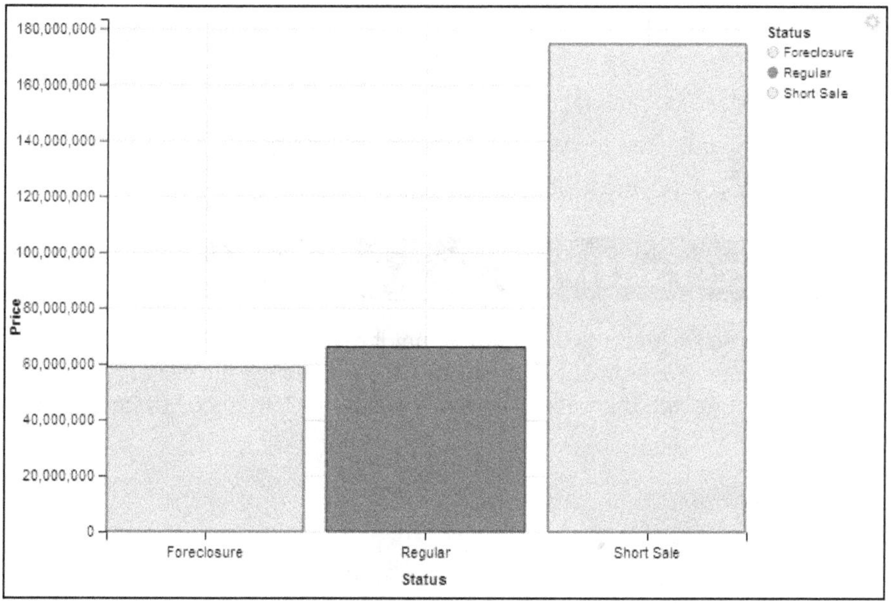

Figure 27. Changing the fill color of the bar plot

How it works...

In the previous recipe, we introduced how to change the axis property with the `add_axis` function. However, the `add_axis` function only modifies the appearance of the axis; it does not alter the mapping data. To scale both the axis and the mapping data, one can use the `scale_*` function.

In this example, we begin by making a bar plot with the `layer_bar` function. We then use `scale_numeric` to replace the linear scale with a power scale. This rescales both the *y* axis and the height of the bars. Besides rescaling continuous values with the `scale_numeric` function, one can change the property of a discrete value with `scale_nominal`. At the end of the previous recipe, we demonstrated how to use `scale_nominal` to refill the bar with a given color vector.

There's more ...

Besides the `scale_nominal` and `scale_numeric` functions, one can use the following function to rescale data on a `ggvis` plot:

- `scale_datetime`: Add a `datetime` scale to a `ggvis` object
- `scale_logical`: Add a logical scale to a `ggvis` object
- `scale_ordinal`: Add an ordinal scale to a `ggvis` object
- `scale_singular`: Add a singular scale to a `ggvis` object

Adding interactivity to a ggvis plot

One of the most attractive features of `ggvis` is that it can be used to create an interactive web form. This allows the user to subset the data, or even change the visual properties of the plot, through interacting with the web form. In this recipe, we introduce how to add interactivity to a `ggvis` object.

Getting ready

Ensure you have installed and loaded `ggvis` into your R session. Also, you need to complete the previous steps by storing `house` in your R environment.

How to do it...

Please perform the following steps to add interactivity to `ggvis`:

1. First, make a bar plot with a drop-down menu and color options:

```
> house %>%
+     ggvis(~Status, ~Price, fill:= input_select(c("red","blue"),
label="Fill Color")) %>%
+     layer_bars() %>%
+     add_axis("x", title = "Status", title_offset = 40) %>%
+     add_axis("y", title = "Price", title_offset = 80)
```

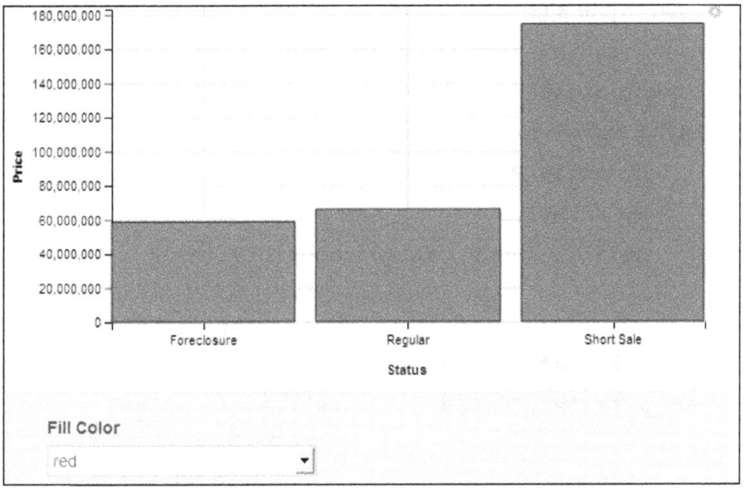

Figure 28: Adding a drop-down menu with color options

2. Create a size slider and a color input text area for a scatterplot:

```
> house %>%
+     ggvis(~Size, ~Price, fill:= input_text("red",label="Fill
Color"), size:=input_slider(1,10)) %>%
+     layer_points() %>%
+     add_axis("x", title = "Size", title_offset = 40) %>%
+     add_axis("y", title = "Price", title_offset = 80)
```

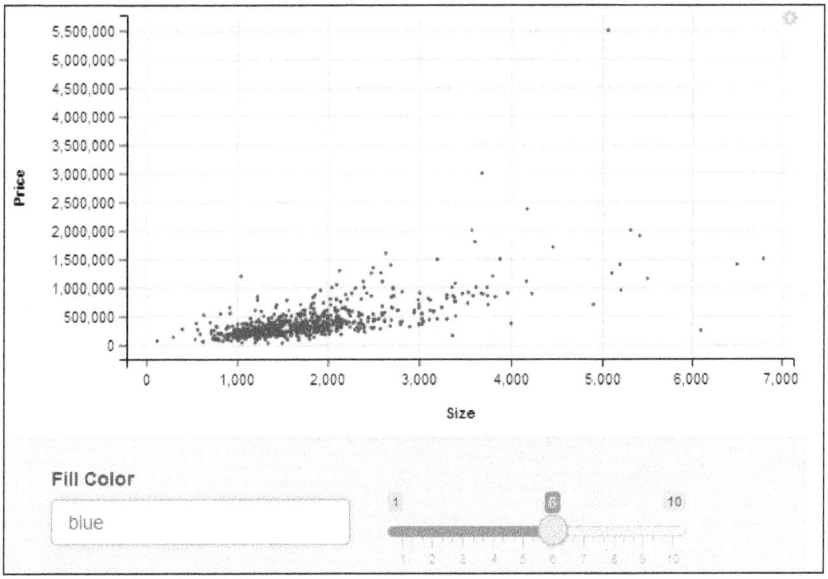

Figure 29: Adding a size slider and a color input text area on a ggvis plot

3. Next, we can create a radio button for users to choose the regression method:

```
> house %>% ggvis(~Size, ~Price) %>%
+     layer_points(fill=~Status)  %>%
+     layer_model_predictions(
+         formula= Price ~ Size,
+         se = TRUE,
+         model = input_radiobuttons(c("loess" = "loess", "Linear" 
= "lm"),
+                                     label = "Model type")) %>%
+     add_axis("x", title = "Size", title_offset = 40) %>%
+     add_axis("y", title = "Price", title_offset = 80)
```

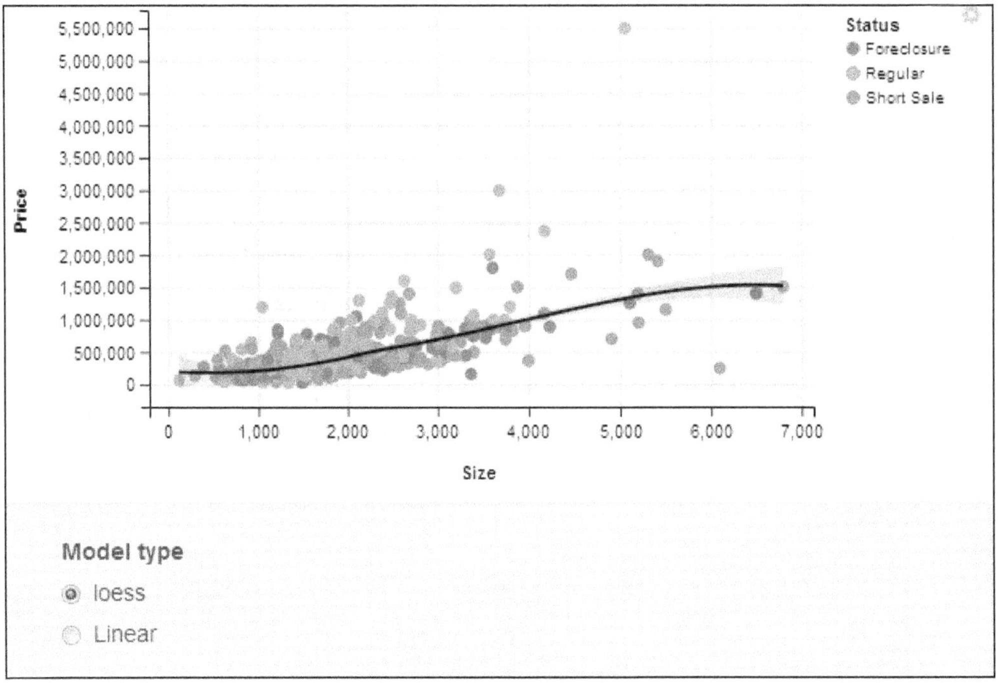

Figure 30: Creating a radio button on a ggvis plot

4. Moving on, we can make a drop-down menu to filter data based on a given sale status:

```
> sale_type <- as.character(unique(house$Status))
> house %>%
+     ggvis(~Size, ~Price, fill=~Status) %>%
+     filter(Status %in% eval(input_select(sale_type)) ) %>%
+     layer_points()  %>%
+     add_axis("x", title = "Size", title_offset = 40) %>%
+     add_axis("y", title = "Price", title_offset = 70)
```

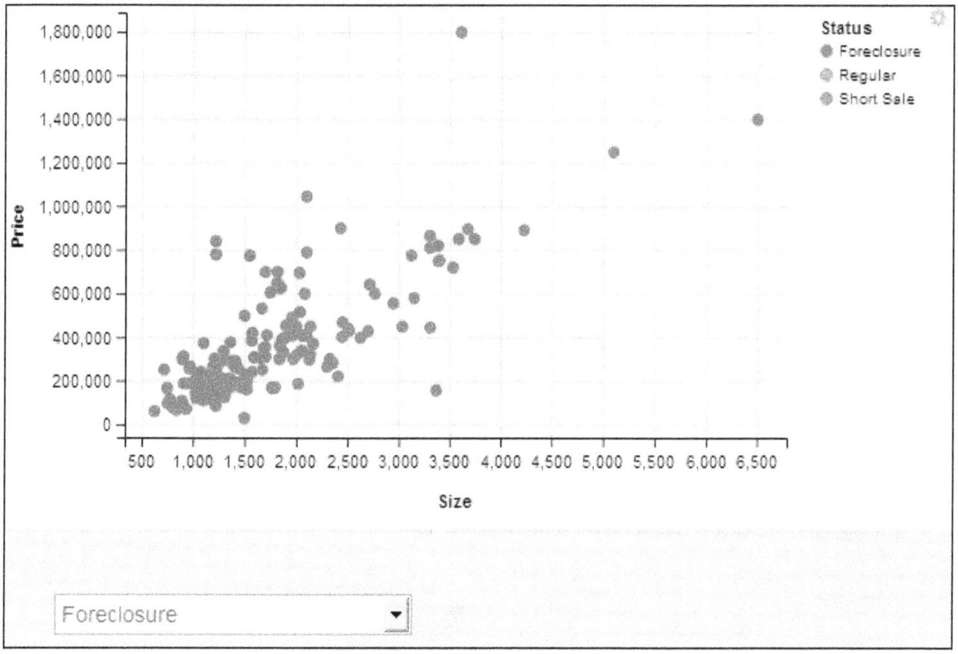

Figure 31: Creating a dropdown menu as a data filter

5. Also, we can subset data based on multiple conditions:

```
> sale_type <- as.character(unique(house$Status))


> house %>%
+     ggvis(~Size, ~Price, fill=~Status) %>%
+     filter(Status %in% eval(input_checkboxgroup(sale_type)) )
%>%
+     layer_points() %>% hide_legend("fill") %>%
+     add_axis("x", title = "Size", title_offset = 40) %>%
+     add_axis("y", title = "Price", title_offset = 70)
```

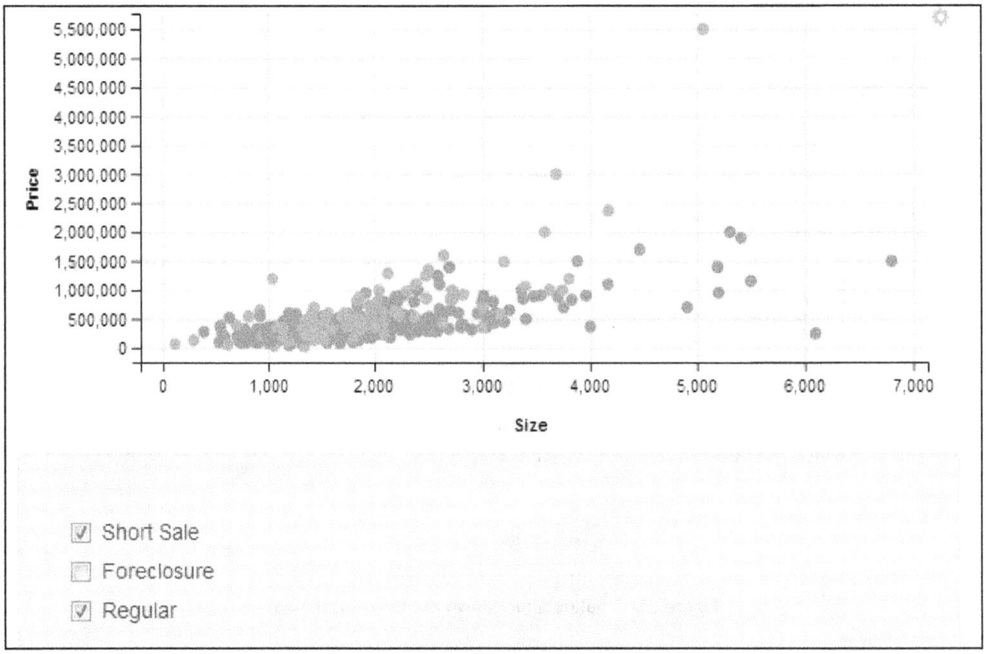

Figure 32: Adding multiple checkboxes onto the ggvis chart

6. Lastly, we can add a radio button for users to choose the attributes of the *x* axis:

```
> house %>%
+     ggvis(x=input_radiobuttons(c("Size", "Price"), map=as.name),
+           y=~Price,
+           fill=~Status) %>%
+     layer_points() %>% hide_legend("fill") %>%
+     add_axis("x", title = "Size", title_offset = 40) %>%
+     add_axis("y", title = "Price", title_offset = 70)
```

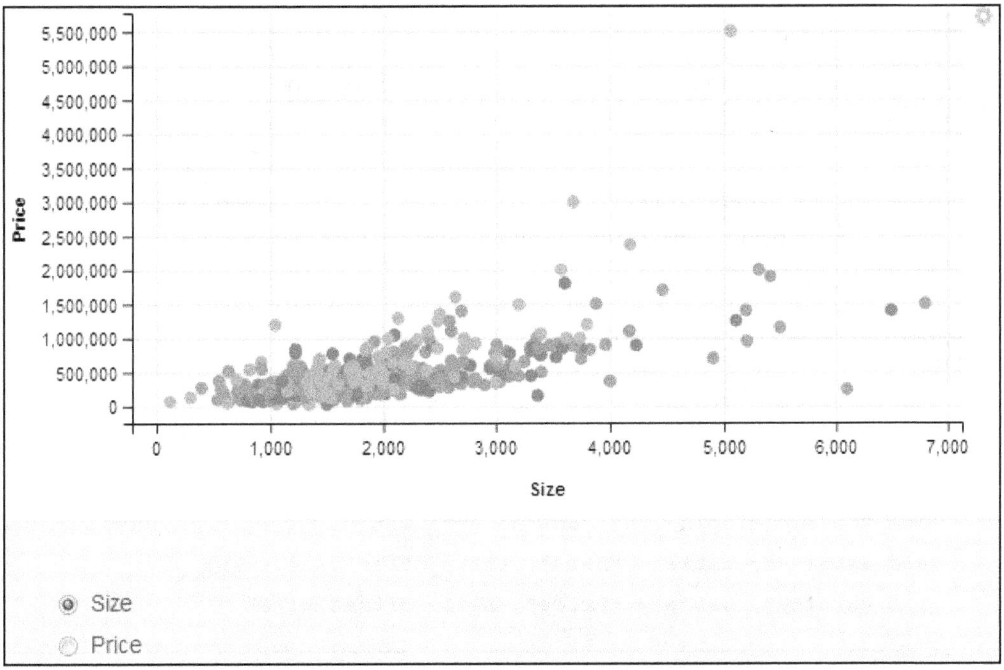

Figure 33: Adding a radio button to control the x axis

How it works...

In the previous steps, we introduced how to add interactivity to a `ggvis` plot. In the first step, we made a bar plot with `layer_bars`. To change the color of the bar plot, we added `input_select` to the fill option. Then, from the rendered plot, a drop-down menu appears with red and blue as the color choices. Besides adding one interactive form on the plot, we can add multiple interactive controls. Thus, in the second step, we add both a drop-down menu and a slider on a scatterplot. This allows the user to adjust both sizes and fill color from the web form. We also made the model property interactive by using a radio button.

Beyond using interactive elements to control visual properties, we can subset the data by variable through form interaction. In step four, we add a drop-down menu to subset the data displayed by housing status. To enable the interactive property, we use an `eval` function to evaluate the expression of the `input_select` function. Furthermore, we can subset data with multiple conditions by adding a checkbox to control the filtering conditions.

Last, we make the *x* position property interactive using a radio button. Therefore, one can choose the variable used in *x* position when rendering the `ggvis` plot.

There's more...

To explore detailed information of each mark in a `ggvis` object, one can use the `add_tooltip` function to add a tooltip to the mark. Moreover, we can control whether the tooltip shows on click or hover by setting the `on` parameter:

```
> display_val <- function(x) {
+     if(is.null(x)) return(NULL)
+     paste0(names(x), ": ", format(x), collapse = "<br />")
+ }
> house %>%
+     ggvis(~Size, ~Price, fill=~Status) %>%
+     filter(Status %in% eval(input_checkboxgroup(sale_type)) ) %>%
+     layer_points() %>%
+     add_tooltip(display_val, on="hover") %>%
+     add_axis("x", title = "Size", title_offset = 40) %>%
+     add_axis("y", title = "Price", title_offset = 70)
```

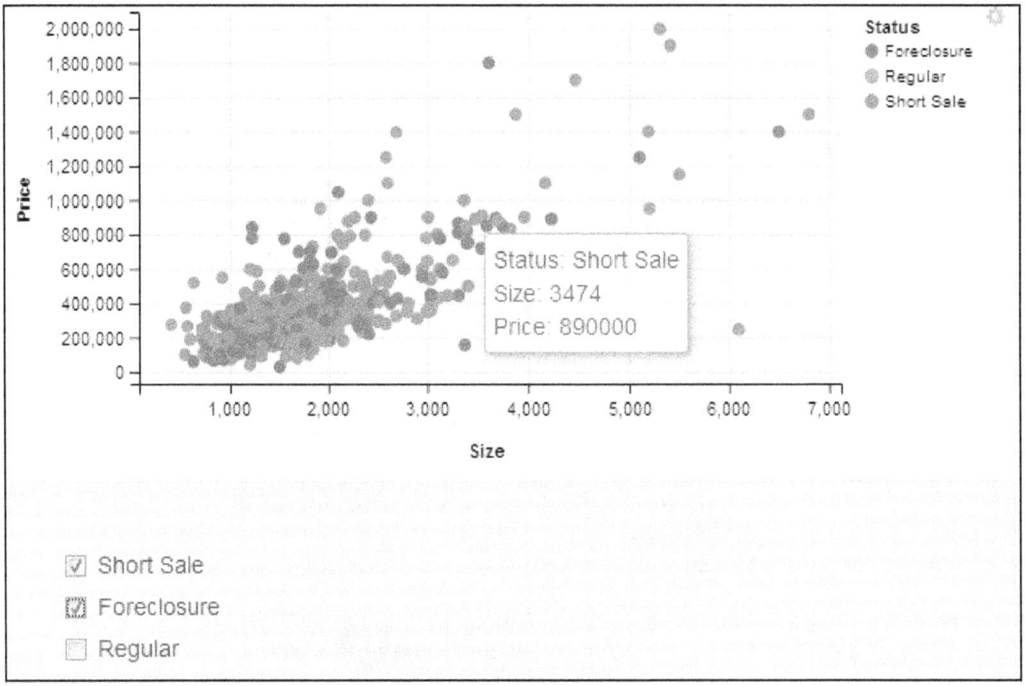

Figure 34. Adding tooltips onto marks

Creating an R Shiny document

Having introduced how to make an R Markdown document and `ggvis` plot, we can now combine these two together in a single report. However, the interactivity property of `ggvis` does not work with a regular R Markdown report. Instead, we can only enable the interactivity of the plot in a Shiny document. In this recipe, we introduce how to create a Shiny document.

Getting ready

Ensure you have installed and loaded `ggvis` into your R session. Also, you need to complete the previous steps by storing `house` in your R environment.

How to do it...

Please perform the following steps to create a Shiny document:

1. First, click **File** | **New File** | **R Markdown** on the menu bar:

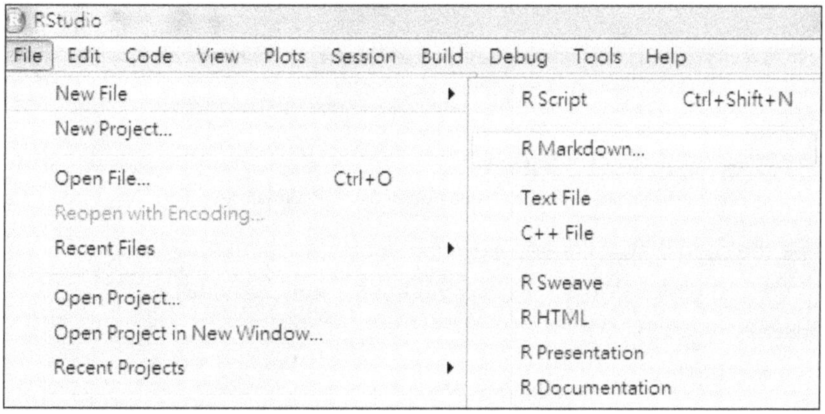

Figure 35: Create an R Markdown file

2. Next, choose **Shiny** as the document type in the left-hand side menu, and then fill in the **Title** and **Author**, and select the output format in the right-hand side menu:

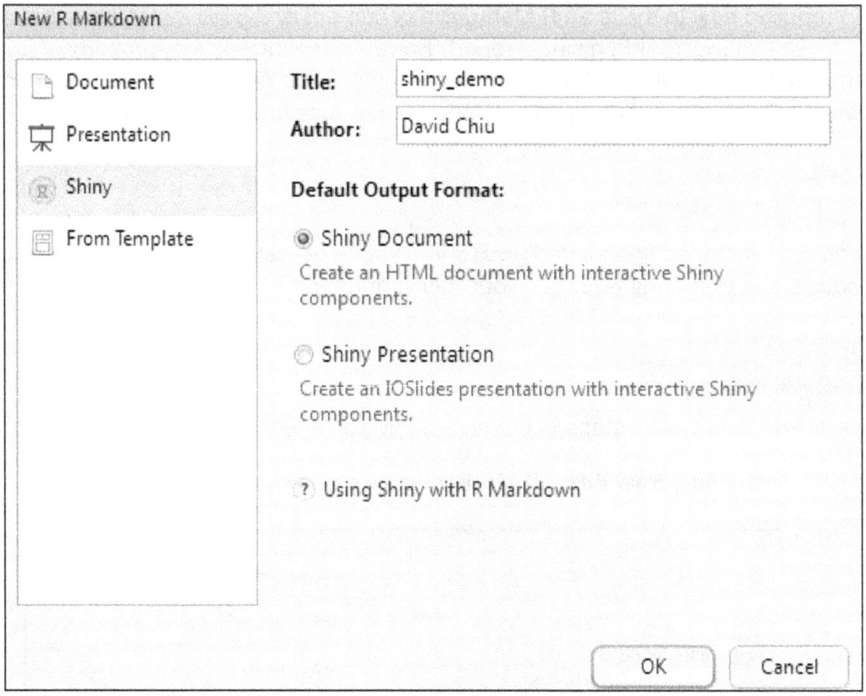

Figure 36: Input title and author of the R Markdown report

3. Type the code that generates a `ggvis` plot into R chunks:

```
```{r, include=TRUE, warning=FALSE, echo=FALSE}
library(ggvis)
library(dplyr)
```


```{r, warning=FALSE}
house <- read.csv('RealEstate.csv', header=TRUE)
housing_status <- as.character(unique(house$Status))
house %>%
  ggvis(~Size, ~Price, fill=~Status) %>%
```
```

````
  filter(Status %in% eval(input_checkboxgroup(housing_status)) )
%>%
  layer_points() %>% hide_legend("fill") %>%
  add_axis("x", title = "Size", title_offset = 40) %>%
  add_axis("y", title = "Price", title_offset = 70)

```
````

4. Click on **Run Document** to compile and render a report:

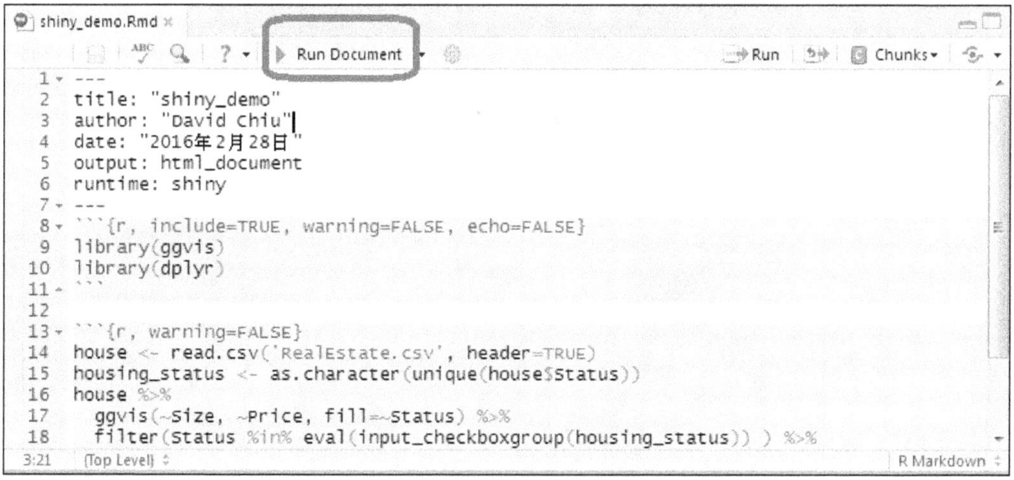

Figure 37: Render the report with Run Document
````

5. The system will request a file name to be entered. Once the file name is given, RStudio will generate a `<filename>.rmd` and `<filename>.html` file and display the formatted report in an open window:

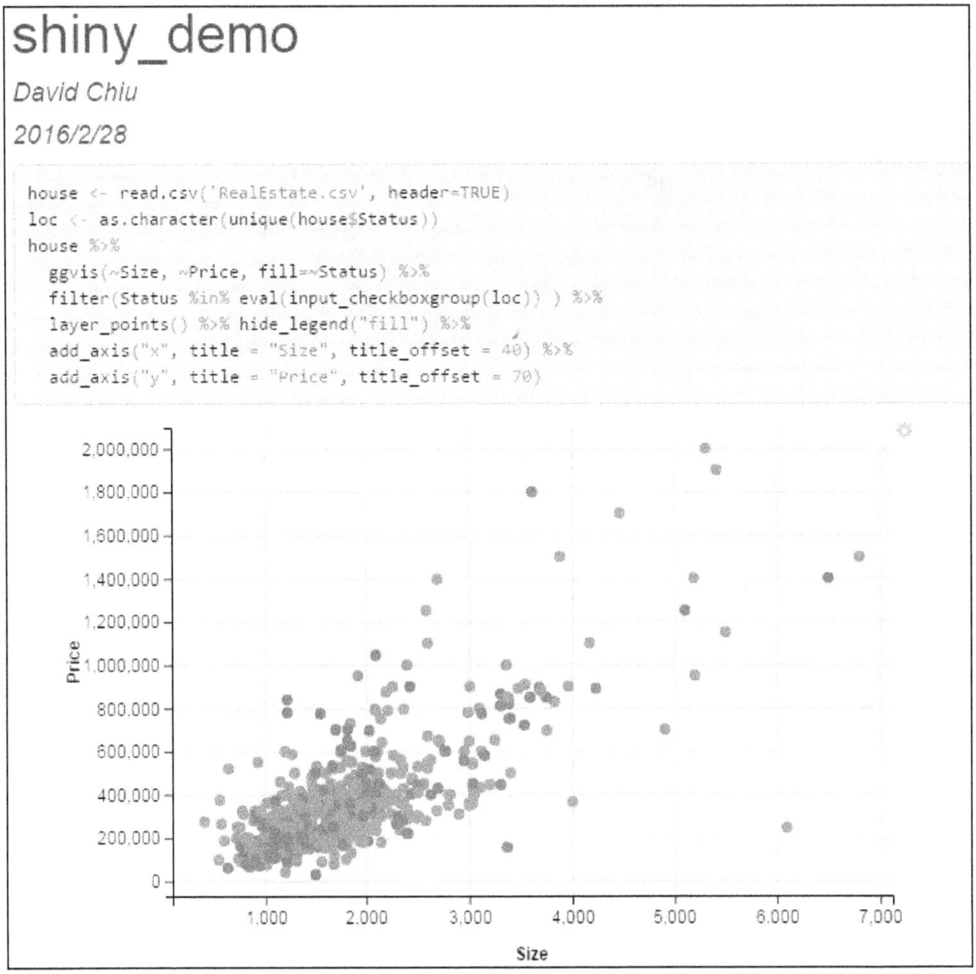
shiny_demo
David Chiu
2016/2/28
house <- read.csv('RealEstate.csv', header=TRUE)
loc <- as.character(unique(house$Status))
house %>%
ggvis(~Size, ~Price, fill=~Status) %>%
filter(Status %in% eval(input_checkboxgroup(loc))) %>%
layer_points() %>% hide_legend("fill") %>%
add_axis("x", title = "Size", title_offset = 40) %>%
add_axis("y", title = "Price", title_offset = 70)
2,000,000
1,800,000
1,600,000
1,400,000
1,200,000
1,000,000
800,000
600,000
400,000
200,000
0
Price
1,000
2,000
3,000
4,000
5,000
6,000
7,000
Size

Figure 38: Create a Shiny report

How it works...

In our first recipe, we introduced how to create an R Markdown report. However, an R Markdown report outputs code chunks and static figures; one cannot perform exploratory data analysis through web interaction. To enable the user to explore the data via a web form, one can build an interactive web page with the Shiny app. In this recipe, we introduced how to create an interactive web report with Shiny.

We began by clicking **R Markdown** in the menu bar to create a new report. We then selected **Shiny** from the left-hand side menu. Next, we needed to fill in the **Title** and **Author**, and choose the output format for our new document. At this point, you will see a new file with R Markdown syntax and simple instructions on how to generate a Shiny document. From the start of the YAML headers, we find a line coded as `runtime: shiny`. With this line, we will compile our report into a Shiny application. On the other hand, if we remove this line, we will compile our report into a regular R Markdown report.

We can now type the code that generates the `ggvis` plot into R chunks. After we have completed our code, we can click **Run Document** in the menu. RStudio will begin compiling the report and render the report in a pop-up window. From the report, one can now interact with the `ggvis` plot using the web form.

There's more...

Besides making an interactive plot with `ggvis`, one can add a Shiny widget to control the rendering. In the following code chunk, we demonstrate how to add a `slideInput` to resize the points of a `ggvis` scatterplot:

````
```{r, include=TRUE, warning=FALSE, echo=FALSE}
library(ggvis)
library(dplyr)
library(shiny)
```


```{r, echo=FALSE}
house <- read.csv('RealEstate.csv', header=TRUE)
sliderInput("size", label = "Point Size:",
            min = 1, max = 100, value = 1, step = 1)
input_size <- reactive(input$size)
house %>%
````
````

```
  ggvis(~Size, ~Price, fill=~Status, size:=input_size) %>%
  layer_points()%>%
  add_axis("x", title = "Size", title_offset = 40) %>%
  add_axis("y", title = "Price", title_offset = 70)
~~~
````

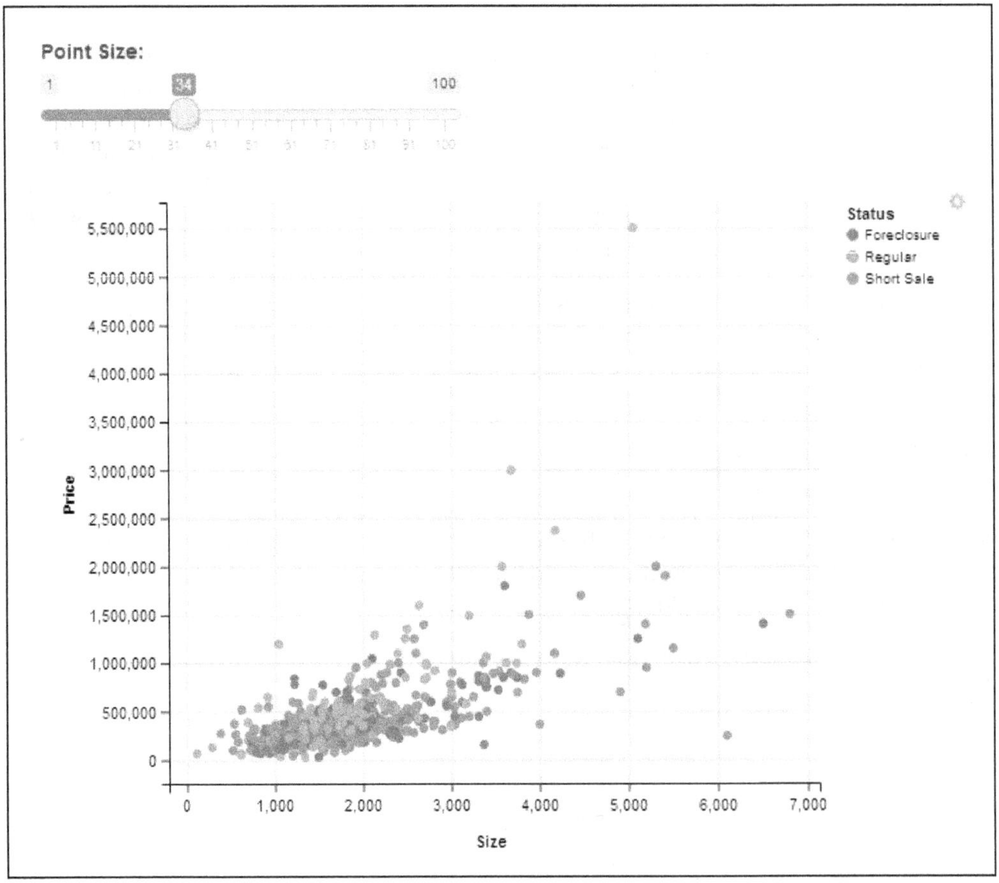

Figure 39: Adding a slideInput to a Shiny report

For more information about the Shiny widget, please use the `help` function to view the related documentation:

```
> help(package="shiny")
```

Publishing an R Shiny report

After creating our interactive report with the Shiny application, we can publish our report online. In this recipe, we introduce how to publish an R Shiny report online.

Getting ready

Ensure you have installed and loaded `ggvis` into your R session. Also, you need to complete the previous steps by storing `house` in your R environment.

How to do it...

Please perform the following steps to publish your Shiny document online:

1. First, click **Publish Document** on the upper-right side of the document:

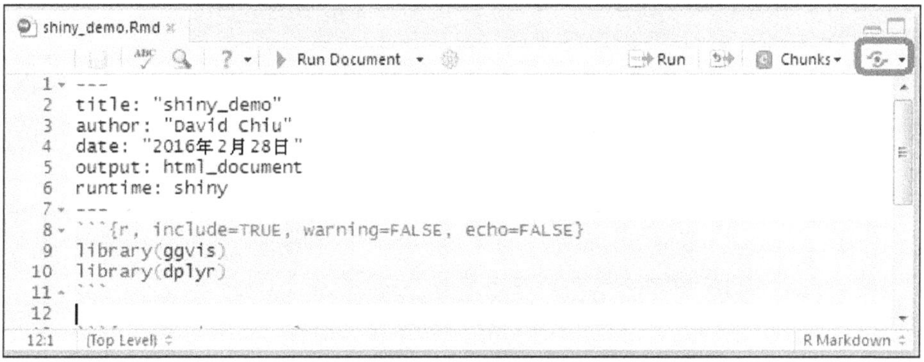
shiny_demo.Rmd
Run Document
Run
Chunks
---
title: "shiny_demo"
author: "David Chiu"
date: "2016年2月28日"
output: html_document
runtime: shiny

```{r, include=TRUE, warning=FALSE, echo=FALSE}
library(ggvis)
library(dplyr)
```
12:1
(Top Level)
R Markdown

Figure 40: Publish the Shiny document

2. If you have not installed the required packages, you should click **Yes** to install the required packages for publishing the Shiny document:

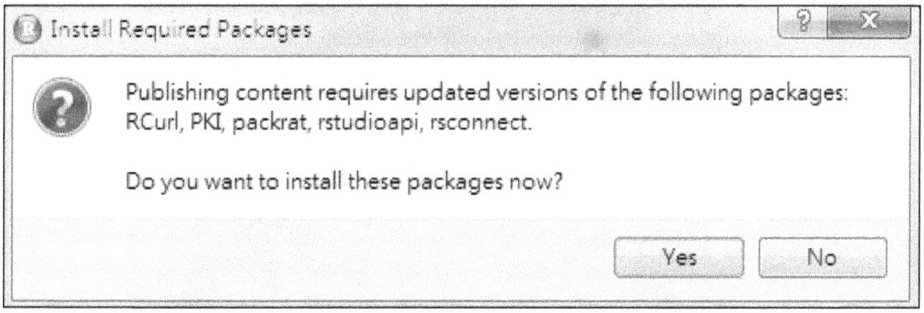
Install Required Packages
Publishing content requires updated versions of the following packages: RCurl, PKI, packrat, rstudioapi, rsconnect.
Do you want to install these packages now?
Yes
No

Figure 41: Install the required packages to publish Shiny app

3. Click on **Publish just this document**:

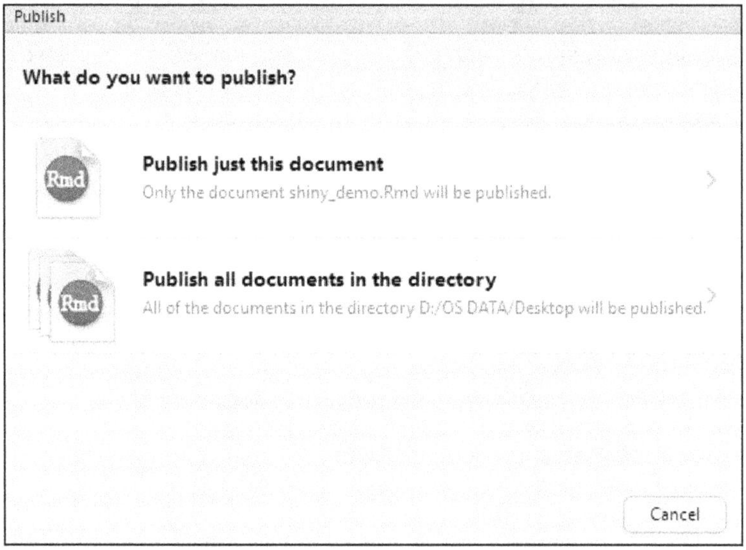

Figure 42: Click on publish just this document

4. At this point, RStudio requires you to connect your ShinyApps account with app tokens. Therefore, you should log in to your account to retrieve the token:

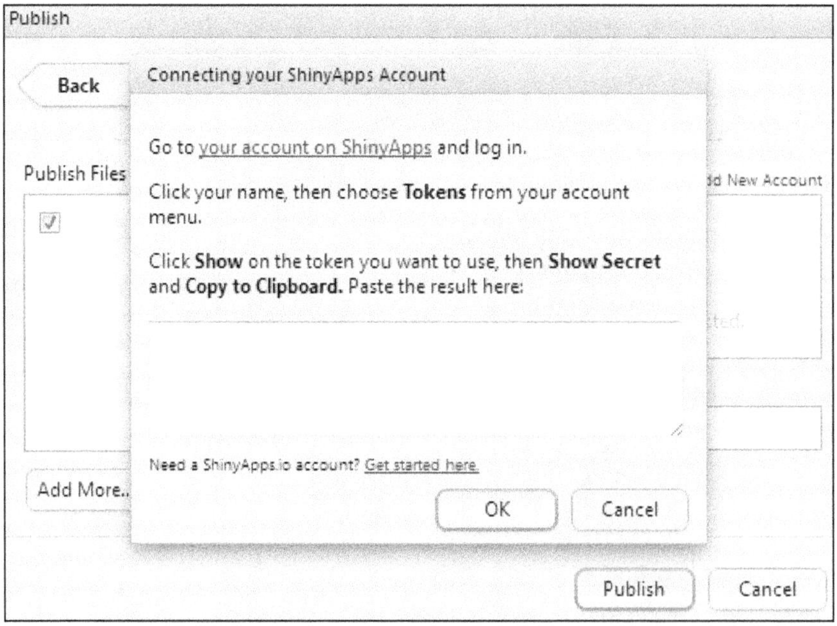

Figure 43: ShinyApps requires a login secret

5. Next, you should create an account and log in to `https://www.shinyapps.io/`. You will be redirected to the dashboard. If this is your first time logging in to the dashboard, you will be asked to set up the account:

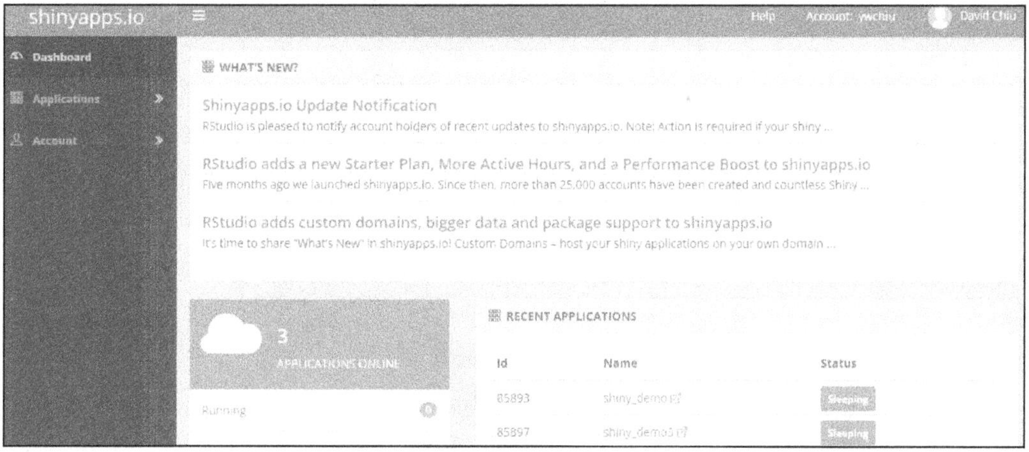

Figure 44: You will be directed to the dashboard of shinyapps.io

6. On the dashboard page, click on **Account** | **Tokens** | **Show** to open the account information window. Click on **Show secret** and **Copy to clipboard**:

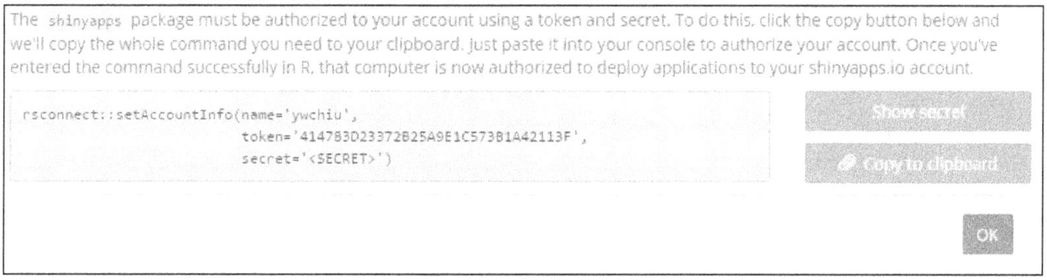

Figure 45: Secret of ShinyApps account

7. Paste what has been copied into the multiple line input box and click on **OK**:

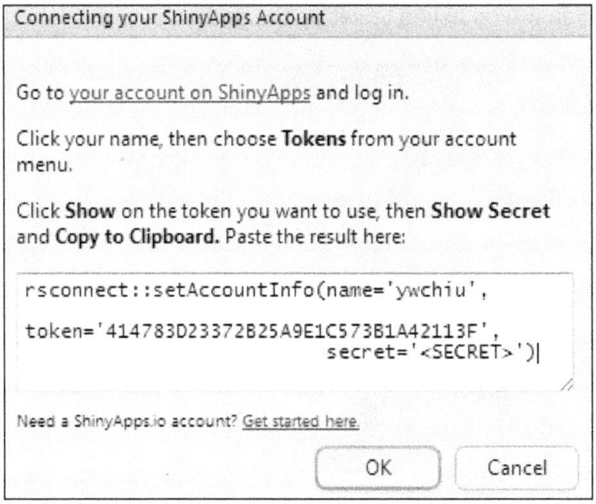

Figure 46: Paste the secret to connect to ShinyApps

8. After setting up the files and destination account, click on **Publish** to publish the Shiny document to your application page:

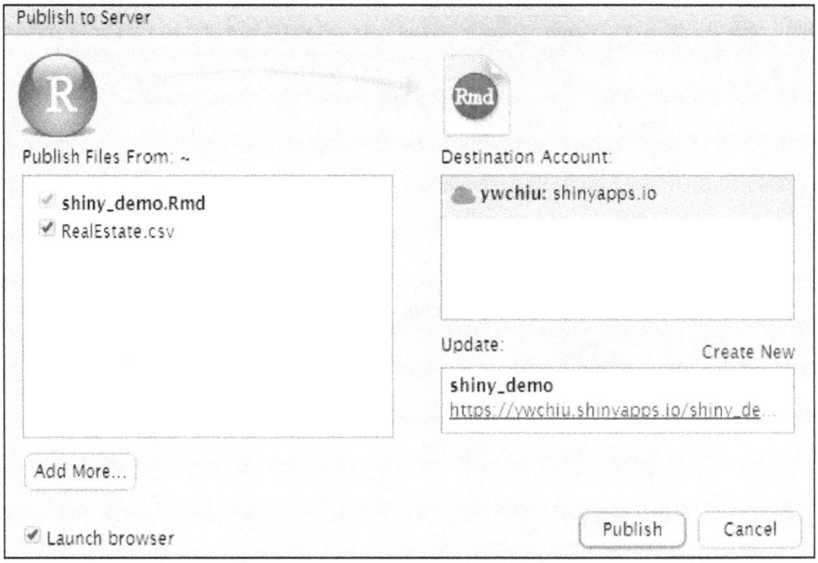

Figure 47: Publish the document to ShinyApps.io

9. Your Shiny document will be published on a webpage hosted by `http://www.shinyapps.io/`:

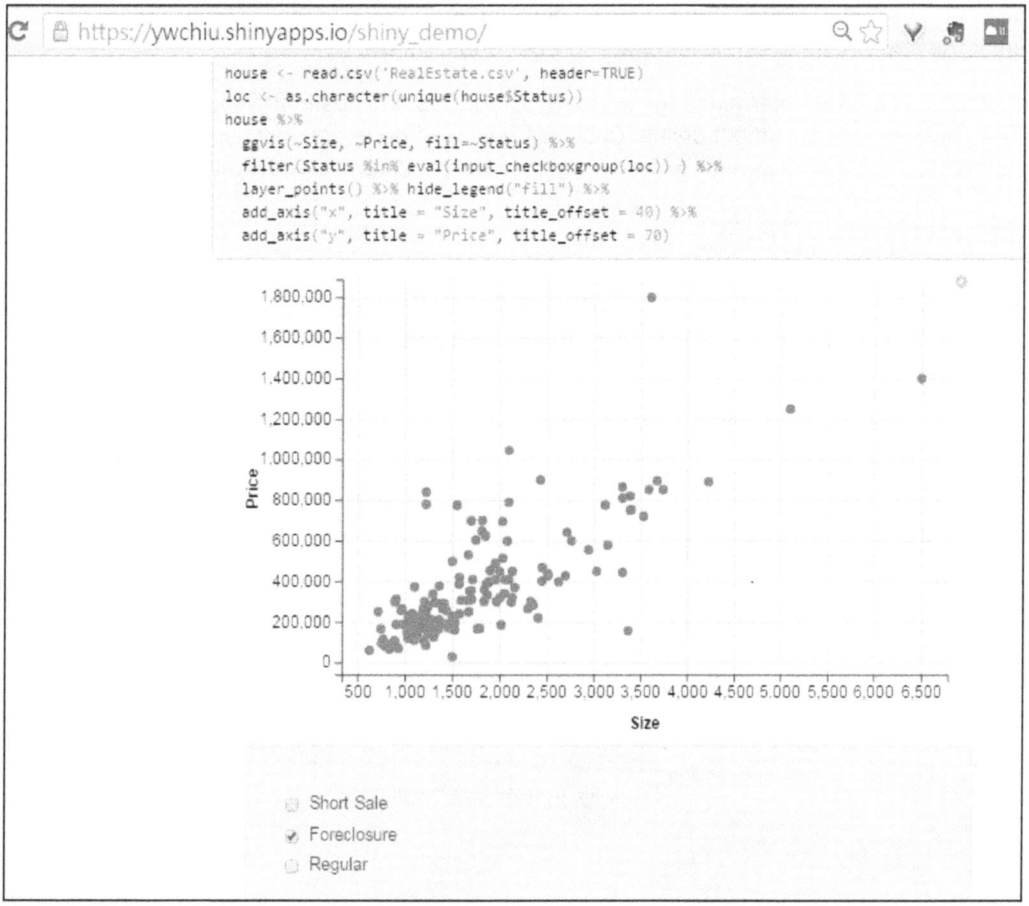
https://ywchiu.shinyapps.io/shiny_demo/
house <- read.csv('RealEstate.csv', header=TRUE)
loc <- as.character(unique(house$Status))
house %>%
ggvis(~Size, ~Price, fill=~Status) %>%
filter(Status %in% eval(input_checkboxgroup(loc))) %>%
layer_points() %>% hide_legend("fill") %>%
add_axis("x", title = "Size", title_offset = 40) %>%
add_axis("y", title = "Price", title_offset = 70)
1,800,000
1,600,000
1,400,000
1,200,000
1,000,000
800,000
600,000
400,000
200,000
0
Price
500 1,000 1,500 2,000 2,500 3,000 3,500 4,000 4,500 5,000 5,500 6,000 6,500
Size
Short Sale
Foreclosure
Regular

Figure 48: Shiny document hosted on shinyapps.io

How it works...

In addition to hosting a Shiny app on a local machine, we can choose to host our Shiny app online. RStudio provides a service, `http://www.shinyapps.io/`, which allows anyone to apply for a free account to upload your Shiny app.

Publishing a Shiny app on shinyapps.io is easy; you just need to click on **Publish Document** on Rstudio. If it is the first time you are publishing a document, you will be asked to install some required packages. Next, after choosing the type of report that you wish to publish, RStudio will open a window and ask you to input a token to connect to your Shiny App account.

If this is your first time publishing a Shiny document online, you are required to apply for an account at `http://www.shinyapps.io/`. Once you have done this and successfully logged in, you can retrieve your login tokens from the admin portal. Here, you need to click on **Show secret** to unmask the secret and copy the account information to the clipboard. You can then enter the account information into the input box and click **OK**.

Last, after you have determined the location of your files and destination account, press **OK** to publish your document online. Once this has been done, you can use the link format https://<account_name>.shinyapps.io/<app_name>/ to view your apps online.

There's more...

In addition to setting account information at `http://www.shinyapps.io/`, you can monitor and control your application online. Click on **Applications** in the left-hand menu of the portal, or use the following link:

`https://www.shinyapps.io/admin/#/applications/all`

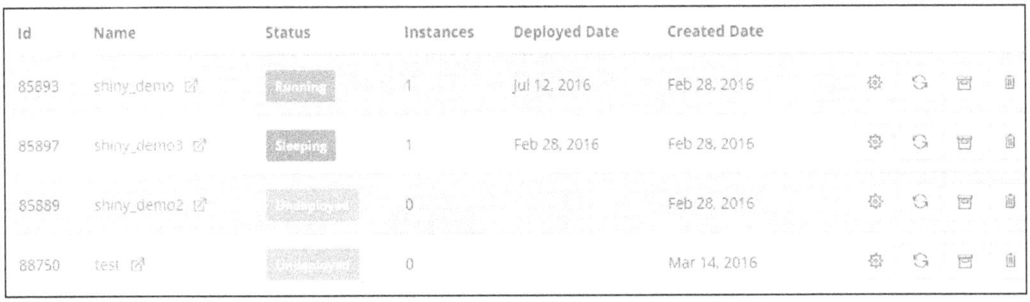
Id
Name
Status
Instances
Deployed Date
Created Date
85893
shiny_demo
Running
1
Jul 12, 2016
Feb 28, 2016
85897
shiny_demo3
Sleeping
1
Feb 28, 2016
Feb 28, 2016
85889
shiny_demo2
0
Feb 28, 2016
88750
test
0
Mar 14, 2016

Figure 49: Monitoring online apps

7
Simulation from Probability Distributions

This chapter covers the following topics:

- Generating random samples
- Understanding uniform distribution
- Generating binomial random variates
- Generating Poisson random variates
- Sampling from normal distribution
- Sampling from chi-squared distribution
- Understanding Student's t-distribution
- Sampling from a dataset
- Simulating the stochastic process

Introduction

To handle the uncertainty of real-world events, we can use probability to measure the likelihood of whether an event will occur. By definition, probability is quantified with a number between 0 and 1; the higher the probability (closer to 1), the more certain we are that an event will occur.

As statistical inference is used to deduce the properties of a given population, knowing the probability distribution of a given population becomes essential. For example, if you find that the data selected for prediction does not follow the exact assumed probability distribution in experiment design, the results should be refuted. In other words, probability provides justification for statistical inference.

In this chapter, we focus on the topic of probability distribution and simulation. We first discuss how to generate random samples, before covering how to use R to generate samples from various distributions such as normal, uniform, Poisson, chi-squared, and Student's t-distribution. We also discuss issues related to data sampling and simulations.

Generating random samples

In this section, we will introduce how to generate random samples from a given population with the `sample` and `sample.int` functions.

Getting ready

In this recipe, you need to prepare your environment with R installed.

How to do it...

Please perform the following steps to generate random samples.

1. First, generate random samples from `1` to `10`:

   ```
   > sample(10)
   ```

2. If you would like to reproduce the same samples, you can set the random seed beforehand:

   ```
   > set.seed(123)
   > sample(10)
    [1]  3  8  4  7  6  1 10  9  2  5
   ```

3. You can then randomly choose two samples from `1` to `10`:

   ```
   > sample(10,2)
   [1] 10  5
   ```

4. If the population and sample size are required arguments, you can also use the `sample.int` function:

   ```
   > sample.int(10,size=2)
   [1] 7 6
   ```

5. For example, one can simulate a lottery game and generate six random samples from a given population with a size of 42:

```
> sample.int(42,6)
[1]  5 37 10  2 13 36
```

6. Alternatively, one can generate random samples with replacements by setting the replace argument to TRUE:

```
> sample(c(1,0), 10, replace=TRUE)
 [1] 0 0 0 0 0 0 0 0 1 1
```

7. Moreover, one can use the sample function to simulate a coin-toss example:

```
> sample(c("HEAD","TAIL"), 10, replace=TRUE)
 [1] "TAIL" "TAIL" "TAIL" "TAIL" "HEAD" "HEAD" "TAIL" "HEAD"
"HEAD" "HEAD"
```

8. Or, one can simulate an example of rolling a fair dice 200 times.

```
> fair <- sample(c(1:6), 200, replace=TRUE)
> table(fair)
fair
 1  2  3  4  5  6
26 39 44 33 30 28
```

9. As an advanced example, to simulate rolling a loaded dice 200 times, one can use the following code:

```
> loaded <- sample(c(1:6), 200, replace=TRUE, prob
=c(0.1,0.1,0.1,0.1,0.1,0.5))
> table(loaded)
loaded
  1   2   3   4   5   6
 22  16  15  17  21 109
```

How it works...

In this recipe, we demonstrate that R can easily generate random samples from a given population. We can generate random samples from 1 to 10 with the sample function. Then to make sure that we can reproduce the same sample sequence, we can set the pseudo-number generator to a value of 123 with the set.seed function. Besides generating all possible samples from the given population, we can specify the value of size in the second argument of the sample function. Thus, we can only choose two samples out of a population from 1 to 10. On the other hand, we can use sample.int instead, if the sample size and population size is known.

If one would like to sample with replacements, one can set `replace` to `TRUE` in the `sample` function. Thus, we can use the `sample` function to generate a random binary sequence with a size of `10` and using ten coin tosses as an example.

Lastly, we can simulate a loaded dice by using the `prob` argument. Here, we first simulate a fair dice and use the `table` function to determine the frequency of each side being drawn, discovering that there is a similar frequency for each side. On the other hand, we can simulate a loaded dice by setting the occurrence probability of `6` to `0.5` in the `prob` argument. By calculating the occurrence frequency of each side in this example, we found that the probability of drawing a `6` is much more than the chance of getting any other side.

There's more...

By default, every R session will produce a different simulation result using a different random seed sequence, which is generated by the current time and process ID. By setting a value to the `set.seed` function, the pseudo-random number generator will assign an integer vector to `.Random.seed`. As a result, all R sessions will then follow the sequence of the `seed` to produce the same results:

```
> set.seed(123)
> sample(10)
 [1]  3  8  4  7  6  1 10  9  2  5
> sample(10)
 [1] 10  5  6  9  1  7  8  4  3  2

> set.seed(123)
> sample(10)
 [1]  3  8  4  7  6  1 10  9  2  5
> sample(10)
 [1] 10  5  6  9  1  7  8  4  3  2
```

Also, one can examine the integer vector by typing `.Random.seed` in the R console:

```
> set.seed(123)
> head(.Random.seed)
[1]        403        624 -983674937  643431772 1162448557 -959247990
```

Understanding uniform distributions

If we roll a fair dice, the likelihood of drawing any side is equal. By plotting the samples in a bar plot, we can see bars with equal heights. This type of distribution is known as **uniform distribution**. In this recipe, we will introduce how to generate samples from a uniform distribution.

Getting ready

In this recipe, you need to prepare your environment with R installed.

How to do it...

Please perform the following steps to generate a sample from uniform distribution:

1. First, we can create samples from uniform distribution by using the `runif` function:

```
> set.seed(123)
> uniform <- runif(n = 1000, min = 0, max = 1)
```

2. We can then make a histogram plot out of the samples from the uniform distribution:

```
> hist(uniform, main = "1,000 random variates from a uniform
distribution")
```

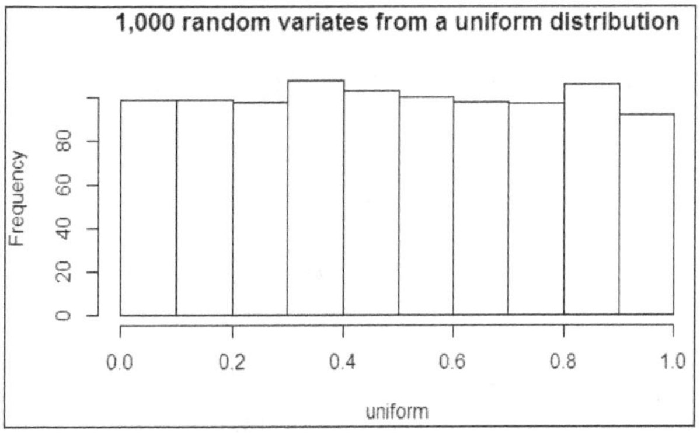

Figure 1: A histogram of 1,000 random variates from a uniform distribution

3. For uniform distribution, we also can find the density at each point equals `1`, and the density function of the distribution is equal to itself:

```
> length(dunif(uniform) == 1)
[1] 1000
> length(punif(uniform) == uniform)
[1] 1000
```

4. Also, we can simulate how to roll a dice with the `sample` function:

```
> fair_dice <- sample(6, 1000, replace=TRUE)
> table(fair_dice)
fair_dice
  1   2   3   4   5   6
169 158 176 179 159 159
```

5. We can then visualize the occurrence of each side in a bar plot:

```
> barplot(table(fair_dice), main="Frequency of each side")
```

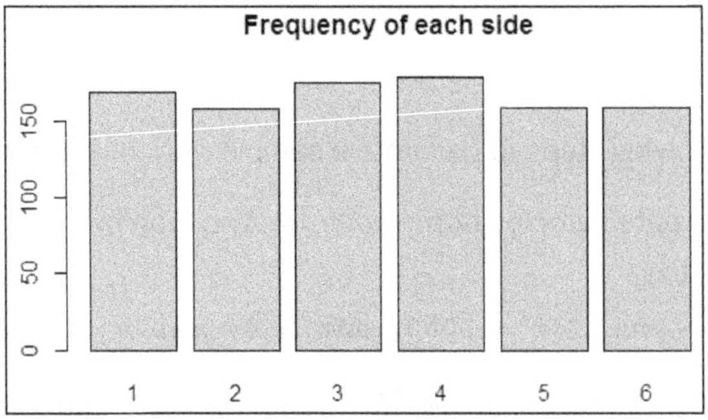

Figure 2: Frequency of each side

How it works...

Uniform distribution can be simply represented by the following formula:

$$f(x) = f(x) = \begin{cases} \frac{1}{b-a}, & for\, a \le x \le b \\ 0, & elsewhere \end{cases}$$

By assigning *a* as the minimum, and *b* as the maximum, the density between *a* and *b* is uniformly distributed.

In this recipe, we demonstrate how to generate samples from a uniform distribution. First, we use `runif` to generate 1,000 samples from a uniform distribution with `min = 0` and `max = 1`. As per *Figure 1*, we can see the frequency height at each value of `x` is alike. We then examine the density and the density function with the `dunif` and `punif` functions. The return of these two functions shows that `dunif` returns one at the different point of uniform distribution. On the other hand, the value obtained from `punif` of the distribution is equal to the distribution value itself all the time.

Lastly, we simulate dice rolling samples with the `sample` function. We can then use the `table` function to obtain the occurrence of each side. We can use `barplot` to visualize the frequency of each side being drawn. As *Figure 2* shows, the dice rolling samples are similar to a uniform distribution.

Generating binomial random variates

To model the success or failure of several independent trials, one can generate samples from binomial distribution. In this recipe, we will discuss how to generate binomial random variates with R.

Getting ready

In this recipe, you need to prepare your environment with R installed.

How to do it...

Please perform the following steps to create a binomial distribution:

1. First, we can use `rbinom` to determine the frequency of drawing a six by rolling a dice 10 times:

```
> set.seed(123)
> rbinom(1, 10, 1/6)
[1] 1
```

2. Next, we can simulate 100 gamblers rolling a dice 10 times, and observe how many times a six is drawn by each gambler:

```
> set.seed(123)
> sim <- rbinom(100,10,1/6)
> table(sim)
sim
 0  1  2  3  4  5
17 36 23 18  4  2
```

3. Additionally, we can simulate 1,000 people tossing a coin 10 times, and compute the number of heads at each tossing:

```
> set.seed(123)
> sim2 <- rbinom(1000,10,1/2)

> table(sim2)
```

```
sim2
   0   1   2   3   4   5   6   7   8   9  10
   2  11  43 126 200 241 213 106  48   9   1
> barplot(table(sim2), main="A simulation of 1,000 people tossing
a coin 10 times")
```

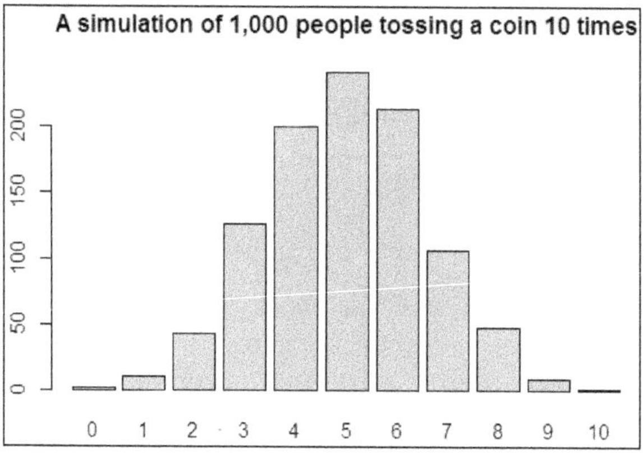

Figure 3: A simulation of 1,000 people tossing a coin 10 times

4. We can then use `dbinom` to retrieve the probability of getting exactly one 6 when rolling a dice 10 times:

```
> dbinom(1, 10, 1/6)
[1] 0.3230112
```

5. If we would like to calculate the probability of getting less than three 6s from 10 dice rolls, we can use the `pbinom` function:

```
> dbinom(0, 10, 1/6) + dbinom(1, 10, 1/6) + dbinom(2, 10, 1/6) +
dbinom(3, 10, 1/6)
[1] 0.9302722
> pbinom(3, 10, 1/6)
[1] 0.9302722
```

How it works...

Binomial distribution models the probability of *k* successes in *n* trials. We can write the formula as follows:

$$P(X=k)=\binom{n}{k}p^k(1-p)^{(n-k)}$$

Having examined how binomial distribution is formulated, we can use R to explore binomial distribution.

First, we use `rbinom` to obtain the frequency of drawing `1` after rolling a dice 10 times. Here, the output shows that we get exactly one `1` during the experiment. We can also simulate 100 gamblers rolling a dice (probability of `1/6`) 10 times to see how many 6s are drawn using the `table` function. In addition to simulating a dice-rolling experiment, we can simulate the number of heads obtained if 1,000 people toss a coin (probability of `1/2`) ten times. Next, we can use a `table` function to determine the number of heads drawn by each person. By using a `barplot` function, we can see the distribution of the experiment tends to normality.

On the other hand, we can use `dbinom` to obtain the probability of getting exactly one 6 when rolling a dice ten times. Furthermore, we can use `pbinom` to calculate the cumulative probability of less than three 6s being drawn during the experiment.

There's more...

Apart from using standard R functions, one can use `gbinom` to plot binomial distribution:

```
> source('https://www.stat.wisc.edu/~larget/R/prob.R')
> gbinom(n=100, p=0.5,scale = T, a=45,b=55)
```

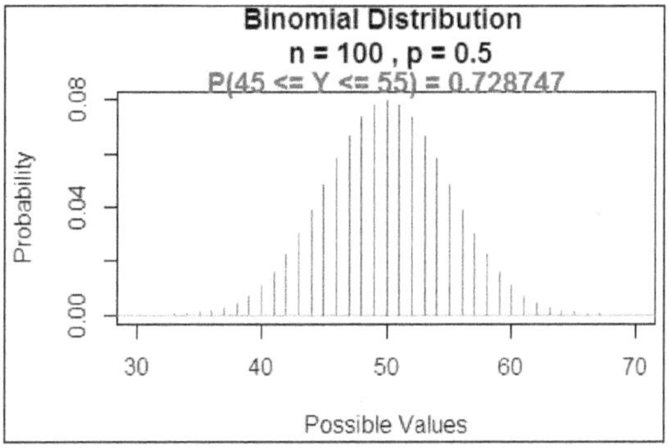

Figure 4: Binomial distribution

Assuming that we would like to visualize the probability of getting 45 and 55 heads in 100 coin tosses, `gbinom` graphs the binomial distribution of 100 trials with 0.5 winning probability and shades the area between 45 and 55 in blue. The blue shade between 45 and 55 covers 72.87% of the area.

Generating Poisson random variates

Poisson distribution is best to use when expressing the probability of events occurring with a fixed time interval. These events are assumed to happen with a known mean rate, λ, and the event of the time is independent of the last event. Poisson distribution can be applied to examples such as incoming calls to customer service. In this recipe, we will demonstrate how to generate samples from Poisson distribution.

Getting ready

In this recipe, you need to prepare your environment with R installed.

How to do it...

Please perform the following steps to generate sample data from Poisson distribution:

1. Similar to normal distribution, we can use `rpois` to generate samples from Poisson distribution:

```
> set.seed(123)
> poisson <- rpois(1000, lambda=3)
```

2. You can then plot sample data from a Poisson distribution into a histogram:

```
> hist(poisson, main="A histogram of a Poisson distribution")
```

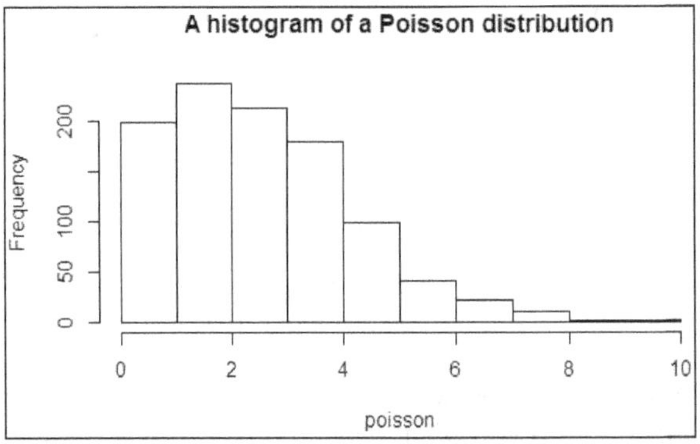

Figure 5: A histogram of a Poisson distribution

3. You can then obtain the height of the distribution function at x=2.

```
> dpois(2,lambda =3)
[1] 0.2240418
```

4. Here, we can use `barplot` to draw samples generated from the `dpois` function:

```
> barplot(dpois(0:10,lambda=3), names = 0:10, main="A barplot of a Poisson distribution")
```

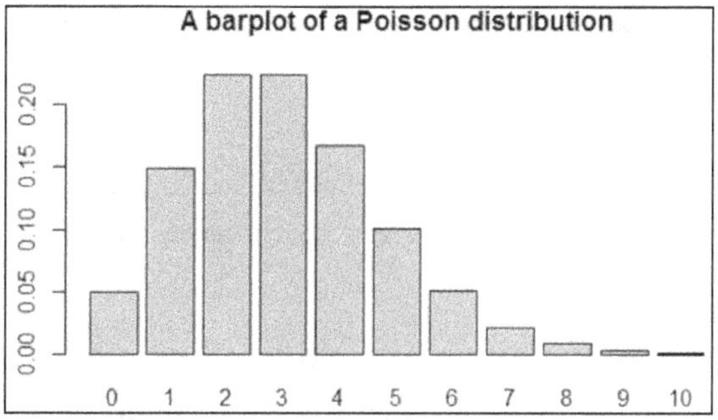

Figure 6: A bar plot of a Poisson distribution

5. Next, you can compute the proportion of area covered from x = *1* to 5 with the `ppois` function:

```
> ppois(5, lambda = 3) - ppois(1, lambda = 3)
```

6. You can also compare the density plot of samples generated using different lambda values:

```
> set.seed(123)
> poisson1 <- rpois(1000, lambda=3)
> hist(poisson1, breaks=seq(0,20,by=1), col=rgb(1,0,0,1), xlim=c(0,15), ylim=c(1,300), main='Histograms with different lambda')
>
> poisson2 <- rpois(1000, lambda=5)
> hist(poisson2, breaks=seq(0,20,by=1), col=rgb(0,0,1,0.6), add=T)
>
> poisson3 <- rpois(1000, lambda=7)
```

```
> hist(poisson3, breaks=seq(0,20,by=1), col=rgb(0,1,0,0.2), add=T)
```

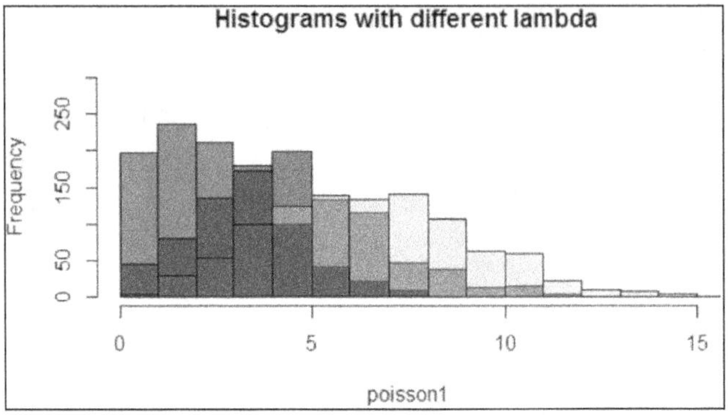

Figure 7: Histograms with different lambda values

How it works...

In a Poisson experiment, we can characterize the mean number of successes that occur in a specified region λ, against the actual number of successes, x, with the following formula:

$$p(x;\lambda)=\frac{e^{-\lambda}\lambda^{x}}{x!}$$

In this recipe, we demonstrate how to generate samples from other distributions. First, we use `rpois` to generate 1,000 samples from the Poisson distribution with lambda (vector of means) equal to 3. As per *Figure 5*, we can see that the data skews toward the right of the chart, and most of the data lies around x=3.

Also, we can use `barplot` to plot the histogram of the Poisson distribution in *Figure 6*. To compute the area under a curve from x=*1* to `5` of a Poisson distribution with a lambda of `3`, we can use the `ppois` function.

In addition, we can compare Poisson distribution with different lambda values. By doing this, we determine that the greater the lambda value, the more the histogram tends to normality (*Figure 7*).

There's more...

Similar to normal distribution, we can use `qpois`, the probability quantile function, to invert cumulative distribution of Poisson distribution:

```
> qpois(ppois(2, lambda=3), lambda =3 )
[1] 2
```

Sampling from a normal distribution

From observing real-world data, we may determine that most data follows a normal distribution. That is to say, when we use a density curve (or histogram) to plot the data, we should see it as bell-shaped. As this kind of distribution is so commonly seen in nature or social science, scientists often use a normal distribution to represent real-value random variables from an unknown distribution. In this recipe, we will demonstrate how R generates samples from a normal distribution.

Getting ready

In this recipe, you need to prepare your environment with R installed.

How to do it...

Please perform the following steps to generate samples from a normal distribution:

1. First, you can use the `rnorm` function to generate `30` and `1000` samples from a normal distribution:

```
> set.seed(123)
> data1 <- rnorm(30)
> data2 <- rnorm(1000)
```

2. We can then use the `hist` function to plot the histogram of `data1` and `data2`:

```
> par(mfrow = c(1,2))
> hist(data1, main="30 samples")
> hist(data2, main="1000 samples ")
```

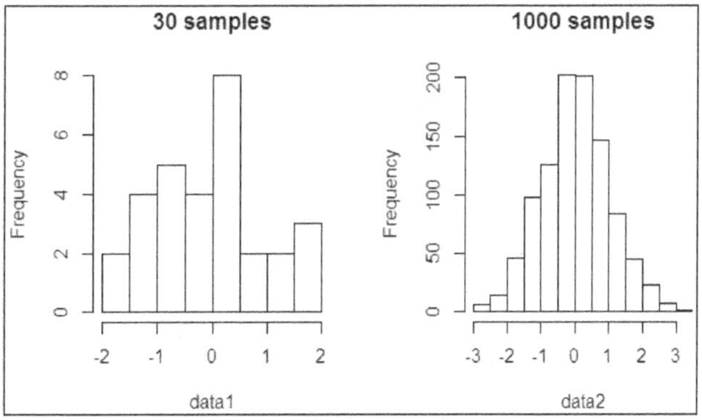

Figure 8: Histogram of 30 and 1,000 samples from a normal distribution

3. Also, you can extract the density property from the sample data, and add a density curve onto the histogram with the `lines` function:

```
> par(mfrow=c(1,1))
> hist(data2, main="Density curve over histogram", freq = FALSE)
> den2 <- density(data2)
> den2
> den2


Call:
density.default(x = data2)

Data: data2 (1000 obs.); Bandwidth 'bw' = 0.2187

       x                  y
 Min.   :-3.4658   Min.   :0.0000206
 1st Qu.:-1.6251   1st Qu.:0.0121127
 Median : 0.2156   Median :0.0776974
 Mean   : 0.2156   Mean   :0.1356826
 3rd Qu.: 2.0564   3rd Qu.:0.2424239
 Max.   : 3.8971   Max.   :0.4192901

> lines(x = den2$x, y = den2$y, col = "red", lwd = 3)
```

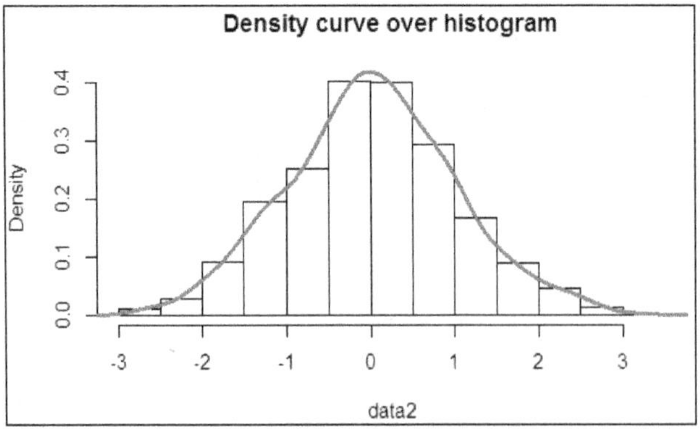

Figure 9: Density curve over histogram

4. On the other hand, you can compare the distribution under different standard deviations in a density plot and histogram:

```
> set.seed(123)
> data3 <- rnorm(1000, sd = 5)
> data4 <- rnorm(1000, sd = 3)
> plot(density(data3), col="blue", ylim=c(0,0.15), main="Density
Plot Under Different sd")
> lines(density(data4), col="red",ylim=c(0,0.15))


> hist(data3, col="blue", main = "Histogram Under Different sd",
ylim=c(0,300))
> hist(data4, col="red", add=TRUE)
```

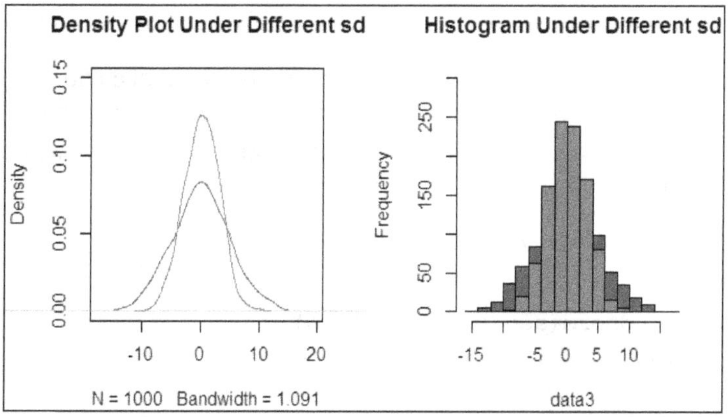

Figure 10: A histogram and density plot of samples under different standard deviations

5. Also, you can compare the distribution under different mean with a density plot and histogram:

```
> set.seed(123)
> data5 <- rnorm(1000, mean = 0)
> plot(density(data5), col="red", main = "Density Plot Under
Different Mean")
>
> data6 <- rnorm(1000, mean = 1)
> lines(density(data6), col="blue")
>
> hist(data5, col="blue", main = "Histogram Under Different Mean")
```

```
> hist(data6, col="red", add=TRUE)
```

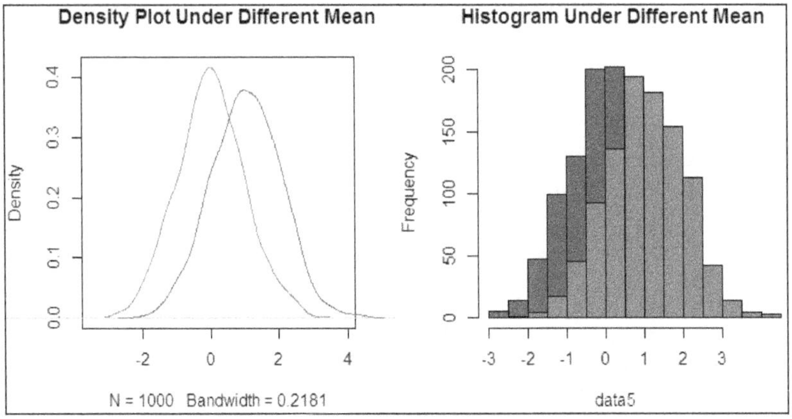

Figure 11: A histogram and density plot of samples under different standard mean

6. On the other hand, if you would like to plot a perfect normal curve, you can use the `curve` function to connect the height of probability distribution from `-3` to `3`:

```
> plot(dnorm, -3,3, main="A normal distribution generated from dnorm function")
```

7. We can also shade the area under a density curve:

```
> x = c(-1,seq(-1,1,0.1),1)
> y = c(0, dnorm(seq(-1,1, 0.1)),0)
> polygon(x,y, col = "blue")
```

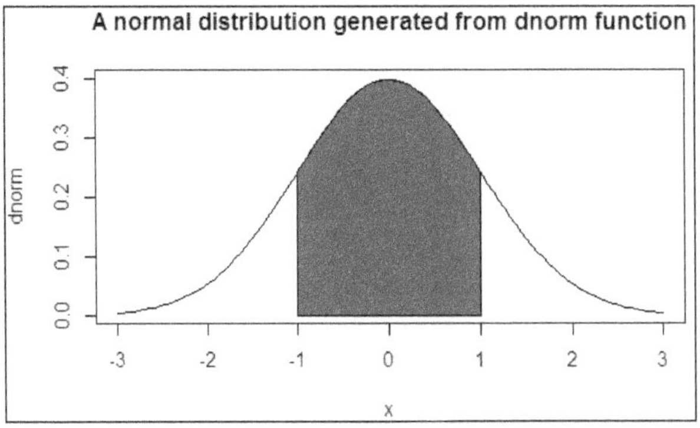

Figure 12: A normal distribution generated from the dnorm function

8. To get the height of probability distribution at `0`, one can use the `dnorm` function:

```
> dnorm(0)
[1] 0.3989423
```

9. Moreover, to compute the area under the curve between x from `-1` to `1`, one can use the cumulative distribution function:

```
> pnorm(1) - pnorm(-1)
[1] 0.6826895
```

10. To plot the graph of `pnorm`, a `curve` function can be used:

```
> curve(pnorm(x), -3,3, main="Cumulative distribution function")
```

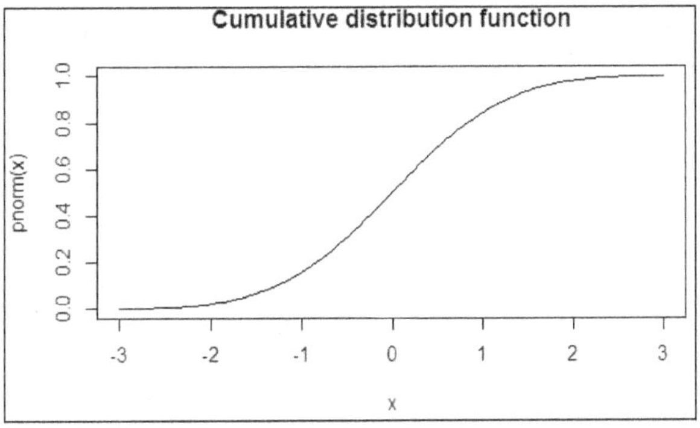

Figure 13: Cumulative distribution function

How it works...

A normal distribution is known as Gaussian distribution, which is defined by a variate *X* with the mean as μ and variance as σ^2. As the distribution curve is shaped liked a bell, some refer to normal distribution as a bell curve. The probability density function can be defined as the following formula:

$$f(x) = \frac{1}{\sqrt{2\pi\sigma}} e\left(-\frac{(x-\mu)^2}{2\sigma^2}\right)$$

In this recipe, we first generate `30` samples from a normal distribution with the `rnorm` function. If we do not adjust `mean` and `sd` when using the `rnorm` function, we will generate data from a standard normal distribution, with the *mean=0* and *sd=1*. Alternatively, we can adjust the `mean` and `sd` of the `rnorm` function so that we generate data from a different normal distribution. We can then use the `hist` function to plot the histogram of the generated dataset. As *Figure 8* shows, the plot is not shaped like a bell. Therefore, while the sample is generated from a normal distribution, the fact that there are insufficient samples means the normal distribution curve is not visible. As a result, we scale up the sample size to 1,000 and then plot the figure with the `hist` function once again. In this example, we can now observe a bell-shaped curve.

Next, we proceed to estimate the kernel density of the data with the `density` function. The `density` function returns the smoothing bandwidth, `bw`, with coordinates as `x` and estimated density as `y`. We can then use the `x` and `y` values to make a density plot. *Figure 9* reveals a bell-shaped curve, which means that the generated data is from a normal distribution.

Moving on, we can make a density curve in regard to different standard deviations and means, and compare them by plotting both on the same figure. As *Figure 10* shows, we determine that as the standard deviation of the normal distribution increases, the density curve appears wider and shorter than data with a lower standard deviation. Also, we can compare a density curve with different means in the same figure. *Figure 11* shows that data with a *mean=1* shifts to the right compared to data with a *mean=0*. Besides comparing different distribution in curves, we can also compare them in histograms.

As an alternative to using `rnorm` to generate samples from a normal distribution, one can use `dnorm` to obtain the height of a distribution function and plot a bell-shaped curve. By default, the `dnorm` function uses `0` as the mean value and `1` as standard deviation. Thus, 99.8% of the distribution lies within three standard deviations. To plot a curve with `dnorm` from `-3` to `3` generates a perfect bell-shaped figure. Also, one can use `dnorm` to obtain the height of the distribution at a specific `x` value. We also demonstrate how to shade the area under a density curve. To achieve this, we first set the *x* and *y* coordinates of the shaded area. We then used the `polygon` function with given *x* and *y* coordinates to shade the density curve in *Figure 12*.

Finally, we can calculate the percentage of area covered by a specific range with the cumulative distribution function, `pnorm`. For example, if we would like to calculate the percentage of area covered within one standard deviation, we can use the value of `pnorm` at *x=1* minus the value of `pnorm` at *x=-1*.

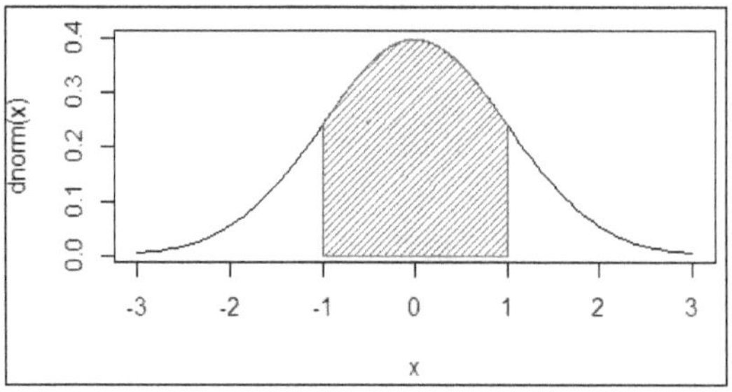

Figure 14: 68% of the observations are between x=-1 and x=1

As a result, we determine that around 68.2% of the area is covered within one standard deviation. Also, we can plot the cumulative distribution function from *x=-3* to *x=3*. As *Figure 13* shows, the height of each point shows the area under the standard normal curve to the left of a particular `x` value.

There's more...

To calculate the point that the distribution accumulates a particular cumulative density, one can use the quantile function, `qnorm`, to inverse `pnorm`. In this sense, we can use `qnorm` to retrieve the approximate value of 97.5 and 2.5 percentile points of a normal distribution:

```
> qnorm(0.975)
[1] 1.959964
> qnorm(0.025)
[1] -1.959964
```

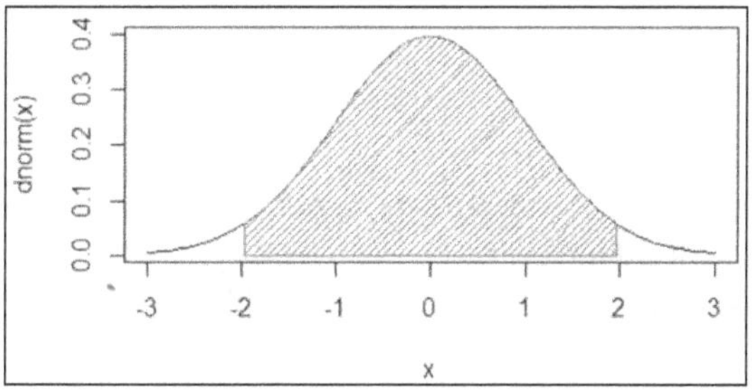

Figure 15: 95% of the observations are between x=-2 and x=2

As per the red-shaded area, we discover that 95% (from point 0.025 to 0.975) of the area under a normal curve lies within approximately 1.96 standard deviations.

Sampling from a chi-squared distribution

Chi-squared distribution is often used by chi-squared tests to inspect the difference between observed value and expected value, or to examine the independence of two variables. In addition, one can infer confidence intervals using chi-squared distribution. In the following recipe, we will discuss how to use R to generate chi-squared distribution further.

Getting ready

In this recipe, you need to prepare your environment with R installed.

How to do it...

Please perform the following steps to generate samples from chi-squared distribution:

1. First, we can use `rchisq` to generate three samples with a degree of freedom equal to `10`:

```
> set.seed(123)
> rchisq(3,df=10)
[1]  6.779170 14.757915  3.259122
```

2. We can then use `dchisq` to obtain the density at *x=3* with a degree of freedom equal to `10`:

```
> dchisq(3,df=10)
[1] 0.02353326
```

3. Also, we can use `pchisq` and `qchisq` to obtain the distribution function and quantile function of the distribution:

```
> pchisq(3,df=10)
[1] 0.01857594
> qchisq(.99,df=10)
[1] 23.20925
```

4. Furthermore, we can increase degrees of freedom and compare the plot under the circumstances with different degrees of freedom:

```
> set.seed(123)
> x = seq(-5,5,0.1)
> op=par(mfrow=c(2,2))
> curve(dchisq(x, 1), 0, qchisq(.99,1), main= "df = 1")
> curve(dchisq(x, 5), 0, qchisq(.99,5), main= "df = 5")
> curve(dchisq(x, 10), 0, qchisq(.99,10), main= "df = 10")
> curve(dchisq(x, 50), 0, qchisq(.99,50), main= "df = 50"
```

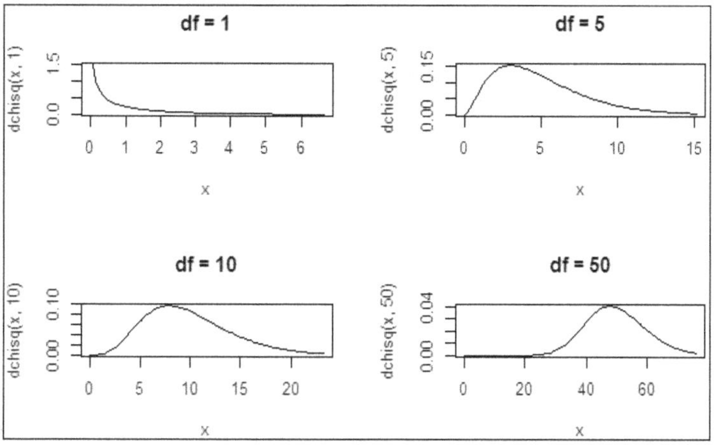

Figure 16: Chi-squared distribution under different degrees of freedom

How it works...

By definition, if $X_1,X_2,\ldots,X_k$ are *k* independent, random variables have the standard normal distribution. Thus, the sum of their square follows a chi-squared distribution with *k* degrees of freedom:

$$Q = \sum_{i=1}^{k} X_i^2$$

This recipe focuses on the actual implementation using R to obtain chi-squared distribution. First, we use `rchisq` to obtain three samples from a chi-squared distribution with a degree of freedom equal to `10`. Next, we can use `dchisq` to obtain the density at *x* =3. Also, we can retrieve the density function and quantile function through the use of `pchisq` and `qchisq`. We can then use `pchisq` to calculate the cumulative probability of `3` in a $X_{10}^{\ 2}$ distribution. Alternatively, using `qchisq` allows us to determine the 99th percentile of $X_{10}^{\ 2}$ distribution.

In the end, we make plots of sample data under different degrees of freedom. As per *Figure 16*, as degrees of freedom increase, figures with larger degrees of freedom tend toward normality.

There's more...

Using the chi-squared table allows us to obtain the probability from chi-squared distribution of a certain degree of freedom (`http://sites.stat.psu.edu/~mga/401/tables/Chi-square-table.pdf`).

Chi-Square Distribution Table

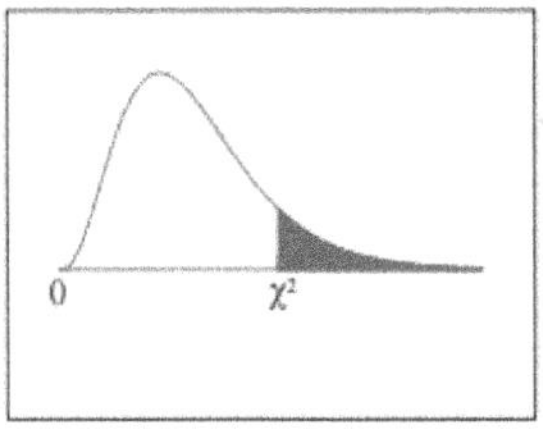

The shaded area is equal to α for $\chi^2 = \chi^2_\alpha$.

df	$\chi^2_{.995}$	$\chi^2_{.990}$	$\chi^2_{.975}$	$\chi^2_{.950}$	$\chi^2_{.900}$	$\chi^2_{.100}$	$\chi^2_{.050}$	$\chi^2_{.025}$	$\chi^2_{.010}$	$\chi^2_{.005}$
1	0.000	0.000	0.001	0.004	0.016	2.706	3.841	5.024	6.635	7.879
2	0.010	0.020	0.051	0.103	0.211	4.605	5.991	7.378	9.210	10.597
3	0.072	0.115	0.216	0.352	0.584	6.251	7.815	9.348	11.345	12.838
4	0.207	0.297	0.484	0.711	1.064	7.779	9.488	11.143	13.277	14.860
5	0.412	0.554	0.831	1.145	1.610	9.236	11.070	12.833	15.086	16.750
6	0.676	0.872	1.237	1.635	2.204	10.645	12.592	14.449	16.812	18.548
7	0.989	1.239	1.690	2.167	2.833	12.017	14.067	16.013	18.475	20.278
8	1.344	1.646	2.180	2.733	3.490	13.362	15.507	17.535	20.090	21.955
9	1.735	2.088	2.700	3.325	4.168	14.684	16.919	19.023	21.666	23.589
10	2.156	2.558	3.247	3.940	4.865	15.987	18.307	20.483	23.209	25.188

Figure 17: Chi-square table

With the help of R, we can easily find the 95th percentile of the chi-squared distribution with six degrees of freedom with the following command:

```
> qchisq(.95,df=6)
[1] 12.59159
```

Understanding Student's t-distribution

Student's t-distribution is also known as **t-distribution**. It is applied to estimate the mean of the population from a normal distribution. Moreover, it is the basis of conducting Student's t-test. In the following recipe, we will introduce how to use R to perform Student's t-distribution.

Getting ready

In this recipe, you need to prepare your environment with R installed.

How to do it...

Please perform the following steps to generate samples from t-distribution:

1. First, we can use `rt` to generate three samples with the degree of freedom equal to `10`:

```
> set.seed(123)
> rt(3, df = 10)
[1] -0.6246844 -1.3782806 -0.1181245
```

2. Then, we use `dt` to obtain the density at *x=3* with a degree of freedom equal to `10`:

```
> dt(3,df=10)
[1] 0.01140055
```

3. Furthermore, we can visualize the Student's t-distribution's degree of freedom equals `1`:

```
> plot(seq(-5,5,0.1), dt(seq(-5,5,0.1), df = 1), type='l',
main="Student's t-distribution of df = 1")
```

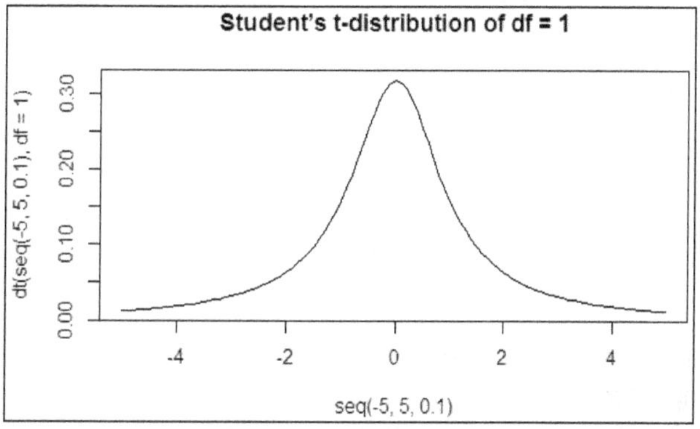

Figure 18: Student's t-distribution of df = 1

4. Also, we can use `pt` to compute the area of Student's t-distribution with degrees of freedom equal to `10` from `x` values between `-2` and `2`:

```
> pt(2,df=1) - pt(-2,df=1)
```

5. Also, we can compare normal and Student's t-distribution on the same plot:

```
> x = seq(-5,5,.1)
> plot(x,dnorm(x), type='l', col='red', main="Student's
t-distribution under different df")
> lines(x, dt(x, df = 1), type='l',col='blue')
```

```
> lines(x, dt(x, df = 5), type='l',col='green')
```

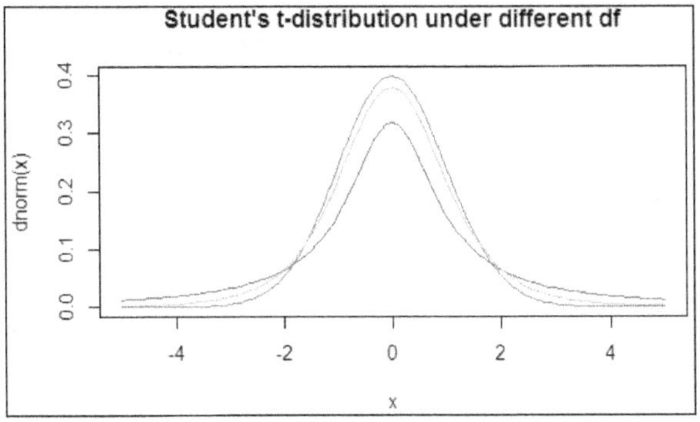

Figure 19: Student's t-distribution under different df

How it works...

For a normal distribution, one should specify its mean and standard deviation to determine its location and shape. However, in reality, the standard deviation is often unknown. Thus, we can use sample variance to substitute standard deviation. As a result, Student's t-distribution can be written as the following formula:

$$T = \frac{Z}{\sqrt{X^2 / k}}$$

In the preceding formula, *Z* is a standard normal distribution, and X_2 is chi-squared value with *k* degrees of freedom.

In this recipe, we explain how to simulate a Student's t-distribution using `rt`, `pt`, and `dt`. First, we create three samples from t-distribution with 10 degrees of freedom using the `rt` function. Then, we can obtain the density at `3` with the `dt` function. Next, we demonstrate how to generate a t-distribution plot from *x* = *-5* to *x=5*. Similar to other distribution functions, we can compute the area of *x=2* to *x= -2* with the `pt` function.

Last, we can make a comparison plot with a normal distribution, t-distribution of a degree of freedom equal to `1` and t-distribution of the degree of freedom equal to `5`. *Figure 19* shows that t-distribution is quite similar to normal distribution; both are shaped like a bell, and the curve is symmetric at *x=0*. However, the tail of t-distribution is heavier than the normal distribution. This means that we should be more concerned about extreme values when examining t-distribution. As the degrees of freedom increase, the curve becomes closer to the standard deviation.

There's more...

Instead of looking up the t-table (`http://www.sjsu.edu/faculty/gerstman/StatPrimer/t-table.pdf`) to find the scores regarding cumulative probability and degrees of freedom, we can use the quantile function `qt` to save some effort and time:

t Table

cum. prob	$t_{.50}$	$t_{.75}$	$t_{.80}$	$t_{.85}$	$t_{.90}$	$t_{.95}$	$t_{.975}$	$t_{.99}$	$t_{.995}$	$t_{.999}$	$t_{.9995}$
one-tail	0.50	0.25	0.20	0.15	0.10	0.05	0.025	0.01	0.005	0.001	0.0005
two-tails	1.00	0.50	0.40	0.30	0.20	0.10	0.05	0.02	0.01	0.002	0.001
df											
1	0.000	1.000	1.376	1.963	3.078	6.314	12.71	31.82	63.66	318.31	636.62
2	0.000	0.816	1.061	1.386	1.886	2.920	4.303	6.965	9.925	22.327	31.599
3	0.000	0.765	0.978	1.250	1.638	2.353	3.182	4.541	5.841	10.215	12.924
4	0.000	0.741	0.941	1.190	1.533	2.132	2.776	3.747	4.604	7.173	8.610
5	0.000	0.727	0.920	1.156	1.476	2.015	2.571	3.365	4.032	5.893	6.869
6	0.000	0.718	0.906	1.134	1.440	1.943	2.447	3.143	3.707	5.208	5.959
7	0.000	0.711	0.896	1.119	1.415	1.895	2.365	2.998	3.499	4.785	5.408
8	0.000	0.706	0.889	1.108	1.397	1.860	2.306	2.896	3.355	4.501	5.041
9	0.000	0.703	0.883	1.100	1.383	1.833	2.262	2.821	3.250	4.297	4.781
10	0.000	0.700	0.879	1.093	1.372	1.812	2.228	2.764	3.169	4.144	4.587

Figure 20: Student's t-table

```
> qt(0.95,5)
[1] 2.015048
```

Sampling from a dataset

In addition to generating random samples with the `sample` function or from any underlying probability distribution, one can sample the subset of a given dataset. In this recipe, we will introduce how to sample a subset from a financial dataset.

Getting ready

In this recipe, you need to prepare your environment with R installed and a computer that can access the Internet.

How to do it...

Please perform the following steps to sample a subset from a given financial dataset:

1. First, install and load the `quantmod` package:

```
> install.packages("quantmod")
> library(quantmod)
```

2. Next, we can use `getSymbols` to obtain the Dow Jones Industrial Average data, and then subset the 2014 data from the population:

```
> getSymbols('^DJI')
> stock = DJI['2014']
```

3. You can then randomly choose `100` samples from the dataset:

```
> set.seed(5)
> sub <- stock[sample(nrow(stock), 100), ]
```

4. Furthermore, you can calculate the daily change from the generated subset:

```
> sub$change =  (sub$DJI.Close - sub$DJI.Open) / sub$DJI.Open
```

5. Make a histogram of the daily price change:

```
> hist(sub$change, main = "Histogram of daily price change")
```

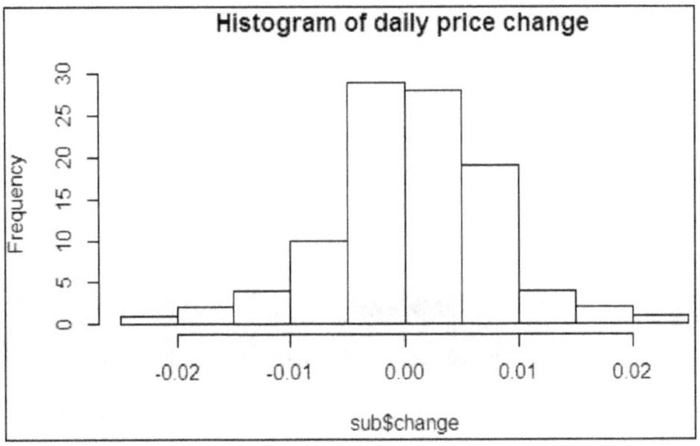

Figure 21: Histogram of daily price change

6. On the other hand, you can employ the `sample` function to separate data into two groups with different proportions:

```
> idx = sample(c(0,1), nrow(stock), prob=c(0.7,0.3), replace=TRUE)
> first_grp = stock[idx == 0, ]
> second_grp = stock[idx == 1, ]
> dim(first_grp)
[1] 183   6
> dim(second_grp)
[1] 69  6
```

How it works...

In this recipe, we demonstrate how to sample the Dow Jones Industrial Average data. First, we install and load `quantmod`, which is a quantitative financial modeling framework. Next, we use the `getSymbols` function to download the Dow Jones Industrial Average data from the Internet. The next step is to subset the 2014 data from the downloaded index. Here, we use the `sample` function to subset 100 samples from the Dow Jones Index in 2014.

Furthermore, we can compute the daily change rate from the subset, and make a histogram of the change rate. As per *Figure 21*, the daily change rate in 2014 is very similar to a normal distribution.

Lastly, we demonstrate that one can also use the `sample` function to separate data into multiple groups with a different proportion. Here, we first generate random samples of either `0` or `1`, and then control the proportion of `0` and `1` by using the `prob` argument. As a result, we can obtain data of different groups by subset data with the generated group index. This technique is especially useful while separating data into training and testing datasets.

There's more...

Besides sampling data randomly, one can sample data by groups or by clusters with the use of the `sampling` package. To learn more about the package, you can install and load the `sampling` package and read the following documentation for more information:

```
> install.packages("sampling")
> library(sampling)
> help(package="sampling")
```

Simulating the stochastic process

The stochastic process, also known as the random process, illustrates the evolution of the system of random values over time. It is the best to use when one or more variables within the model are random. In this recipe, we will introduce how to simulate a random stock trading process with R.

Getting ready

In this recipe, you need to prepare your environment with R installed and a computer that can access the Internet. You should complete the previous step and have `quantmod` or DJI data loaded in an R session.

How to do it...

Please perform the following steps to simulate trading stocks:

1. First, subset the 2014 data from the Dow Jones Industrial Average Index:

```
> DJI.2014 = DJI['2014']
```

2. We can then use the sample function to generate samples of `0` and `1`:

```
> set.seed(123)
> randIDX = sample(c(0,1), nrow(DJI.2014), replace=TRUE)
```

3. Next, we subset the DJI close price with the index equal to `1` only, which denotes that we either buy or sell stock on that day:

```
> stock = DJI.2014[which(randIDX==1),]$DJI.Close
```

4. Now, we can calculate the return of each trade:

```
> stock$lag_close = lag(stock$DJI.Close, -1)
> stock$returns = (stock$lag_close - stock$DJI.Close) / stock$DJI.
Close
```

5. By using $10,000 as our initial investment, we then compute our assets after each trade:

```
> stock$asset = 0
> fund = 10000
> for(i in 1:nrow(stock)){
+      fund = (1 + as.numeric(stock[i,]$returns))* fund;
+      stock[i,]$asset = fund
+ }
```

6. Lastly, we use the `plot` function to visualize our returns:

```
> plot(stock$asset[which(!is.na(stock$asset))], main="Returns on
random trading")
```

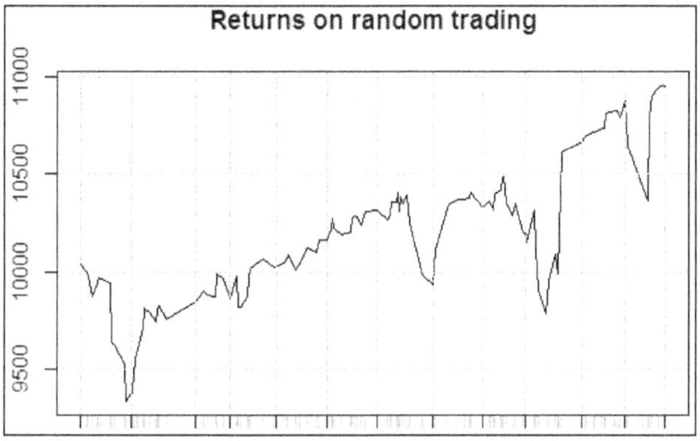

Figure 22: Returns on random trading

How it works...

In this recipe, we test what will happen if we randomly buy and sell the Dow Jones Index at different times. As stock prices go up or down, it is quite unlikely anyone can give you a certain answer that you will profit or not. However, one can answer this kind of question through a simulation. In this case, we simulate a stochastic process to calculate our earnings if we randomly trade at different times.

In this recipe, we first subset the 2014 data of the DJI index. Next, we need to determine whether to buy or sell the index. Thus, we randomly generate some samples of `0` and `1`. Here, a sample of `1` denotes buying or selling, while a sample of `0` means that we do not do anything. We can then subset the DJI data from a sample of index `1`. As we have extracted the time that we buy or sell stocks, we can then use the `lag` function to shift the close price of the next trade forward. This allows us to calculate the return of each trade.

We assume our initial investment is $10,000, and perform an iteration to calculate the change of our assets after each trade. We visualize our return with the `plot` function. *Figure 22* reveals that if we randomly bought or sold in 2014, we would very likely profit. This example shows how to compare the returns of your trading strategy by using random trading. Thus, we can easily determine if the trading strategy will help us profit from the market.

There's more...

Another famous stochastic process is Brownian motion. The following code shows how to simulate price changes as Brownian motion in R:

```
> set.seed(123)
> brown <- cumsum(rnorm(100,0,1))
> brown <- brown - brown[1]
```

```
> plot(brown, type = "l", main="Brownian motion")
```

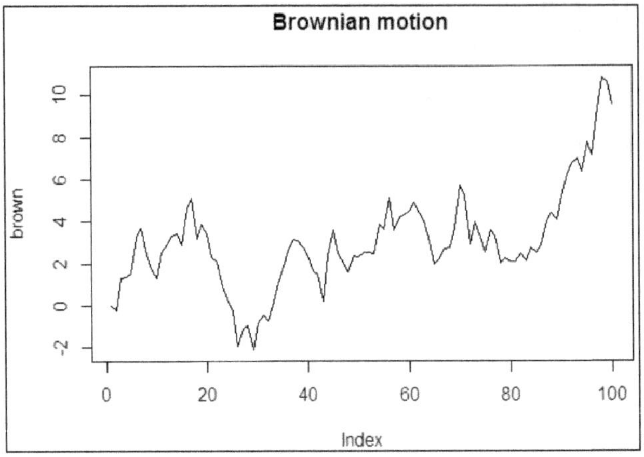

Figure 23: Brownian motion

In this recipe, we covered a simple method to simulate stock trading and calculate the return of each trade. However, for those who need a more complete tool to measure performance, one can consider using `PerformanceAnalytics`. The following URL contains more information:

https://cran.r-project.org/web/packages/PerformanceAnalytics/index.html

8

Statistical Inference in R

This chapter covers the following topics:

- Getting confidence intervals
- Performing Z-tests
- Performing student's T-tests
- Conducting exact binomial tests
- Performing the Kolmogorov-Smirnov tests
- Working with the Pearson's chi-squared tests
- Understanding the Wilcoxon Rank Sum and Signed Rank test
- Performing one-way ANOVA
- Performing two-way ANOVA

Introduction

The most prominent feature of R is that it implements a wide variety of statistical packages. Using these packages, it is easy to obtain descriptive statistics about a dataset or infer the distribution of a population from sample data. Moreover, with R's plotting capabilities, we can easily display data in a variety of charts.

To apply statistical methods in R, the user can categorize the method of implementation into descriptive statistics and inferential statistics, described as follows:

- **Descriptive statistics**: These are used to summarize the characteristics of data. The user can use mean and standard deviation to describe numerical data, and they can use frequency and percentages to describe categorical data.
- **Inferential statistics**: This is when, based on patterns within sample data, the user can infer the characteristics of the population. Methods relating to inferential statistics include hypothesis testing, data estimation, data correlation, and relationship modeling. Inference can be further extended to forecasting and estimation of unobserved values either in or associated with the population being studied.

We cover the following topics in the following recipes: confidence intervals, Z-tests, student's T-tests, skewness identification, exact binomial tests, Kolmogorov-Smirnov tests, the Pearson's chi-squared test, the Wilcoxon Rank Sum and Signed Rank test, one-way ANOVA, two-way ANOVA, and linear regression.

Getting confidence intervals

Using confidence intervals allows us to estimate the interval range of unknown parameters in the data. In this recipe, we will teach you methods that can help obtain confidence intervals in R.

Getting ready

Ensure that you installed R on your operating system.

How to do it...

Perform the following steps to obtain confidence intervals:

1. Let's first generate a normal distribution using the `rnorm` function:

```
>set.seed(123)
>population<- rnorm(1000, mean = 10, sd = 3)
>plot(dens, col="red", main="A density plot of normal
distribution")
```

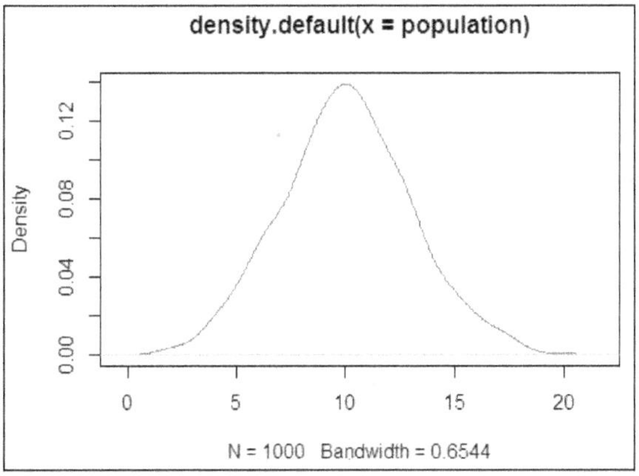

Figure 1: A density plot of normal distribution

2. Next, we will sample 100 samples out of the population:

```
>samp<- sample(population, 100)
>mean(samp)
[1] 10.32479
>sd(samp)
[1] 3.167692
```

3. At this point, we can obtain the Z-score at a confidence of 99%:

```
> 1 - 0.01 / 2
[1] 0.995

>qnorm(0.995)
[1] 2.575829
```

4. We can now compute the standard deviation error and estimate the upper and lower bounds of the population mean:

```
>sde<-(sd(samp)/sqrt(100))
>sde
[1] 0.3167692

>upper<- mean(samp) + qnorm(0.995) * sde
>upper
```

```
[1] 11.14073

>lower<- mean(samp) - qnorm(0.995) * sde
>lower
[1] 9.508846
```

5. Shade the area between the lower and upper bound of the density plot:

```
>polygon(c(lower, dens$x[dens$x>lower &dens$x< upper], upper),
+         c(0, dens$y[dens$x>=lower &dens$x<= upper], 0),
+         col="red", density = c(10, 50) ,angle = c(-45, 45))
```

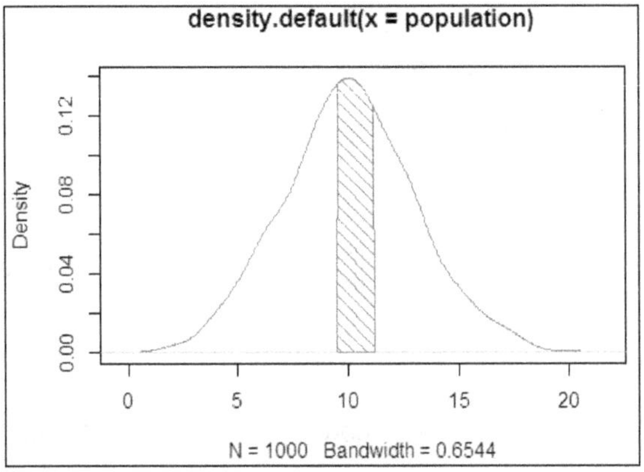

Figure 2: Shade the 99% confidence interval in red

6. Now, we can find the number *z* that the area under the standard normal curve between −*z* and *z* is `0.95`:

```
> 1 - 0.05 / 2
[1] 0.975

>qnorm(0.975)
[1] 1.959964
```

7. Let's now check what happens if we compute the range of confidence at 95% level:

```
>upper2<- mean(samp) + qnorm(0.975) * sde
>upper2
[1] 10.94565

>lower2<- mean(samp) - qnorm(0.975) * sde
>lower2
[1] 9.703933
```

8. We can then color the 95% confidence interval in blue:

```
>polygon(c(lower2, dens$x[dens$x>lower2&dens$x<upper2], upper2),
+                 c(0, dens$y[dens$x>=lower2&dens$x<= upper2],
0),
+                 col="blue")
```

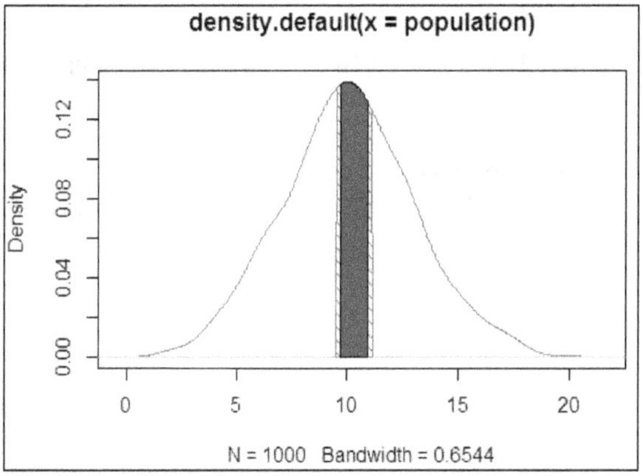

Figure 3: Shade the 95% confidence interval in blue

9. Moving on, let's take a look at how to obtain the confidence intervals of a Poisson distribution:

```
>set.seed(123)
>population2<- rpois(1000, lambda = 3)
>mean(population2)
[1] 2.967
>hist(population2, main= "A histogram of A Poisson distribution")
```

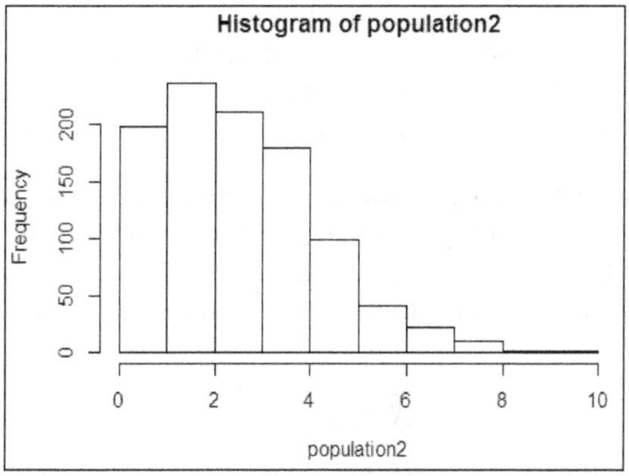

Figure 4: A histogram of a Poisson distribution

10. Here, we sample 100 instances from the population:

```
>sample2<-sample(population2, 100)
```

11. We continue to sample these samples 1,000 times to generate 1,000 sample means:

```
>sample_mean<- rep(NA, 1000)
>for(i in 1:1000){
+    sample_mean[i] <- mean(sample(sample2, replace=TRUE))
+ }
```

12. Finally, we can obtain the 95% confidence interval using a 2.5% sample quantile as the lower bound, and the 97.5% sample quantile as the upper bound:

```
>upper3<-  quantile(sample_mean, 0.975)
>upper3
97.5%
 3.55

>lower3<- quantile(sample_mean, 0.025)
>lower3
    2.5%
2.76975
```

How it works...

As an example, assume that you are a quality assurance engineer who wants to ensure that all manufactured screws are of a similar size. As it is very unlikely that you can check all the screws that are produced, one possible method is to collect a few samples and estimate the mean of the population from the samples. The method used to calculate the means of the samples is known as point estimation. This is a simple estimation method, but it fails to address issues with uncertainty. As an alternative, you can use confidence intervals to estimate a range of values that are likely to contain the population mean.

Confidence intervals are constructed by confidence level; a 95 % confidence interval means that 95 % of the interval will include the population mean. On the other hand, a 95% confidence interval can be stated as $1-\alpha$, where α is the significance level. You can formulate the confidence interval calculation in the following equation:

$$\bar{X} \pm Z_{1-\alpha/2}\frac{\sigma}{\sqrt{n}}$$

Here, $\overline{X}$ is the sample means, $Z_{1-\alpha/2}$ is the critical value of significance level α, σ is the known standard deviation, and *n* is the number of population. In R, we can use the `qnorm` function to obtain the $Z_{1-\alpha/2}$ critical value with the α given significance level.

Estimating confidence intervals is closely related to parametric hypothesis tests, but both have different goals and methods. The purpose of using a confidence interval is mainly to estimate the interval of a population. On the other hand, in hypothesis testing, we first make an assumption about unknown parameters and then determine whether the observed sample leads to a rejection of the null hypothesis at a certain significance level.

In this recipe, we first generate a population of 1,000 samples under normal distribution with the `rnorm` function. We then make a density plot from the population. Next, we sample 100 data points from the population. We can now compute a 99% confidence interval with the `qnorm` function. As the significance level under the 99% confidence interval equals *1 - 0.99 = 0.01*, the confidence interval should, therefore, lie between *qnorm(0.01 / 2) = -2.575829* and *qnorm(1- (1- 0.99) / 2) = 2.572829*, as shown in the following plot:

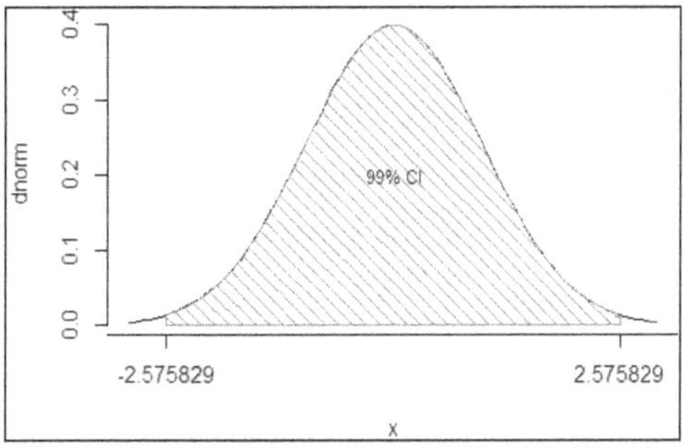

Figure 5: The 99% confidence interval

Next, we can compute the standard error, which is the standard deviation of the sampling distribution. We can also calculate the margin error by timing standard error to Z-score. At this point, we can obtain the 99% confidence upper and lower bound of the population mean. By overlaying the 99% confidence interval on the density plot, we can see the mean of the population lies between the x-axis range of the red-shaded area.

Moreover, we can compute the upper and lower bound of the 95% confidence interval. The area covered in between the 95% confidence interval is slightly narrower than the confidence bound of the 99% confidence interval. We can compare the difference by overlaying the area of 99% confidence in blue on the same density plot. After doing this, we observe a narrower blue region above the red-shaded area.

In the preceding example, we demonstrated how to obtain confidence intervals from a population of normal distribution. However, for a population that belongs to a different category, such as Poisson distribution, we can perform bootstrapping to obtain confidence intervals. The concept is to resample *n* samples from a sample of size *n*, and repeatedly calculate the mean value of the resampled samples. In this case, we generate a Poisson distribution, and then sample 100 data points out of the population. We then continue to resample the samples 1,000 times, and assign the sample mean of each iteration in a vector named `sample_mean`. Finally, we can use the `quantile` function to retrieve the sample mean at the 2.5% and 97.5% quantiles from `sample_mean`, which is the 95% confidence lower and upper bound of the population mean. Here, we discovered that the population mean from Poisson distribution, 2.967, lies between 2.76975 and 3.55.

There's more...

Besides using the bootstrapping method, we can also use the `boot.ci` function from the boot package to obtain the confidence intervals. This is how it works:

1. First, load the `boot` package:

```
>library(boot)
```

2. We can now resample the data repeatedly using the `boot` function:

```
>boot.mean<- function(x,i){boot.mean<- mean(x[i])}
> z <- boot(sample2, boot.mean, R = 2000)
```

3. Finally, we can use `boot.ci` to obtain the bootstrap confidence interval:

```
>boot.ci(z)
BOOTSTRAP CONFIDENCE INTERVAL CALCULATIONS
Based on 2,000 bootstrap replicates

CALL :
boot.ci(boot.out = z)

Intervals :
Level      Normal              Basic
95%   ( 2.786,  3.546 )   ( 2.780,  3.550 )

Level     Percentile            BCa
95%   ( 2.77,  3.54 )   ( 2.78,  3.55 )
Calculations and Intervals on Original Scale
```

In this sample, we first loaded the boot package. We then resampled the sample mean of `sample2` 2,000 times. Finally, we used the `boot.ci` function to obtain the 95% confidence interval under different measurements.

Performing Z-tests

When making decisions, it is important to know whether decision error can be controlled or measured. In other words, we want to prove that the hypothesis formed is unlikely to have occurred by chance, and it is statistically significant. In hypothesis testing, there are two types of hypothesis: null hypothesis and alternative hypothesis (research hypothesis). The purpose of hypothesis testing is to validate whether the experiment results are significant. However, to validate whether the alternative hypothesis is acceptable, the alternative hypothesis is deemed to be true if the null hypothesis is rejected.

A Z-test is a parametric hypothesis method that can determine whether the observed sample is statistically significantly different from a population with known standard deviation, based on standard normal distribution.

Getting ready

Ensure that you installed R on your operating system.

How to do it...

Perform the following steps to calculate the Z-score:

1. First, collect the volume (in milliliters) of 20 bottles of soft drink. Let's assume that this data comes from a normally-distributed population with a mean equal to 300 and standard deviation equal to 10:

```
>pop_mean<- 300
>pop_sd<- 10
>soft_drink<-c(278,289,291,291,291,285,295,278,304,287,291,287,288,
300,309,280,294,283,292,306)
```

2. We can then calculate the standard deviation error and Z-score of the 20 bottles of soft drink:

```
>sde<- pop_sd / sqrt(length(soft_drink))
> z   <- (mean(soft_drink) - pop_mean) / sde
>z
[1] -4.047283
```

3. We can now compute the one-tailed p-value from Z-score:

```
>pnorm(z)
[1] 2.590779e-05
> 1- pnorm(z)
[1] 0.9999741
```

4. Alternatively, we can compute the two-tailed p-value from Z-score:

```
> p   <- (1 - pnorm(abs(z))) * 2
>p
[1] 5.181557e-05
```

5. We can then create a function that can perform a Z-test and calculate the p-value of either one side or two-sided:

```
>z.test<- function(x, pop_mean, pop_sd, side="twoside"){
+    sde<- pop_sd / sqrt(length(x))
+    z   <- (mean(x) - pop_mean) / sde
+
+    switch(side,
+           twoside={
+             p   <- (1 - pnorm(abs(z))) * 2
+           },
+           less={
+             p   <- pnorm(z)
+           },
+           {
+             p   <- 1- pnorm(z)
+           }
+    )
+    return(list(z = z , p = p))
+ }
```

6. We can now calculate the one-tailed p-value and two-tailed p-value from Z-score:

```
>z.test(a, 300,10)
$z
[1] -4.047283


$p
```

```
[1] 5.181557e-05

>z.test(a, 300,10, "l")
$z
[1] -4.047283

$p
[1] 2.590779e-05
```

How it works...

A Z-test is where the test statistics follow a standard normal distribution with a known standard deviation (or variance). Mostly, we perform Z-tests to compare whether the two mean values are significantly different.

The procedure to perform a Z-test takes the following steps:

1. Make a null hypothesis, $H_1 : \mu \neq \mu_1$, and an alternative hypothesis, $H_0 : \mu = \mu_1$.
2. Offer a significant level α (for example, 0.05).
3. Compute $Z = \frac{\overline{x} - \mu_0}{\sigma / \sqrt{n}}$ where $\overline{x}$ is the sample mean, μ_0 is the population mean, σ is the population standard deviation, and *n* is the length of the sample.
4. Calculate the corresponding p-value.
5. Determine whether we can reject the null hypothesis.

In this recipe, we used the example of determining whether the mean volume of produced soft drink is equal to 300 ml. To answer our question, we first randomly sampled 20 bottles of soft drink volumes from the population and listed the 20 samples in a vector named `soft_drink`. The volume of the population follows a standard normal distribution, with a known standard deviation of 10. Thus, we can then perform the Z-test to determine whether the volume of soft drink is equal to 300 ml.

Here, we first calculated the standard error of the sample and compute the Z-score from calculated standard error. Next, we computed the one-tailed p-value from the Z-score with the `pnorm` function. From the resulting p-value, we discovered that it is highly likely (99.99741%) that the mean value of produced soft drink is less than 300 ml.

However, as our goal is to check whether the mean volume of the produced soft drink is equal to 300 ml, we need to perform a two-tailed T-test instead. Therefore, we calculated the two-tailed p-value from the Z-score and find the output p-value is equal to 5.181557e-05, which is significantly lower than the significant level (0.05 in this case). As a result, we can refute the null hypothesis and conclude that the volume of produced soft drink is very unlikely to equal 300 ml.

Moreover, besides calculating Z-score and p-value step by step, we can make the procedure into a `z.test` function. Thus, we can easily calculate the Z-score and p-value of any sample with a given mean value, standard deviation, and the type of test.

There's more...

Besides making our own `z.test` function, we can also find a `z.test` function in the `BSDA` package:

```
>install.packages("BSDA")
>library(BSDA)
> ?z.test
```

Also, you can find a Z-test in the `PairedData` package (`https://cran.r-project.org/web/packages/PairedData/index.html`) and the `PASWR` package (`https://cran.r-project.org/web/packages/PASWR/index.html`).

Performing student's T-tests

In a Z-test, we can determine whether two mean values are significantly different if the standard deviation (or variance) is known. However, if the standard deviation is unknown and the sample size is fairly small (less than 30), we can perform a student's T-test instead. A one sample T-test allows us to test whether two means are significantly different; a two sample T-test allows us to test whether the means of two independent groups are different. In this recipe, we will discuss how to conduct a one sample T-test and a two sample T-test using R.

Getting ready

Ensure that you installed R on your operating system.

How to do it...

Perform the following steps to calculate the t-value:

1. First, we visualize a sample weight vector in a boxplot:

```
> weight <- c(84.12,85.17,62.18,83.97,76.29,76.89,61.37,70.38,90.9
8,85.71,89.33,74.56,82.01,75.19,80.97,93.82,78.97,73.58,85.86,76.4
4)
>boxplot(weight, main="A boxplot of weight")
```

```
>abline(h=70,lwd=2, col="red")
```

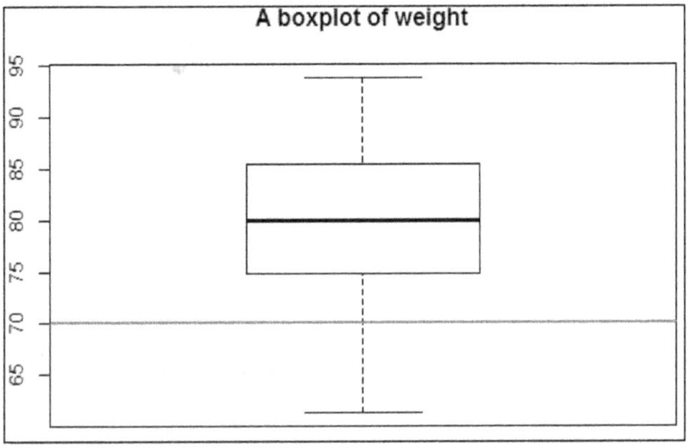

Figure 6: A boxplot of weight

2. We then perform a statistical procedure to validate whether the average weight of the sample is above 70:

```
>t.test(weight, mu = 70)

  One Sample t-test

data:  weight
t = 4.873, df = 19, p-value = 0.0001056
alternative hypothesis: true mean is not equal to 70
95 percent confidence interval:
 75.35658 83.42242
sample estimates:
mean of x
  79.3895
```

3. Next, we visualize two weight samples in one boxplot:

```
>weight2<- c(69.35,63.21,71.57,73.23,65.26,60.32,66.96,59.78,69.71
,76.88,81.39,64.9,75.53,65.05,77.21,64.9,71.93,75.04,74.29,77.53)
>boxplot(list(weight, weight2), main="A boxplot of two weight
samples")
```

```
>abline(h=mean(weight),lwd=2, col="blue")
>abline(h=mean(weight2),lwd=2, col="red")
```

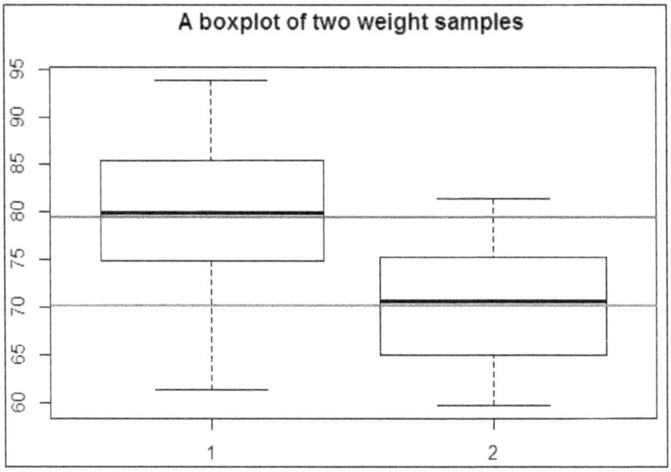

Figure 7: A boxplot of two weight samples

4. We then perform a two-sample T-test to determine whether the means of the two samples are the same:

```
>t.test(weight, weight2)

  Welch Two Sample t-test

data:  weight and weight2
t = 3.8717, df = 34.497, p-value = 0.0004592
alternative hypothesis: true difference in means is not equal to 0
95 percent confidence interval:
  4.367613 14.007387
sample estimates:
mean of x mean of y
  79.3895   70.2020
```

How it works...

A student's T-test is where the test statistics follow a normal distribution (student's t-distribution) if the null hypothesis is true. You can use this to determine whether there is a difference between two independent datasets. A student's T-test is best used with the problems associated with inference based on small samples with an unknown standard deviation.

The procedure to perform one-sample student's T-test is as follows:

1. Make a null hypothesis $H_0 : \mu = \mu_1$, and an alternative hypothesis $H_1 : \mu \neq \mu_1$.
2. Offer significant level α (for example, 0.05).
3. Compute $t = \frac{\overline{x} - \mu_0}{s/\sqrt{n}}$, which follows a t-distribution with degree of freedom of *n-1*, where $\overline{x}$ is the sample mean, μ_0 is the population mean, *n* is the length of the sample, and s is the ration of sample standard deviation over population standard deviation.
4. Calculate the corresponding p-value.
5. Determine whether we can reject the null hypothesis.

On the other hand, the procedure to perform a two-sample student's T-test is as follows:

1. Make a null hypothesis $H_0 : \mu = \mu_1$, and an alternative hypothesis $H_1 : \mu \neq \mu_1$.
2. Offer significant level α (for example, 0.05).
3. Compute $t = \frac{(\overline{x} - \overline{y}) - (\mu_1 - \mu_2)}{\sqrt{\frac{S_x^2}{m} \times \frac{S_y^2}{n}}}$, where $\overline{x}$ and $\overline{y}$ are the sample means, μ_1 and μ_2 are the population means, S_x^2 is the sample variance of sample x, and S_y^2 is the sample variance of sample y.
4. Calculate the corresponding p-value.
5. Determine whether we can reject the null hypothesis.

In this recipe, we discussed a one sample student's T-test and a two-sample student's T-test. In the one sample student's T-test, a research question is often formulated as, "Is the mean of the population different from the null hypothesis?" Thus, to test whether the average weight of city residents is higher than the overall average weight of a nation, we first used a boxplot to view the differences between populations without making any assumptions. From the figure, it is clear that the mean weight of our sample is higher than the average weight (red line) of the overall population. When we applied a one sample T-test, the low p-value of `0.0001056` (< 0.05) suggests that we should reject the null hypothesis that the mean weight for city residents is higher than the average weight of the overall population.

As a one sample T-test enables us to test whether two means are significantly different, a two sample T-test allows us to test whether the means of two independent groups are different. Here, we perform a T-test to determine whether there is a weight difference between city residents and rural residents. Similarly to the one sample T-test, we first use a boxplot to see differences between populations and then apply a two-sample T-test. The test results show the p-value = `0. 0004592` (p< 0.05). In other words, the test provides evidence to reject the null hypothesis, which shows the mean weight of city residents is different from rural residents.

There's more...

To read more about the usage of the student's T-test, please use the `help` function to view related documents:

```
>?t.test
```

Conducting exact binomial tests

To perform parametric testing, one must assume that the data follows a specific distribution. However, in most cases, we do not know how the data is distributed. Thus, we can perform a nonparametric (that is, distribution-free) test instead. In the following recipes, we will show you how to perform nonparametric tests in R. First, we will cover how to conduct an exact binomial test in R.

Getting ready

In this recipe, we will use the `binom.test` function from the stat package.

How to do it...

Perform the following steps to conduct an exact binomial test:

1. Let's assume there is a game where a gambler can win by rolling the number six on a dice. As part of the rules, the gambler can bring their own dice. If the gambler tried to cheat in the game, they would use a loaded dice to increase their chances of winning. Therefore, if we observe that the gambler won 92 games out of 315, we could determine whether the dice was likely fair by conducting an exact binomial test:

```
>binom.test(x=92, n=315, p=1/6)

	Exact binomial test

data:  92 and 315
number of successes = 92, number of trials = 315, p-value =
3.458e-08
alternative hypothesis: true probability of success is not equal
to 0.1666667
95 percent confidence interval:
 0.2424273 0.3456598
sample estimates:
probability of success
             0.2920635
```

How it works...

The binomial test uses binomial distribution to determine whether the true success rate is likely to be *P* for *n* trials with a binary outcome. The formula of probability *P* can be defined in the following equation:

$$P(X=k)=\binom{n}{k}p^k q^{n-k}$$

Here, *X* denotes the random variable counting the number of outcomes of interest, *n* denotes the number of trials, *k* indicates the number of successes, *p* indicates the probability of success, and *q* denotes the probability of failure.

After we compute the probability *P*, we can then perform a sign test to determine whether the probability of success is in line with our expectations. If the probability is not equal to what we expected, we can reject the null hypothesis.

By definition, the null hypothesis is a skeptical perspective or a statement about the population parameter that will be tested. We denote the null hypothesis using H0. A range of population values that are not included in the null hypothesis represents an alternative hypothesis. The alternative hypothesis is denoted by H1. In this case, the null hypothesis and alternative hypothesis are illustrated as follows:

- **H0 (null hypothesis)**: True probability of success is equal to what we expected
- **H1 (alternative hypothesis)**: True probability of success is not equal to what we expected

In this example, we demonstrated how to use a binomial test to determine the number of times the dice is rolled, the frequency of rolling the number six, and the probability of rolling a six with unbiased dice. The result of the T-test shows a p-value is equal to `3.458e-08` (lower than 0.05). For significance at the 5% level, the null hypothesis (the dice is unbiased) is rejected as too many sixes were rolled (probability of success is equal to `0.2920635`).

There's more...

To read more about the usage of the exact binomial test, please use the `help` function to view related documents:

```
>?binom.test
```

Performing Kolmogorov-Smirnov tests

We use a one-sample Kolmogorov-Smirnov test to compare a sample with reference probability. A two-sample Kolmogorov-Smirnov test compares the cumulative distributions of two datasets. In this recipe, we will demonstrate how to perform a Kolmogorov-Smirnov test with R.

Getting ready

In this recipe, we will use the `ks.test` function from the stat package.

How to do it...

Perform the following steps to conduct a Kolmogorov-Smirnov test:

1. Validate whether the *x* dataset (generated with the `rnorm` function) is distributed normally with a one-sample Kolmogorov-Smirnov test:

```
>set.seed(123)
> x <-rnorm(50)
>ks.test(x,"pnorm")

  One-sample
  Kolmogorov-Smirnov test

data:  x
D = 0.073034, p-value =
0.9347
alternative hypothesis: two-sided
```

2. Next, we can generate uniformly distributed sample data:

```
>set.seed(123)
> x <- runif(n=20, min=0, max=20)

> y <- runif(n=20, min=0, max=20)
```

3. We first plot the ECDF of two generated data samples:

```
>plot(ecdf(x), do.points = FALSE, verticals=T, xlim=c(0, 20))
>lines(ecdf(y), lty=3, do.points = FALSE, verticals=T)
```

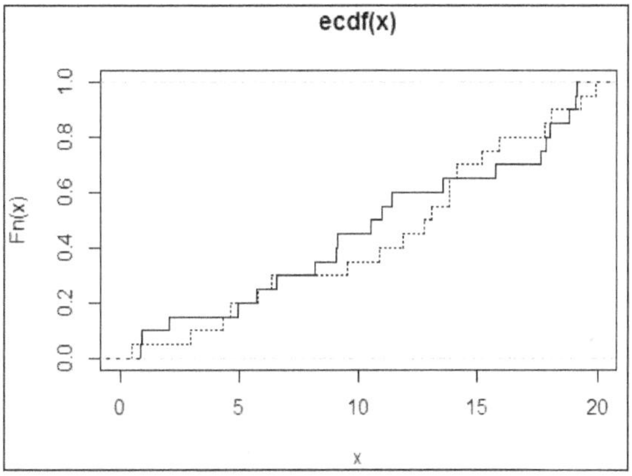

Figure 8: The ECDF plot

4. Finally, we apply a two-sample Kolmogorov-Smirnov test on two groups of data:

```
>ks.test(x,y)

  Two-sample Kolmogorov-Smirnov test

data:  x and y
D = 0.2, p-value = 0.832
alternative hypothesis: two-sided
```

How it works...

The **Kolmogorov-Smirnov test** (**K-S test**) is a nonparametric statistical test for the equality of continuous probability distributions. We can use it to compare a sample to a reference probability distribution (one-sample K-S test), or we can use it to directly compare two samples (two-sample K-S test). The test is based on the **empirical cumulative distribution function** (**ECDF**). Let $x_1, x_2 \ldots x_n$ be a random sample of size *n*; the $F_n(x)$ empirical distribution function is defined as follows:

$$F_n(x) = \frac{1}{n}\sum_{i=1}^{n} I\{x_i \le x\}$$

Here, $I\{x_i \le x\}$ is the indicator function. If $x_i \le x$ the function equals *1*. Otherwise, the function equals *0*.

The Kolmogorov-Smirnov statistic (D) is based on the greatest (where *supx* denotes the supremum) vertical difference between *F(x)* and *Fn(x)*. It is defined as follows:

$$D_n = \sup_x \left| F_n(x) - F(x) \right|$$

- **H0**: The sample follows the specified distribution
- **H1**: The sample does not follow the specified distribution

If *Dn* is greater than the critical value obtained from a table, then we reject *H0* at the level of significance α.

We first tested whether a generated random number from a normal distribution is normally distributed. At the 5% significance level, the p-value of `0.9347` indicates the input is normally distributed.

We then plotted an empirical cumulative distribution function (`ecdf`) plot to show how a two-sample test calculates maximum distance D (showing 0.2), and applied the two-sample Kolmogorov-Smirnov test to discover whether the two input datasets possibly come from the same distribution.

The p-value is `0.832` (above 0.05), which does not reject the null hypothesis. In other words, it means the two datasets could be from the same distribution.

There's more...

To read more about the usage of the Kolmogorov-Smirnov test, use the `help` function to view related documents:

```
>?ks.test
```

For the definition of the empirical cumulative distribution function, refer to the ECDF help page:

```
>?ecdf
```

Working with the Pearson's chi-squared tests

In this recipe, we introduced Pearson's chi-squared test, which is used to examine whether the distribution of categorical variables of two groups differ. We will discuss how to conduct Pearson's chi-squared Test in R.

Getting ready

In this recipe, we will use the `chisq.test` function that originated from the `stat` package.

How to do it...

Perform the following steps to conduct a Pearson's chi-squared test:

1. First, build a matrix containing the number of male and female smokers and nonsmokers:

   ```
   >mat<- matrix(c(2047, 2522, 3512, 1919), nrow = 2, dimnames =
   list(c("smoke","non-smoke"), c("male","female")))
   >mat
   malefemale
   smoke2047    3512
   non-smoke 2522    1919
   ```

2. Then, plot the portion of male and female smokers and nonsmokers in a mosaic plot:

   ```
   >mosaicplot(mat, main="Portion of male and female smokers/non-
   smokers", color = TRUE)
   ```

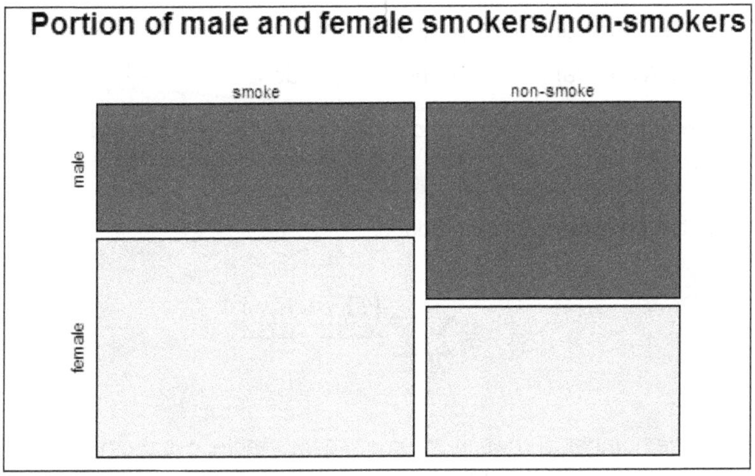

Figure 9: The mosaic plot

3. Next, perform a Pearson's chi-squared test on the contingency table to test whether the factor of gender and smoking habit is independent:

```
>chisq.test(mat)


  Pearson's Chi-squared test with Yates'
  continuity correction


data:  mat
X-squared = 395.79, df = 1, p-value <2.2e-16
```

How it works...

Pearson's chi-squared test is a statistical test used to discover whether there is a relationship between two categorical variables. It is best used for unpaired data from large samples. If you want to conduct a Pearson's chi-squared test, you need to make sure that the input samples satisfy two assumptions: the two input variables should be categorical, and the variable should include two or more independent groups.

In a Pearson's chi-squared test with the assumption that we have two variables, A and B, we can illustrate the null and alternative hypothesis in the following statements:

- **H0**: Variable A and Variable B are independent
- **H1**: Variable A and Variable B are not independent

To test whether the null hypothesis is correct or incorrect, the chi-squared test takes the following three steps:

1. Calculate the chi-squared test statistic, X^2:

$$X^2 = \sum_{i=1}^{r} \sum_{j=1}^{c} \frac{\left(O_{i,j} - E_{i,j}\right)^2}{E_{i,j}}$$

   Here, *r* is the number of rows in the contingency table, c is the number of columns in the contingency table, *Oi, j* is the observed frequency count, and *Ei, j* is the expected frequency count.

2. Determine the degrees of freedom, *df*, of this statistic. The degree of freedom is equal to the following:

$$df = (r-1) \times (c-1)$$

Here, *r* is the number of levels of one variable and c is the number of levels for another variable.

3. Compare X^2 to the critical value from the chi-squared distribution with degrees of freedom.

In this recipe, we used a contingency table and mosaic plot to illustrate the differences in count numbers. It is obvious that the portion of female smokers is much smaller than male smokers.

Then, we performed a Pearson's chi-squared test on the contingency table to determine if the factor of gender and smoking habit is independent. The output p-value = `2.2e-16` (< 0.05) refutes the null hypothesis and shows that the smoking habit is related to gender.

There's more...

To read more about the usage of Pearson's chi-squared test, use the `help` function to view related documents:

```
>?chisq.test
```

Understanding the Wilcoxon Rank Sum and Signed Rank tests

The **Wilcoxon Rank Sum and Signed Rank test** (**Mann-Whitney-Wilcoxon**) is a nonparametric test of the null hypothesis that two different groups come from the same population without assuming the two groups are normally distributed. This recipe will show you how to conduct a Wilcoxon Rank Sum and Signed Rank test in R.

Getting ready

In this recipe, we will use the `wilcox.test` function that originated from the `stat` package.

How to do it...

Perform the following steps to conduct a Wilcoxon Rank Sum and Signed Rank test:

1. First, prepare the Facebook likes of a fan page:

```
> likes <- c(17,40,57,30,51,35,59,64,37,49,39,41,17,53,21,28,46,23
,14,13,11,17,15,21,9,17,10,11,13,16,18,17,27,11,12,5,8,4,12,7,11,8
,4,8,7,3,9,9,9,12,17,6,10)
```

2. Then, plot the *Facebook Likes* data into a histogram:

```
>hist(likes)
```

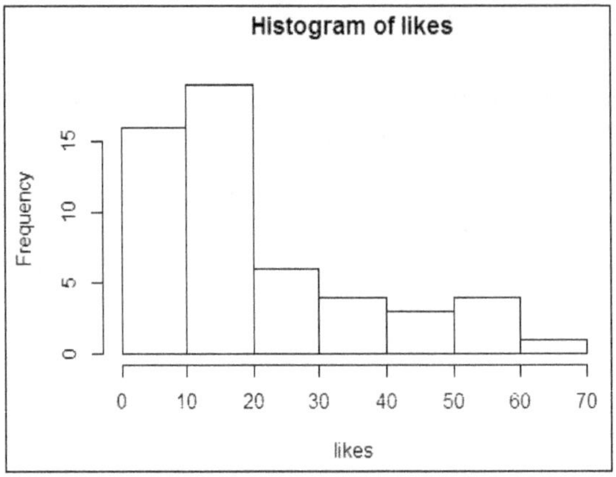

Figure 10: The histogram of Facebook likes of a fan page

3. Now, perform a one-sample Wilcoxon signed rank test to determine whether the median of the input dataset is equal to 30:

```
>wilcox.test(likes, mu = 30)

  Wilcoxon signed rank test with continuity
  correction

data:  likes
V = 314.5, p-value = 0.0006551
alternative hypothesis: true location is not equal to 30
```

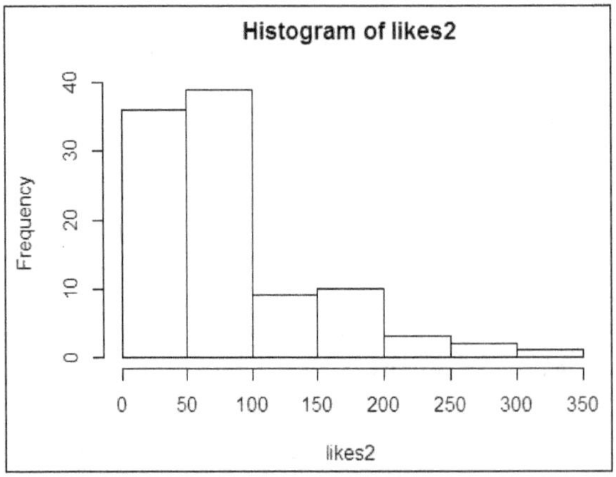

Figure 11: The histogram of Facebook likes of another fan page

4. Then, input and plot likes data from another fanpage into a histogram:

```
>likes2<- c(28,152,197,25,62,39,32,202,85,74,125,32,67,29,37,297,1
01,45,24,63,17,92,46,60,317,85,46,61,56,59,91,54,133,87,200,28,97,
28,30,103,77,78,80,159,39,46,151,278,75,124,213,35,145,68,30,71,58
,52,36,61,48,31,165,93,74,30,86,88,145,21,47,167,63,55,36,215,52,8
4,24,189,65,44,101,36,39,98,140,32,65,33,84,61,45,40,160,64,65,41,
36,165)
>hist(likes2)
```

5. We can also place the likes data of two fanpages in the same boxplot:

```
>boxplot(list(likes, likes2))
```

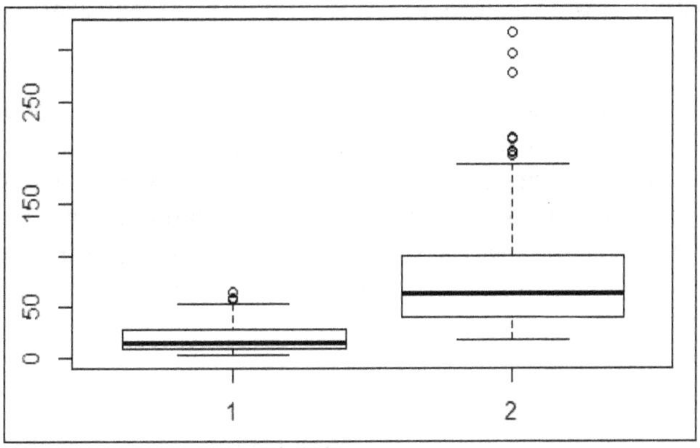

Figure 12: The boxplot of two Facebook likes page

6. Finally, perform a two-sample Wilcoxon rank sum test to validate whether the likes median of the two fanpages is identical:

```
>wilcox.test(likes, likes2)

    Wilcoxon rank sum test with continuity
    correction

data:  likes and likes2
W = 426, p-value <2.2e-16
alternative hypothesis: true location shift is not equal to 0
```

How it works...

In this recipe, we discussed a nonparametric test method—the Wilcoxon Rank Sum test (also known as the Mann-Whitney U-test). For a student's T-test, it is assumed that the differences between the two samples are normally distributed (it also works best when the two samples are normally distributed). However, when the normality assumption is uncertain, we can adopt the Wilcoxon Rank Sum test for hypothesis testing.

Here, we first plotted the histogram of fanpage likes data. From the plot, the fanpage likes appear similar to a Poisson distribution. We then performed a one-sample Wilcoxon Rank Sum test to determine whether the median of fanpage likes is equal to 30 of the likes-number median of the competitor's fanpage. From the test result, the p-value = `0.0006551` (< 0.05) rejects the null hypothesis, which shows the median of the fanpage likes is not equal to 30.

Next, we wanted to compare whether the likes from two groups come from the same population. We first made the boxplot of each fanpage and put them in one figure. Here, we discovered that the likes median of the first fanpage is lower than the likes median of the second fanpage. We then performed a two-sample Wilcoxon Rank Sum test to determine whether the two sets of fanpage likes data come from the same population. Here, we obtained a p-value of `2.2e-16` (< 0.05), which rejected the null hypothesis. We can conclude that the two sets of sample likes data do not come from the same population.

There's more...

To read more about the usage of the Wilcoxon Rank Sum and Signed Rank test, use the `help` function to view related documents:

```
>?wilcox.test
```

Conducting one-way ANOVA

Analysis of variance (**ANOVA**) investigates the relationship between categorical independent variables and continuous dependent variables. You can use it to test whether the means of several groups are equal. If there is only one categorical variable as an independent variable, you can perform a one-way ANOVA. On the other hand, if there are more than two categorical variables, you should perform a two-way ANOVA. In this recipe, we discuss how to conduct one-way ANOVA with R.

Getting ready

In this recipe, we will use the `oneway.test` and `TukeyHSD` functions.

How to do it...

Perform the following steps to perform a one-way ANOVA:

1. We begin by visualizing data with a boxplot:

```
>data_scientist<- c(95694,82465,85001,74721,73923,94552,96723,9079
5,103834,120751,82634,55362,105086,79361,79679,105383,85728,71689,
92719,87916)

>software_eng<- c(78069,82623,73552,85732,75354,81981,91162,83222,
74088,91785,89922,84580,80864,70465,94327,70796,104247,96391,75171
,65682)

>bi_eng<- c(62895,72568,67533,66524,60483,69549,62150,53320,66197,
79189,64246,76079,53821,69444,75194,73011,71056,63592,61502,59758)

>salary<- c(data_scientist, software_eng, bi_eng)

>profession<- c(rep("Data Scientist",20), rep("Software
Engineer",20), rep("BI Engineer",20))

>boxplot(salary ~ profession,xlab='Profession', ylab = "Salary",
main="Salary boxplots of different profession")
```

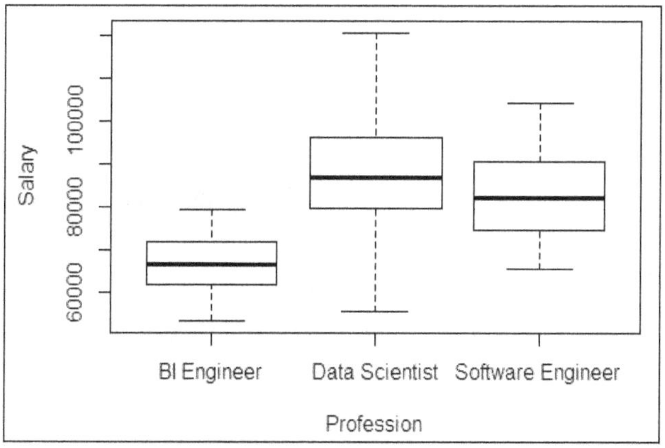

Figure 13: The salary boxplots of different professions

2. Next, we conduct a one-way ANOVA to examine whether engineers with different professions have different salary means. We use the `oneway.test` function:

```
>oneway.test(salary ~ profession)

  One-way analysis of means (not assumingequal variances)

data:  salary and profession
F = 27.972, numdf = 2.000, denomdf =
35.291, p-value = 5.263e-08
```

3. In addition to `oneway.test`, there is an `aov` standard function for ANOVA analysis:

```
>salary.aov<- aov(salary ~ profession)
>summary(salary.aov)
Df     Sum Sq    Mean Sq F value    Pr(>F)
profession    2 5.111e+092.555e+09    21.23 1.29e-07
Residuals    57 6.862e+091.204e+08

profession   ***
Residuals
---
Signif.codes:
0 '***' 0.001 '**' 0.01 '*' 0.05 '.' 0.1 ' ' 1
```

4. The model generated by the `aov` function can also generate a summary as a fitted table:

```
>model.tables(salary.aov, "means")
Tables of means
Grand mean

79035.67

profession
profession
      BI Engineer    Data Scientist
            66406             88201
Software Engineer
            82501
```

5. For the `salary.aov` model, we can use `TukeyHSD` for a post-hoc comparison test:

```
>salary.posthoc<-TukeyHSD(salary.aov)
>salary.posthoc
Tukey multiple comparisons of means
    95% family-wise confidence level

Fit: aov(formula = salary ~ profession)

$profession
diff
Data Scientist-BI Engineer          21795.25
Software Engineer-BI Engineer       16095.10
Software Engineer-Data Scientist -5700.15
lwr
Data Scientist-BI Engineer           13445.609
Software Engineer-BI Engineer         7745.459
Software Engineer-Data Scientist -14049.791
upr
Data Scientist-BI Engineer          30144.891
Software Engineer-BI Engineer       24444.741
Software Engineer-Data Scientist   2649.491
padj
Data Scientist-BI Engineer          0.0000001
Software Engineer-BI Engineer       0.0000616
Software Engineer-Data Scientist 0.2362768
```

How it works...

To understand whether engineers with different professions have different salary means, we first plot salary by profession as a boxplot. This offers a simple indication of whether the salary means of engineers vary by profession. We then perform the most basic form of ANOVA, a one-way ANOVA, to test whether the populations have different means.

The procedure to perform a one-way ANOVA takes the following steps:

1. Make a null hypothesis $_0 \ \mu_1 = \mu_2 = \quad \mu$ and an alternative hypothesis $H_1 : \mu_1, \mu_2, \cdots \mu_r$. The means are not all equal.
2. Offer a significant level α (for example, 0.05).

3. Compute the summary table of ANOVA:

Sources	Sum of Squares	Degree of Freedom	Mean Square
Between Group	SS_b	r-1	$MS_b = SS_b/(r-1)$
Within Group	SS_w	n-r	$MS_w = SS_w/(n-r)$
Total	SS_{w+b}	n	

4. Compute $f = \frac{MS_b}{MS_w}$.
5. Calculate the corresponding p-value.
6. Determine whether we can reject the null hypothesis.

In R, there are two functions to perform the ANOVA test: `oneway.test` and `aov`. The advantage of `oneway.test` is that the function applies a Welch correction to address heterogeneity of variance. However, it does not provide as much information as `aov`, and it does not offer a post-hoc test. Next, we perform `oneway.test` and `aov` on the independent variable, profession, in regard to the dependent variable, salary. Both test results show a low p-value, which rejects the null hypothesis that engineers with different professions have the same salary means.

As the results of the ANOVA only suggest that there is a significant difference in means within the overall population, you may not know which two populations differ. Therefore, we apply the `TukeyHSD` post-hoc comparison test on our ANOVA model. The results show the largest variance in salaries is between data scientists and business intelligence engineers, as their confidence interval is furthest to the right within the plot.

There's more...

ANOVA relies on F-distribution as the basis of probability distribution. The F-score is obtained by dividing the between-group variance by in-group variance. If the overall F test were significant, you could conduct a post-hoc test (or multiple comparison tests) to measure the differences between groups. The most commonly used post-hoc tests are Scheffé's method, the Tukey-Kramer method, and Bonferroni correction.

To interpret the output of ANOVA, you need to have a basic understanding of multiple terms, including degrees of freedom, sum of the square total, sum of squares groups, sum of squares errors, mean square error, and F statistic. If you require more information about these terms, you may refer to *Using Multivariate Statistics* by *Barbara G. Tabachnick, Linda S. Fidell, Allyn and Bacon (2006)* or the Wikipedia entry *Analysis of variance* at `http://en.wikipedia.org/wiki/Analysis_of_variance#cite_ref-31`.

Performing two-way ANOVA

Two-way ANOVA can be viewed as an extension of one-way ANOVA because the analysis covers more than two categorical variables rather than just one. In this recipe, we will discuss how to conduct two-way ANOVA in R.

Getting ready

Download the GDP dataset from the following link and ensure that you have installed R on your operating system: `https://github.com/ywchiu/rcookbook/raw/master/chapter5/engineer.csv`.

How to do it...

Perform the following steps to perform two-way ANOVA:

1. First, load the engineer's salary data from engineer.csv:

```
>engineer<-read.csv("engineer.csv", header = TRUE)
```

2. Plot the two boxplots of the salary factor in regard to profession and region:

```
>par(mfrow=c(1,2))
>boxplot(Salary~Profession, data = engineer,xlab='Profession',
ylab = "Salary",main='Salary v.s. Profession')
>boxplot(Salary~Region, data = engineer,xlab='Region', ylab =
"Salary",main='Salary v.s. Region')
```

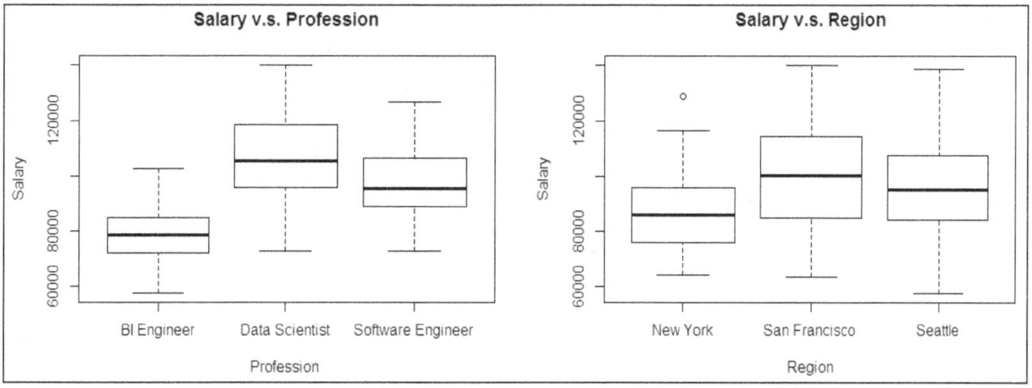

Figure 14: A boxplot of Salary versus Profession and Salary versus Region

3. Also, you can produce a boxplot of salary by a number of `Profession * Region` with the use of the `*` operation in the boxplot function:

```
>boxplot(Salary~Profession * Region,
+          data = engineer,xlab='Profession * Region',
+          ylab = "Salary",
+          cex.axis=0.7,
+          main='Salary v.s. Profession * Region')
```

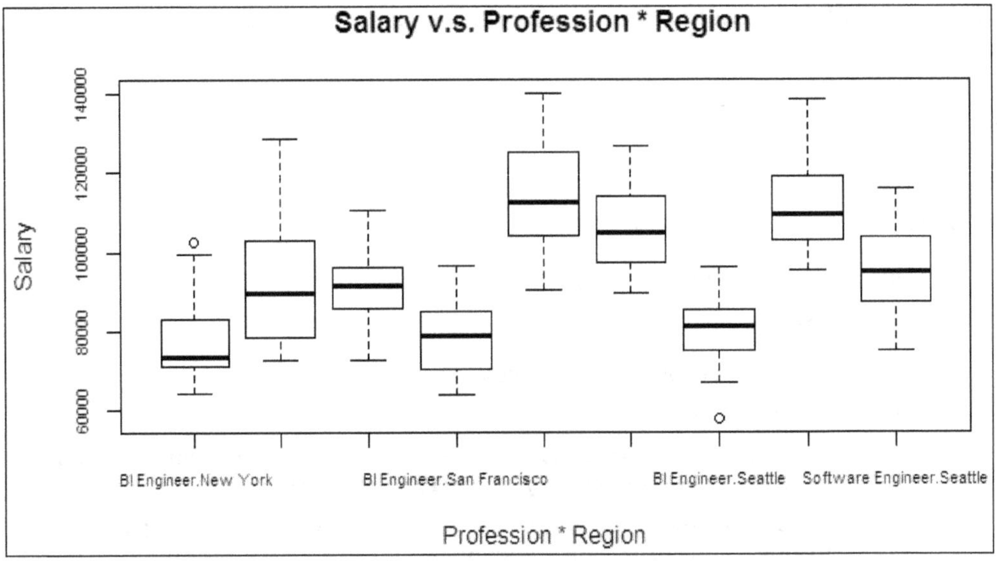

Figure 15: A boxplot of Salary versus. combinations of Profession and Region

4. Next, we use an interaction plot to characterize the relationship between variables:

```
>interaction.plot(engineer$Region, engineer$Profession,
engineer$Salary,
+                    type="b", col=c(1:3),leg.bty="o",
+                    leg.bg="beige", lwd=2,
+                    pch=c(18,24,22),
+                    xlab="Region", ylab="Salary", main="Interaction
Plot")
```

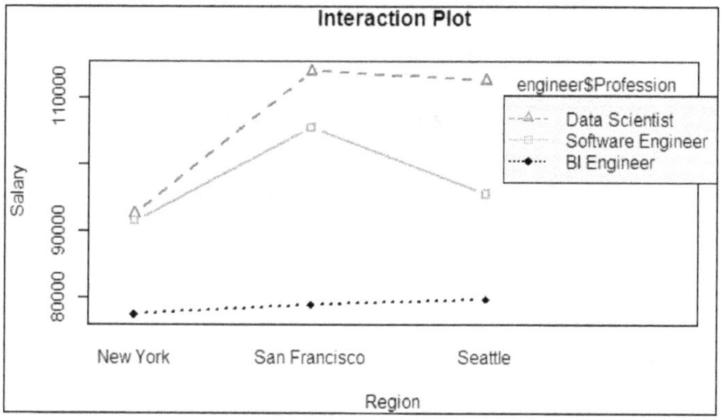

Figure 16: The interaction plot

5. We then perform a two-way ANOVA on salary with a combination of the profession and region factors:

```
>salary.anova<-aov(Salary~Profession * Region, data=engineer)
>summary(salary.anova)
Df      Sum Sq    Mean Sq F value     Pr(>F)
Profession              2 2.386e+101.193e+10  86.098<2e-16 ***
Region                  2 4.750e+092.375e+09  17.1431.64e-07 ***
Profession:Region       4 3.037e+097.593e+08   5.481 0.000355 ***
Residuals             171 2.369e+101.385e+08
---
Signif.codes:  0 '***' 0.001 '**' 0.01 '*' 0.05 '.' 0.1 ' ' 1
```

6. In a similar manner to one-way ANOVA, we can perform a post-hoc comparison test to the results of the two-way ANOVA model:

```
>TukeyHSD(salary.anova)
```

7. We then visualize the differences in mean level with plot function:

```
>plot(TukeyHSD(salary.anova))
```

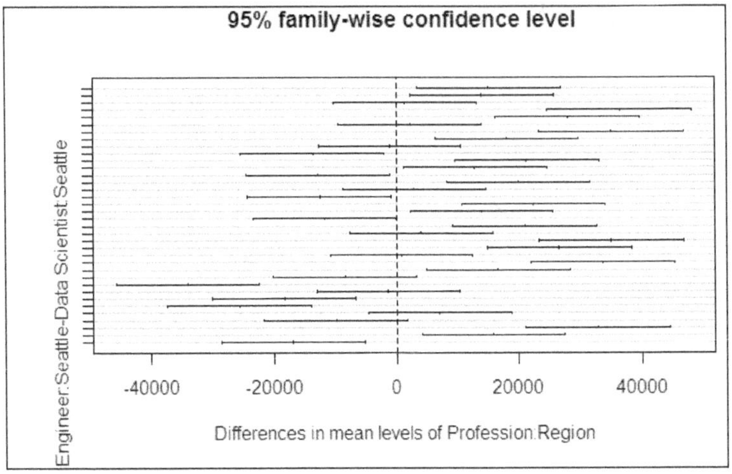

Figure 17: The 95% family-wise confidence level

How it works...

You can look at a two-way ANOVA as an extension of one-way ANOVA. There are two independent variables, and this is why it is known as two-way.

The procedure to perform two-way ANOVA takes the following steps:

1. Make a null hypothesis $H_0(A): \mu_1 = \mu_2 = \cdots \mu_a$, $H_0(B): \mu_1 = \mu_2 = \cdots \mu_b$ and alternative hypothesis $H_1(A): \mu_1, \mu_2, \cdots \mu_a$, $H_1(B): \mu_1, \mu_2, \cdots \mu_b$, the means are not all equal.
2. Offer significant level α (for example, 0.05).
3. Compute the summary table of ANOVA:

Sources	**Sum of Squares**	**Degree of Freedom**	**Mean Square**	**F**
Effect A	SS_A	a-1	$MS_a = SS_a/(a-1)$	MS_a/MS_w
Effect B	SS_B	b-1	$MS_b = SS_b/(b-1)$	MS_b/MS_w
Within Group	SS_w	(a-1)(b-1)	$MS_w = SS_w/(a-1)(b-1)$	

Sources	Sum of Squares	Degree of Freedom	Mean Square	F
Total	SS_{w+b}	abn-1		

4. Calculate correspondent p-value, p_a, p_b.
5. Determine whether we can reject the null hypothesis. If $p_a > \alpha$, effect A is not significant; otherwise, effect A is significant. On the other hand, if $p_b > \alpha$, effect B is not significant; otherwise, effect B is significant.

In this recipe, we performed a two-way ANOVA to examine the influence of the independent variables, `Profession` and `Region`, on dependent variable `Salary`. We first used a boxplot to examine the mean of `Salary` by `Region` and `Profession`. Secondly, we applied an interaction plot to visualize the change of salary in regard to different regions and professions.

The resulting plot shows us that `Profession` and `Region` do have an effect on the mean of `Salary`. Thirdly, we performed a two-way ANOVA with the `aov` function. The output shows the p-value of factor `Profession` and `Region`, and the combination of these two factors rejects the null hypothesis. In other words, both Profession and Region affect the average salary of an engineer. Finally, to examine which two populations have the largest differences, we performed a post-hoc analysis, which revealed that a data scientist from San Francisco has a much higher salary than a BI engineer in New York.

There's more...

For multivariate analysis of variance, you can use the `manova` function to examine the effect of multiple independent variables on multiple dependent variables. Further information about MANOVA is included in the `help` function in R:

```
>?MANOVA
```

9
Rule and Pattern Mining with R

This chapter covers the following topics:

- Transforming data into transactions
- Displaying transactions and associations
- Mining associations with the Apriori rule
- Pruning redundant rules
- Visualizing association rules
- Mining frequent itemsets with Eclat
- Creating transactions with temporal information
- Mining frequent sequential patterns with cSPADE

Introduction

The majority of readers will be familiar with Wal-Mart moving beer next to diapers in its stores because it found that the purchase of both products is highly correlated. This is one example of what data mining is about; it can help us find how items are associated in a transaction dataset. Using this skill, a business can explore the relationship between items, allowing it to sell correlated items together to increase sales.

As an alternative to identifying correlated items with association mining, another popular application of data mining is to discover frequent sequential patterns from transaction datasets with temporal information. This can be used in a number of applications, including predicting customer shopping sequence order, web click streams and biological sequences.

The recipes in this chapter cover creating and inspecting transaction datasets, performing association analysis with the Apriori algorithm, visualizing associations in various graph formats, and finding frequent itemsets using the Eclat algorithm. Last, we create transactions with temporal information and use the cSPADE algorithm to discover frequent sequential patterns.

Transforming data into transactions

Before using any rule mining algorithm, we need to transform data from the data frame format into transactions. In this example, we demonstrate how to transform a purchase order dataset into transactions with the `arules` package.

Getting ready

Download the `purchase_order.RData` dataset from the `https://github.com/ywchiu/rcookbook/raw/master/chapter9/product_by_user.RData` GitHub link.

How to do it...

Perform the following steps to create transactions:

1. First, install and load the `arules` package:

   ```
   > install.packages("arules")
   > library(arules)
   ```

2. Use the `load` function to load purchase orders by user into an R session:

   ```
   > load("product_by_user.RData")
   ```

3. Last, convert the `data.table` (or `data.frame`) into transactions with the `as` function:

   ```
   > trans = as(product_by_user $Product, "transactions")
   > trans
   transactions in sparse format with
    32539 transactions (rows) and
    20054 items (columns)
   ```

How it works...

Before mining a frequent item set or association rule, it is important to prepare the dataset for the class of transactions. In this recipe, we demonstrate how to transform a dataset from a `data.table` (or `data.frame`) into transactions.

We first install and load the `arules` package into an R session. Next, we need to load the `purchase_order.RData` into an R session with `load` function. As we have a transaction dataset, we can transform the data into transactions using the `as` function, which results in 32,539 transactions and 20,054 items.

There's more...

In addition to converting the data of a `data.frame` (or `data.table`) class into transactions, you can transform the data of a list or matrix class into transactions by following these steps:

1. Make a list with three vectors containing purchase records:

```
> tr_list = list(c("Apple", "Bread", "Cake"),
+                c("Apple", "Bread", "Milk"),
+                c("Bread", "Cake", "Milk"))
> names(tr_list) = paste("Tr",c(1:3), sep = "")
```

2. Next, use the `as` function to transform the data frame into transactions:

```
> trans = as(tr_list, "transactions")
> trans
transactions in sparse format with
 3 transactions (rows) and
 4 items (columns)
```

3. You can also transform matrix format data into transactions:

```
> tr_matrix = matrix(
+   c(1,1,1,0,
+     1,1,0,1,
+     0,1,1,1), ncol = 4)
> dimnames(tr_matrix) =  list(
+   paste("Tr",c(1:3), sep = ""),
+   c("Apple","Bread","Cake", "Milk")
+   )
```

```
> trans2 =  as(tr_matrix, "transactions")
> trans2
transactions in sparse format with
 3 transactions (rows) and
 4 items (columns)
```

Displaying transactions and associations

The `arules` package uses its `transactions` class to store transaction data. As such, we must use the generic function provided by `arules` to display transactions and association rules. In this recipe, we illustrate how to plot transactions and association rules with various functions in the `arules` package.

Getting ready

Ensure you have completed the previous recipe by generating transactions and storing these in a variable named `trans`.

How to do it...

Perform the following steps to display transactions and associations:

1. First, obtain a `LIST` representation of the transaction data:

```
> head(LIST(trans),3)
$'00001'
[1] "P0014520085"

$'00002'
[1] "P0018800250"

$'00003'
[1] "P0003926850034" "P0013344760004" "P0013834251"
"P0014251480003"
```

2. Next, use the `summary` function to show a summary of the statistics and details of the transactions:

```
> summary(trans)
transactions as itemMatrix in sparse format with
```

```
 32539 rows (elements/itemsets/transactions) and
 20054 columns (items) and a density of 7.720787e-05

most frequent items:
P0005772981 P0024239865 P0004607050 P0003425855 P0006323656
(Other)
        611         610         489         462         327
47882

element (itemset/transaction) length distribution:
sizes
    1     2     3     4     5     6     7     8     9    10    11
12    13    14
23822  5283  1668   707   355   221   132    88    62    49    29
20    23    19
   15    16    17    18    19    20    21    22    23    24    27
29    30    32
    9     8     7     6     1     6     5     3     3     2     4
1     2     1
   35    44
    2     1

   Min. 1st Qu.  Median    Mean 3rd Qu.    Max.
  1.000   1.000   1.000   1.548   2.000  44.000

includes extended item information - examples:
       labels
1 P0000005913
2 P0000006020
3 P0000006591

includes extended transaction information - examples:
  transactionID
1         00001
2         00002
3         00003
```

3. You can then display transactions using the `inspect` function:

```
> inspect(trans[1:3])
  items
transactionID
1 {P0014520085}                                                       00001
2 {P0018800250}                                                       00002
3 {P0003926850034,P0013344760004,P0013834251,P0014251480003} 00003
```

4. In addition, you can filter transactions by size:

```
> filter_trans = trans[size(trans) >=3]
> inspect(filter_trans[1:3])
  items
transactionID
3 {P0003926850034,P0013344760004,P0013834251,P0014251480003} 00003
5 {P0018474750044,P0023729451,P0024077600013,P0024236730}     00005
7 {P0003169854,P0008070856,P0020005801,P0024629850}            00007
```

5. You can also use the `image` function to inspect transactions visually:

```
> image(trans[1:300,1:300])
```

6. To visually show frequency/support bar plot, use `itemFrequencyPlot`:

```
> itemFrequencyPlot(trans, topN=10, type="absolute")
```

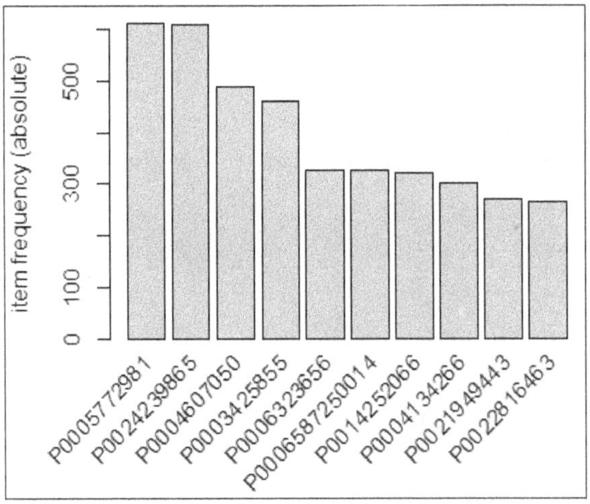

Figure 1: The item frequency bar plot

How it works...

After creating transactions, we can display the association to gain insights into how relationships are built. The `arules` package provides a variety of methods to inspect transactions. First, we use the `LIST` function to obtain a list representation of the transaction data. We can then use the `summary` function to view information such as basic descriptions, most frequent items, and transaction length distribution.

Next, we use the `inspect` function to display the first three transactions. Besides showing all transactions, one can filter transactions by size and then reveal associations by using the `inspect` function. Furthermore, we can use the `image` function to visually inspect transactions. Last, we illustrate how to use a frequency/support bar plot to display the top 10 most frequent items of the transaction dataset.

There's more...

Besides using `itemFrequencyPlot` to show the frequency/bar plot, one can use the `itemFrequency` function to show support distribution. For more detail, please use the `help` function to view the following document:

```
> help(itemFrequency)
```

Mining associations with the Apriori rule

Association mining is a technique that can discover interesting relationships hidden in a transaction dataset. This approach first finds all frequent itemsets and generates strong association rules from frequent itemsets. In this recipe, we will introduce how to perform association analysis using the Apriori rule.

Getting ready

Ensure you have completed the previous recipe by generating transactions and storing these in a variable, `trans`.

How to do it...

Please perform the following steps to analyze association rules:

1. Use `apriori` to discover rules with support over `0.001` and confidence over `0.1`:

   ```
   > rules <- apriori(trans, parameter = list(supp = 0.001, conf =
   0.1, target= "rules"))
   > summary(rules)
   ```

```
set of 6 rules

rule length distribution (lhs + rhs):sizes
2
6

   Min. 1st Qu.  Median    Mean 3rd Qu.    Max.
      2       2       2       2       2       2

summary of quality measures:
    support           confidence          lift
 Min.   :0.001137   Min.   :0.1131   Min.   :16.01
 1st Qu.:0.001183   1st Qu.:0.1351   1st Qu.:17.72
 Median :0.001321   Median :0.1503   Median :22.85
 Mean   :0.001281   Mean   :0.1740   Mean   :22.06
 3rd Qu.:0.001368   3rd Qu.:0.2098   3rd Qu.:26.21
 Max.   :0.001383   Max.   :0.2704   Max.   :27.33

mining info:
  data ntransactions support confidence
 trans         32539   0.001        0.1
```

2. We can then use `inspect` to display all rules:

```
> inspect(rules)
  lhs                  rhs                 support     confidence lift
1 {P0014252070}     => {P0014252066}     0.001321491 0.2704403
27.32874
2 {P0014252066}     => {P0014252070}     0.001321491 0.1335404
27.32874
3 {P0006587250003} => {P0006587250014} 0.001137097 0.1608696
16.00775
4 {P0006587250014} => {P0006587250003} 0.001137097 0.1131498
16.00775
5 {P0014252055}     => {P0014252066}     0.001382956 0.2261307
22.85113
6 {P0014252066}     => {P0014252055}     0.001382956 0.1397516
22.85113
```

3. Last, you can sort rules by confidence and then inspect all rules:

```
> rules <- sort(rules, by="confidence", decreasing=TRUE)
> inspect(rules)
  lhs                 rhs                 support     confidence lift
1 {P0014252070}    => {P0014252066}    0.001321491 0.2704403
27.32874
5 {P0014252055}    => {P0014252066}    0.001382956 0.2261307
22.85113
3 {P0006587250003} => {P0006587250014} 0.001137097 0.1608696
16.00775
6 {P0014252066}    => {P0014252055}    0.001382956 0.1397516
22.85113
2 {P0014252066}    => {P0014252070}    0.001321491 0.1335404
27.32874
4 {P0006587250014} => {P0006587250003} 0.001137097 0.1131498
16.00775
```

How it works...

The purpose of association mining is to discover relationships between items from a transactional database. Typically, the process of association mining finds itemsets that have support greater than the minimum support. Next, the process uses the frequent itemsets to generate strong rules (for example, *Milk => Bread*, a customer who buys milk is likely to buy bread) that have confidence greater than minimum confidence. By definition, an association rule can be expressed in the form of *X=>Y*, where *X* and *Y* are disjointed itemsets. We can measure the strength of associations using two terms: **support** and **confidence**. Support shows the percentage of a rule that is applicable within a dataset, while confidence indicates the probability of both *X* and *Y* appearing in the same transaction:

$$Support = \frac{\sigma(X \cup Y)}{N}$$

$$Confidence = \frac{\sigma(X \cup Y)}{\sigma(X)}$$

Where σ refers to the frequency of a particular itemset, and *N* denotes the population.

As support and confidence are metrics for rule strength only, one may still obtain many redundant rules with high support and confidence. Thus, we can use the third measure, lift, to evaluate the quality (ranking) of the rule. By definition, lift indicates the strength of a rule over the random co-occurrence of *X* and *Y*. We can use lift in the following form:

$$Lift = \frac{\sigma(X \cup Y)}{\sigma(X) \times \sigma(Y)}$$

Apriori is the best-known algorithm for mining associations, which performs a level-wise, breadth-first algorithm to count candidate itemsets. The process of Apriori starts by finding frequent itemsets (sets of items that have minimum support) level-wise. For example, the process starts by finding frequent 1 itemsets. It then continues by using frequent 1 itemsets to find frequent 2 itemsets. The process iteratively discovers new frequent k+1 itemsets from frequent *k* itemsets until no frequent itemsets are found. Finally, the process utilizes frequent itemsets to generate association rules:

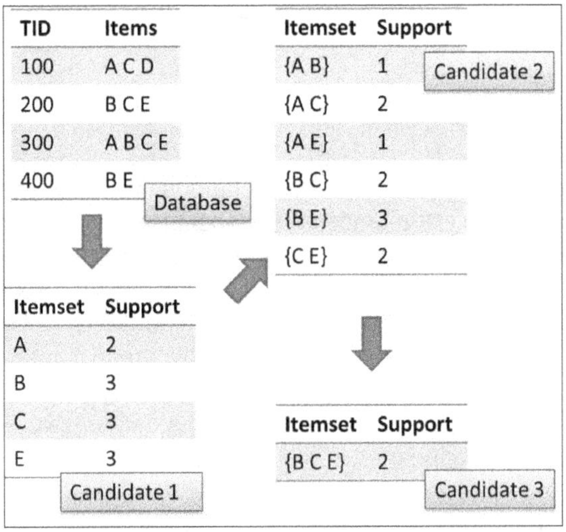

Figure 2: The process of the Apriori algorithm

In this recipe, we use the Apriori algorithm to find association rules within transactions. First, we apply the Apriori algorithm to search for rules with support over 0.001 and confidence over 0.1. We then use the `summary` function to inspect detailed information on the generated rules. From the output summary, we find the Apriori algorithm produces six rules. Furthermore, we can find rule length distribution, the summary of quality measurement, and mining information. In the summary of quality measurement, we find descriptive statistics of three measurements, support, confidence, and lift. Support is the proportion of transactions containing a certain itemset. Confidence is the correctness percentage of the rule. Lift is the response target association rule divided by the average response.

To explore some generated rules, we can use the `inspect` function to view all rules. Last, we can sort rules by confidence and list rules with the most confidence. Thus, we find that `P0014252070` associated to `P0014252066` is the most confident rule, with support equal to `0.001321491`, confidence equal to `0.2704403` and lift equal to `27.32874`.

There's more...

For those interested in research results using the `Groceries` dataset, and how support, confidence, and lift measurement are defined, refer to the following papers:

- *Implications of Probabilistic Data Modeling for Mining Association Rules* by *Michael Hahsler, Kurt Hornik, and Thomas Reutterer, Springer (2006)*
- *Modeling Techniques in Predictive Analysis* by *In M. Spiliopoulou, R. Kruse, C. Borgelt, A*
- *From Data and Information Analysis to Knowledge Engineering, Studies in Classification, Data Analysis, and Knowledge Organization* by *Nuernberger, and W. Gaul, and editors, pages 598–605. Springer-Verlag.*

Also, in addition to using the `summary` and `inspect` functions to inspect association rules, one can use `interestMeasure` to obtain additional interest measures:

```
> head(interestMeasure(rules, c("support", "chiSquare", "confidence",
"conviction","cosine", "coverage", "leverage", "lift","oddsRatio"),
Groceries))
```

Pruning redundant rules

Among generated rules, we sometimes find repeated or redundant rules (for instance, one rule is the super rule of another rule). In this recipe, we will show how to prune (or remove) repeated or redundant rules.

Getting ready

In this recipe, one has to have completed the previous recipe by generating rules and having these stored in a variable named `rules`.

How to do it...

Perform the following steps to prune redundant rules:

1. First, you need to identify the redundant rules:

   ```
   > rules.sorted = sort(rules, by="lift")
   ```

   ```
   > subset.matrix = is.subset(rules.sorted, rules.sorted)
   ```

```
> subset.matrix[lower.tri(subset.matrix, diag=T)] = NA
> redundant = colSums(subset.matrix, na.rm=T) >= 1
```

2. You can then remove the redundant rules:

```
> rules.pruned = rules.sorted[!redundant]
> inspect(rules.pruned)
  lhs                  rhs                 support     confidence lift
1 {P0014252070}    => {P0014252066}      0.001321491 0.2704403
27.32874
5 {P0014252055}    => {P0014252066}      0.001382956 0.2261307
22.85113
3 {P0006587250003} => {P0006587250014} 0.001137097 0.1608696
16.00775
```

How it works...

Choosing between support and confidence are the main constraints in association mining. For example, if one uses a high support threshold, one might remove rare item rules without considering whether these rules have a high confidence value. On the other hand, if one chooses to use a low support threshold, association mining can produce large sets of redundant association rules, which make these rules difficult to utilize and analyze. We therefore need to prune redundant rules so we can discover meaningful information from the generated rules.

In this recipe, we demonstrate how to prune redundant rules. First, we search for redundant rules. We sort the rules by lift measure, and then find subsets of sorted rules using the `is.subset` function, which will generate an `itemMatrix` object. We can then set the lower triangle of the matrix to NA. Last, we compute the `colSums` of the generated matrix, of which *colSums >=1* indicates that the specific rule is redundant.

After we have found redundant rules, we can prune these rules from the sorted rules. Last, we can examine the pruned rules using the `inspect` function.

There's more...

To find a subset or superset of rules, one can use the `is.superset` and `is.subset` functions on association rules. These two methods may generate an `itemMatrix` object to show which rule is the superset of a subset of other rules. You can refer to the `help` function for more information:

```
> help(is.superset)
> help(is.subset)
```

Visualizing association rules

To explore the relationship between items, one can visualize the association rules. In the following recipe, we introduce how to use the `arulesViz` package to visualize association rules.

Getting ready

In this recipe, one has to have completed the previous recipe by generating rules and have these stored in a variable named `rules.pruned`.

How to do it...

Please perform the following steps to visualize association rules:

1. First, install and load the `arulesViz` package:

```
> install.packages("arulesViz")
> library(arulesViz)
```

2. You can then make a scatterplot from the pruned rules:

```
> plot(rules.pruned)
```

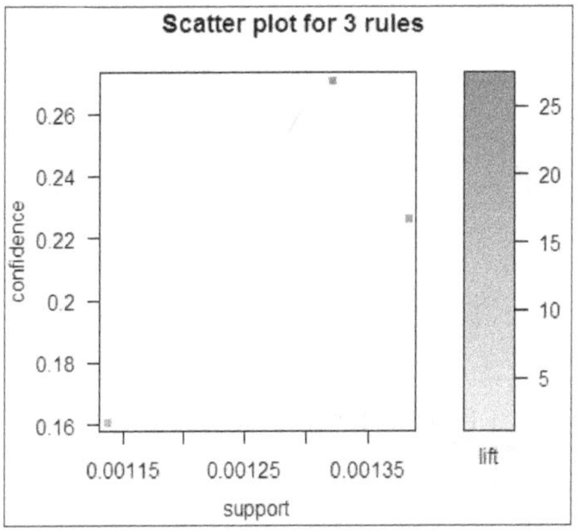

Figure 3: The scatterplot of pruned rules

3. We can also present the rules in a grouped matrix:

```
> plot(rules.pruned,method="grouped")
```

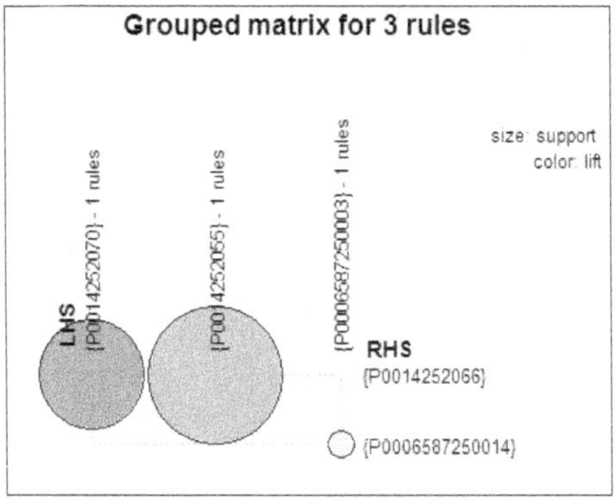

Figure 4: The grouped matrix for three rules

4. Alternatively, we can use a graph to present the rules:

```
> plot(rules.pruned,method="graph")
```

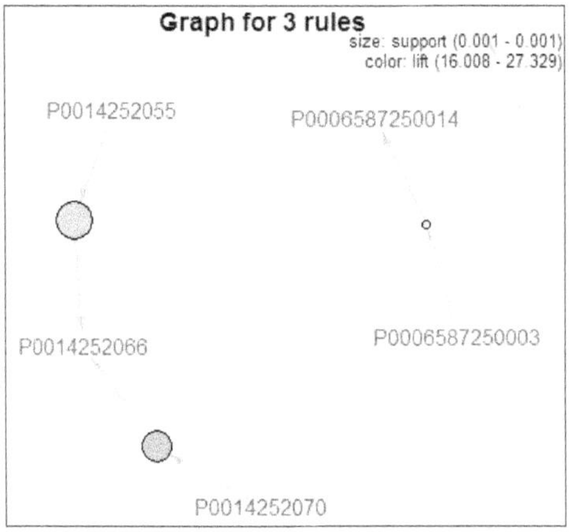

Figure 5: The graph for three rules

How it works...

As an alternative to presenting association rules as text, we can use `arulesViz` to visualize association rules. The `arulesViz` package is an `arules` extension package, which provides many visualization techniques to explore association rules. To start using `arulesViz`, we first need to install and load the `arulesViz` package. We then use the pruned rules generated in the previous recipe to make a scatterplot. The rules are shown as points on the scatterplot, with the *x* axis as support and the *y* axis as confidence. The shade of the plot shows the lift of the rule.

In addition to making a scatterplot, `arulesViz` enables you to plot rules in a grouped matrix or graph. From the grouped matrix, the left-hand rule is shown as column labels and the right-hand rule is shown as row labels. The size of balloons in the grouped matrix illustrates the support of the rule and the color of the balloon indicates the lift of the rule. Furthermore, we can discover the relationship between items in a graph representation; every itemset is presented in a vertex, and their relationship is shown as an edge. Similar to a grouped matrix, the size of the balloons in the grouped matrix shows the support of the rule and the color of the balloon indicates the lift of the rule.

See also

In addition to generating a static plot, one can create an interactive plot by setting `interactive` equal to `TRUE` at the following step:

```
> plot(rules.pruned,interactive=TRUE)
```

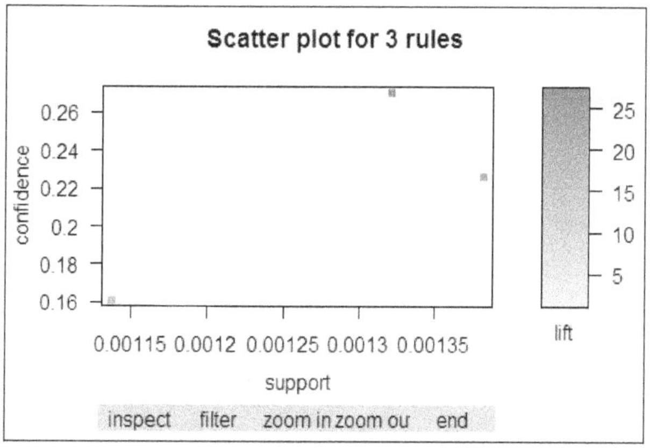

Figure 6: The interactive scatterplot for three rules

Mining frequent itemsets with Eclat

As the Apriori algorithm performs a breadth-first search to scan the complete database, support counting is rather time-consuming. Alternatively, if the database fits into memory, one can use the Eclat algorithm, which performs a depth-first search to count supports. The Eclat algorithm, therefore, runs much more quickly than the Apriori algorithm. In this recipe, we introduce how to use the Eclat algorithm to generate a frequent itemset.

Getting ready

In this recipe, one has to have completed the previous recipe by generating rules and have these stored in a variable named `rules`.

How to do it...

Please perform the following steps to generate a frequent itemset using the Eclat algorithm:

1. Similar to the Apriori method, we can use the `eclat` function to generate a frequent itemset:

```
> frequentsets=eclat(trans,parameter=list(support=0.01,maxlen=10))
```

2. We can then obtain the summary information from the generated frequent itemset:

```
> summary(frequentsets)
set of 6 itemsets

most frequent items:
   P0003425855     P0004607050     P0005772981     P0006323656
P0006587250014
             1               1               1               1
1
       (Other)
             1

element (itemset/transaction) length distribution:sizes
1
6

   Min. 1st Qu.  Median    Mean 3rd Qu.    Max.
      1       1       1       1       1       1

summary of quality measures:
    support
 Min.   :0.01005
 1st Qu.:0.01109
 Median :0.01461
 Mean   :0.01447
 3rd Qu.:0.01782
 Max.   :0.01878

includes transaction ID lists: FALSE

mining info:
  data ntransactions support
 trans         32539    0.01
```

3. Last, we can examine the frequent itemset:

```
> inspect(sort(frequentsets,by="support"))
  items              support
1 {P0005772981}      0.01877747
2 {P0024239865}      0.01874673
4 {P0004607050}      0.01502812
3 {P0003425855}      0.01419835
5 {P0006587250014} 0.01004948
6 {P0006323656}      0.01004948
```

How it works...

In this recipe, we introduce another algorithm, Eclat, to perform frequent itemset generation. While Apriori is a straightforward and easy-to-understand association mining method, the algorithm has two main disadvantages:

- It generates large candidate sets
- It performs inefficiently in support counting, as it requires multiple scans of databases

In contrast to Apriori, Eclat uses equivalence classes, depth-first searching, and set intersection, which greatly improves the speed of support counting.

In Apriori, the algorithm uses a horizontal data layout to store transactions. In contrast, Eclat uses a vertical data layout to store a list of transaction IDs (`tid`) for each item. Eclat then determines support of any *k+1* itemset by intersecting `tid` lists of two *k* itemsets, and it finally utilizes frequent itemsets to generate association rules:

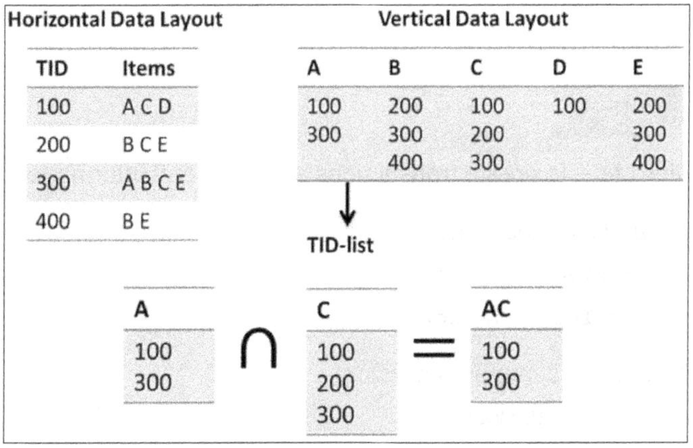

Figure 7: The Eclat algorithm

Similar to the recipe using the Apriori algorithm, we can use the `eclat` function to generate a frequent itemset with a given support and maximum length. We can then use the `summary` function to obtain summary statistics, which include most frequent items, itemset length distributions, a summary of quality measures, and mining info. Last, we can sort the frequent itemset by the support and inspect the frequent itemsets.

There's more...

Besides Apriori and Eclat, another popular association mining algorithm is FP-growth. Similar to Eclat, this takes a depth-first search to count supports. However, there is no existing R package that you can download from CRAN that contains this algorithm. If you are interested in how to apply the FP-growth algorithm to your transaction dataset, you can refer to Christian Borgelt's web page (`http://www.borgelt.net/fpgrowth.html`) for more information.

Creating transactions with temporal information

In addition to mining interesting associations within the transaction database, we can mine interesting sequential patterns using transactions with temporal information. In the following recipe, we demonstrate how to create transactions with temporal information from a web traffic dataset.

Getting ready

Download the `web_traffic.csv` dataset from the `https://github.com/ywchiu/rcookbook/raw/master/chapter9/traffic.RData` GitHub link.

We can then generate transactions from the loaded dataset for frequent sequential pattern mining.

How to do it...

Perform the following steps to create transactions with temporal information:

1. First, install and load the `arulesSequences` package:

   ```
   > install.packages("arulesSequences")
   > library(arulesSequences)
   ```

2. Load web traffic data into an R session:

   ```
   > load('traffic.RData')
   ```

3. Create the transaction data with temporal information:

```
> traffic_data<-data.frame(item=traffic$Page)
> traffic.tran<-as(traffic_data,"transactions")
> transactionInfo(traffic.tran)$sequenceID <- traffic$sequence
> transactionInfo(traffic.tran)$eventID<-traffic$Timestamp
> traffic.tran
transactions in sparse format with
 17 transactions (rows) and
 6 items (columns)
```

4. Use the `inspect` function to inspect the transactions:

```
> inspect(head(traffic.tran))
   items              transactionID sequenceID eventID
 1 {item=/}           1             1          1458565800
 2 {item=/login}      2             1          1458565803
 3 {item=/profile}    3             1          1458565811
 4 {item=/shop_list}  4             1          1458565814
 5 {item=/}           5             2          1458565802
 6 {item=/login}      6             2          1458565808
```

5. You can then obtain summary information about the transactions with temporal information:

```
> summary(traffic.tran)

 transactions as itemMatrix in sparse format with
  17 rows (elements/itemsets/transactions) and
  6 columns (items) and a density of 0.1666667
 most frequent items:
          item=/      item=/login   item=/profile item=/shop_list
               4                4               4               3
   item=/contact          (Other)
               1                1
 element (itemset/transaction) length distribution:
 sizes
  1
 17
    Min. 1st Qu.  Median    Mean 3rd Qu.    Max.
       1       1       1       1       1       1
```

```
includes extended item information - examples:
        labels variables    levels
1        item=/       item         /
2 item=/contact       item /contact
3   item=/login       item   /login
includes extended transaction information - examples:
  transactionID sequenceID     eventID
1             1          1 1458565800
2             2          1 1458565803
3             3          1 1458565811
```

How it works...

Before mining frequent sequential patterns, you are required to create transactions with temporal information. In this recipe, we introduce how to create transactions with temporal information from a data frame. First, we load `web_traffic.csv` as a data frame into an R session. The web traffic dataset contains the source user IP, access timestamp, access URL, and status code. In this example, we use the IP of origin, timestamp, and URL to create transactions.

We convert the source IP to a sequence, and then sort the dataset by user sequence and event timestamp. We can use the `as` function to transform list data into a transaction dataset. We then add `eventID` and `sequenceID` as temporal information; `sequenceID` is the sequence that the event belongs to, and `eventID` indicates when the event occurred. After generating transactions with temporal information, one can use this dataset for frequent sequential pattern mining.

There's more...

In addition to creating your own transactions with temporal information, if you already have data stored in a text file, you can use the `read_basket` function from `arulesSequences` to read transaction data into basket format. We can also read the transaction dataset for further frequent sequential pattern mining:

```
> zaki=read_baskets(con = system.file("misc", "zaki.txt", package =
"arulesSequences"), info = c("sequenceID","eventID","SIZE"))
> as(zaki, "data.frame")
   transactionID.sequenceID transactionID.eventID transactionID.SIZE
items
1                         1                    10                  2
{C,D}
2                         1                    15                  3
{A,B,C}
```

3 {A,B,F}	1	20	3
4 {A,C,D,F}	1	25	4
5 {A,B,F}	2	15	3
6 {E}	2	20	1
7 {A,B,F}	3	10	3
8 {D,G,H}	4	10	3
9 {B,F}	4	20	2
10 {A,G,H}	4	25	3

Mining frequent sequential patterns with cSPADE

One of the most famous frequent sequential pattern mining algorithms is the **SPADE** (**Sequential PAttern Discovery using Equivalence classes**) algorithm, which employs characteristics of the vertical database to perform intersection on ID-list with efficient lattice search and allows us to place constraints on mined sequences. In this recipe, we will demonstrate how to use cSPADE to mine frequent sequential patterns.

Getting ready

In this recipe, one has to have completed the previous recipe by generating transactions with temporal information and have it stored in a variable named `traffic.tran`.

How to do it...

Please perform the following steps to mine frequent sequential patterns:

1. First, use the `cspade` function to generate frequent sequential patterns:

```
> frequent_pattern <-cspade(traffic.tran,parameter = list(support
= 0.50))
> inspect(frequent_pattern)
    items                              support
  1 <{item=/}>                      1.00
```

```
2 <{item=/login}>           1.00
3 <{item=/profile}>         0.75
4 <{item=/shop_list}>     0.75
5 <{item=/},
   {item=/shop_list}>       0.75
6 <{item=/login},
   {item=/shop_list}>       0.75
7 <{item=/},
   {item=/login},
   {item=/shop_list}>       0.75
8 <{item=/},
   {item=/profile}>           0.75
9 <{item=/login},
   {item=/profile}>           0.75
10 <{item=/},
   {item=/login},
   {item=/profile}>           0.75
11 <{item=/},
   {item=/login}>             1.00
```

2. You can then examine the summary of frequent sequential patterns:

```
> summary(frequent_pattern)
set of 11 sequences with

most frequent items:
         item=/     item=/login   item=/profile item=/shop_list
(Other)
              6               6               4               4
14

most frequent elements:
         {item=/}     {item=/login}   {item=/profile} {item=/shop_
list}
              6                 6                 4
4
         (Other)
              14
```

```
element (sequence) size distribution:
sizes
1 2 3
4 5 2

sequence length distribution:
lengths
1 2 3
4 5 2

summary of quality measures:
    support
 Min.   :0.7500
 1st Qu.:0.7500
 Median :0.7500
 Mean   :0.8182
 3rd Qu.:0.8750
 Max.   :1.0000

includes transaction ID lists: FALSE

mining info:
         data ntransactions nsequences support
 traffic.tran            17          4     0.5
```

3. Transform generated sequence format data back to the data frame:

```
> as(frequent_pattern, "data.frame")
                                                              sequence
support
1                                                          <{item=/}>
1.00
2                                                     <{item=/login}>
1.00
3                                                   <{item=/profile}>
0.75
4                                                  <{item=/shop_list}>
0.75
```

```
5                             <{item=/},{item=/shop_list}>    0.75
6                        <{item=/login},{item=/shop_list}>    0.75
7  <{item=/},{item=/login},{item=/shop_list}>    0.75
8                               <{item=/},{item=/profile}>    0.75
9                          <{item=/login},{item=/profile}>    0.75
10  <{item=/},{item=/login},{item=/profile}>    0.75
11                                <{item=/},{item=/login}>    1.00
```

How it works...

The object of sequential pattern mining is to discover sequential relationships or patterns in transactions. You can use the pattern mining result to predict future events, or recommend items to users.

One popular method of sequential pattern mining is SPADE (**S**equential **PA**ttern **D**iscovery using **E**quivalence classes). SPADE uses a vertical data layout to store a list of IDs, in which each input sequence in the database is called **SID**, and each event in a given input sequence is called **EID**. The process of SPADE is performed by generating patterns level-wise by Apriori candidate generation. SPADE generates subsequent *n* sequences from joining (*n-1*) sequences from the intersection of ID lists. If the number of sequences is greater than the minimum support (`minsup`), we can consider the sequence to be frequent enough. The algorithm stops when the process cannot find more frequent sequences:

Database

SID	EID	Items
1	100	C D
1	150	A B C
1	200	A B F
1	250	A C D F
2	150	A B F
2	200	E
3	100	A B F
4	100	D G H
4	200	B F
4	250	A G H

Frequent Sequences (minsup =2)

Frequent 1-Sequences	
A	4
B	4
D	2
F	4

Frequent 2-Sequences	
AB	3
AF	3
B->A	2
BF	4
D->A	2
D->B	2
D->F	2
F->A	2

Frequent 3-Sequences	
ABF	3
BF->A	2
D->BF	2
D->B->A	2
D->F->A	2

Frequent 4-Sequences	
D->BF->A	2

Figure 8: The SPADE algorithm

In this recipe, we illustrate how to use a frequent sequential pattern mining algorithm, cSPADE, to mine frequent sequential patterns. First, as we have transactions with temporal information loaded in variable `trans`, we can use the `cspade` function with support over `0.50` to generate frequent sequential patterns in sequences format. We can then obtain summary information such as the most frequent items, sequence size distribution, the summary of quality measures, and mining info. We can transform the generated **sequence** information back to data frame format, so we can examine the sequence and support of frequent sequential patterns with support over `0.50`. Last, we may discover the access pattern of a user from the generated patterns. A web designer may inspect the patterns to see how to improve the access flow of a website.

See also

If you are interested in the concept and design of the SPADE algorithm, you can refer to the original published paper: *SPADE: An Efficient Algorithm for Mining Frequent Sequences* by *M. J. Zaki, Machine Learning Journal, pages 42, 31–60 (2001).*

10
Time Series Mining with R

This chapter covers the following topics:

- Creating time series data
- Plotting a time series object
- Decomposing time series
- Smoothing time series
- Forecasting time series
- Selecting an ARIMA model
- Creating an ARIMA model
- Forecasting with an ARIMA model
- Predicting stock prices with an ARIMA model

Introduction

The first example of time series analysis in human history occurred in ancient Egypt. The ancient Egyptians recorded the inundation (rise) and relinquishment (fall) of the Nile river every day, noting when fertile silt and moisture occurred. Based on these records, they found that the inundation period began when the sun rose at the same time as the Sirius star system became visible. By being able to predict the inundation period, the ancient Egyptians were able to make sophisticated agricultural decisions, greatly improving the yield of their farming activities.

As demonstrated by the ancient Egyptian inundation period example, time series analysis is a method that can extract patterns or meaningful statistics from data with temporal information. It allows us to forecast future values based on observed results. One can apply time series analysis to any data that has temporal information. For example, an economist can perform time series analysis to predict the GDP growth rate of a country. Also, a website administrator can create a network indicator to predict the growth rate of website visitors or number of page views.

In this chapter, we cover recipes that involve creating and plotting time series objects. We also introduce how to decompose and smooth time series objects, and how to forecast future values based on time series analysis. Moving on, we discuss how the ARIMA model works with time series objects. Last, we introduce how to predict the future value of a stock with the ARIMA model.

Creating time series data

To begin time series analysis, we need to create a time series object from a numeric vector or matrix. In this recipe, we introduce how to create a time series object from the financial report of Taiwan Semiconductor (2330.TW) with the `ts` function.

Getting ready

Download the `tw2330_finance.csv` dataset from the following GitHub link:

`https://github.com/ywchiu/rcookbook/raw/master/chapter10/tw2330_finance.csv`

How to do it...

Please perform the following steps to create time series data:

1. First, read Taiwan Semiconductor's financial report into an R session:

```
> tw2330 = read.csv('tw2330_finance.csv', header=TRUE)
> str(tw2330)
'data.frame': 32 obs. of  5 variables:
 $ Time            : Factor w/ 32 levels "2008Q1","2008Q2",..: 1 2 3 4 5 6 7 8 9 10 ...
 $ Total.Income    : int  875 881 930 646 395 742 899 921 922 1050 ...
 $ Gross.Sales     : num  382 402 431 202 74.8 343 429 447 442 519 ...
```

```
 $ Operating.Income: num  291 304 329 120 12.1 251 320 336 341 405
...
 $ EPS              : num  1.1 1.12 1.17 0.48 0.06 0.95 1.18 1.26
1.3 1.55 ...
```

2. Use the `ts` function to transform the finance data into a time series object:

```
> m <- ts(tw2330$Total.Income, frequency= 4, start=c(2008, 1))
```

3. Use the `class` function to determine the data type:

```
> class(m)
[1] "ts"
```

4. Print the contents of the time series object:

```
> m
     Qtr1 Qtr2 Qtr3 Qtr4
2008  875  881  930  646
2009  395  742  899  921
2010  922 1050 1122 1101
2011 1054 1105 1065 1047
2012 1055 1281 1414 1313
2013 1328 1559 1626 1458
2014 1482 1830 2090 2225
2015 2220 2054 2125 2035
```

5. We can use `window` to subset the time series object:

```
> m2 <- window(m, start=c(2012, 2), end=c(2014, 2))
> m2
     Qtr1 Qtr2 Qtr3 Qtr4
2012      1281 1414 1313
2013 1328 1559 1626 1458
2014 1482 1830
```

6. Alternatively, we can convert the entire `data.frame` into a `ts` object:

```
> m_ts <- ts(tw2330[,-1], start=c(2008,01), frequency = 4)
> class(m_ts)
[1] "mts"    "ts"     "matrix"
> head(m_ts)
     Total.Income Gross.Sales Operating.Income  EPS
[1,]          875       382.0            291.0 1.10
```

```
[2,]            881        402.0                304.0 1.12
[3,]            930        431.0                329.0 1.17
[4,]            646        202.0                120.0 0.48
[5,]            395         74.8                 12.1 0.06
[6,]            742        343.0                251.0 0.95
```

7. Finally, we can print the EPS information from the time series object:

```
> m_ts[,"EPS"]
     Qtr1 Qtr2 Qtr3 Qtr4
2008 1.10 1.12 1.17 0.48
2009 0.06 0.95 1.18 1.26
2010 1.30 1.55 1.81 1.57
2011 1.40 1.39 1.17 1.22
2012 1.29 1.61 1.90 1.60
2013 1.53 2.00 2.00 1.73
2014 1.85 2.30 2.94 3.09
2015 3.05 3.06 2.91 2.81
```

How it works...

A time series is a sequence of random variables that are indexed in time. One can present GDP, CPI, stock index, website indicators, or any data with temporal information as a time series object. In R, we can use the `ts` function to create time series data from a given vector or matrix.

In this recipe, we first download `tw2330_finance.csv` from GitHub. This dataset contains the time, total income, gross sales, operational income, and EPS of Taiwan's largest company, Taiwan Semiconductor (2330.TW). Next, we read the dataset into an R session as an R `data.frame`. In the following steps, we convert this dataset into a time series object.

Moving on, we transform Taiwan Semiconductor's total income into time series data with the `ts` function. As it is a quarterly dataset, we set the frequency equal to `4`; we also configure the time of the first observation, which starts at the first quarter of 2008. Next, we examine the data type of the time series data and print the contents. From the contents, we separate the dataset by quarter and year. Moreover, to subset the dataset, we use the `window` function to subset data with a given time interval.

Besides creating data from a given vector, we can create time series data from a data.frame. Here, we transform all columns except the time column into time series data. This enables us to examine the EPS information from the time series data.

There's more...

Besides using the `ts` function to create a time series object, you can use the `xts` package to create an extensible time series object. This package provides much simpler and more useful APIs to process time series data:

1. First, install and load the `xts` package:

   ```
   > install.packages("xts")
   > library(xts)
   ```

2. Convert the time series data to an xts object with the `as.xts` function:

   ```
   > m.xts <- as.xts(m)
   ```

3. Use the `head` function to examine the first six rows:

   ```
   > head(m.xts)
           [,1]
   2008 Q1  875
   2008 Q2  881
   2008 Q3  930
   2008 Q4  646
   2009 Q1  395
   2009 Q2  742
   ```

4. We can subset an `xts` series with the `window` function:

   ```
   > sub.m.xts <- window(m.xts, start = "2010 Q1", end = "2012 Q1")
   > sub.m.xts
           [,1]
   2010 Q1  922
   2010 Q2 1050
   2010 Q3 1122
   2010 Q4 1101
   2011 Q1 1054
   2011 Q2 1105
   2011 Q3 1065
   2011 Q4 1047
   2012 Q1 1055
   ```

Plotting a time series object

Plotting a time series object will make trends and seasonal composition clearly visible. In this recipe, we introduce how to plot time series data with the `plot.ts` function.

Getting ready

Ensure you have completed the previous recipe by generating a time series object and storing it in two variables: `m` and `m_ts`.

How to do it...

Please perform the following steps to plot time series data:

1. First, use the `plot.ts` function to plot time series data, `m`:

```
> plot.ts(m)
```

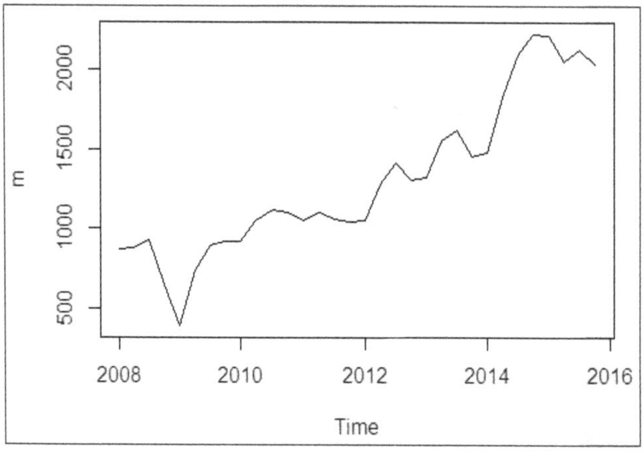

Figure 1: A time series plot of single time series data

2. Also, if the dataset contains multiple time series objects, you can plot multiple time series data in a separate sub-figure:

```
> plot.ts(m_ts, plot.type = "multiple",)
```

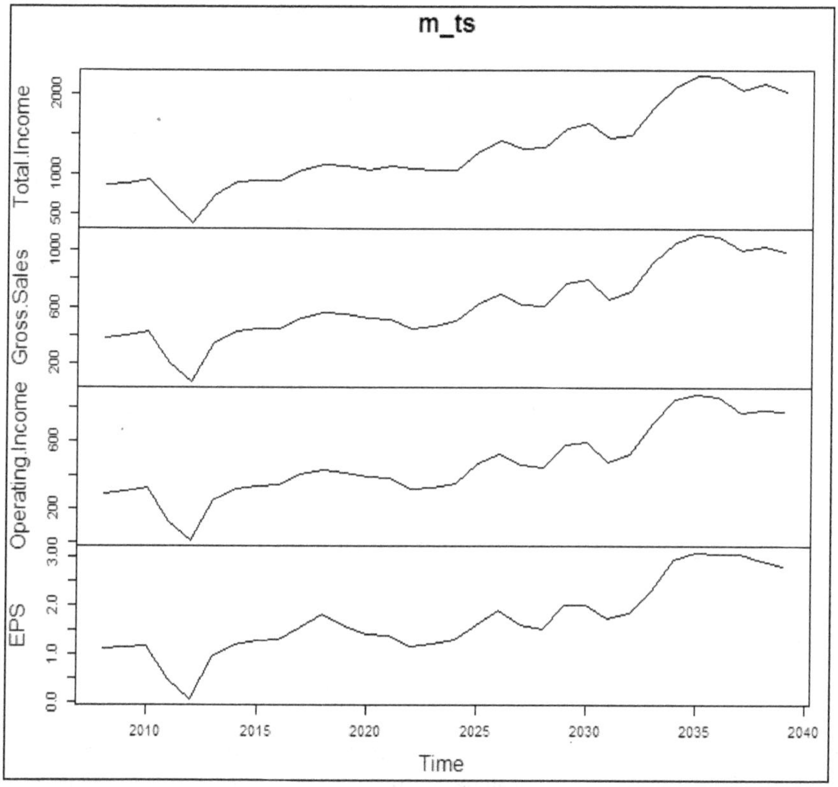

Figure 2: A time series plot of multiple time series data

3. Alternatively, you can plot all four time series objects in a single figure:

```
> plot.ts(m_ts, plot.type = "single", col=c("red","green","blue",
"orange"))
```

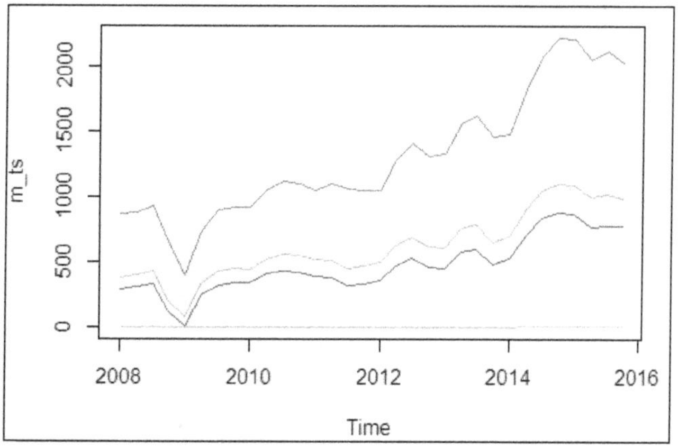

Figure 3: A multiple time series plot in different colors

4. Moreover, you can exclusively plot the EPS time series information:

```
> plot.ts(m_ts[,"EPS"])
```

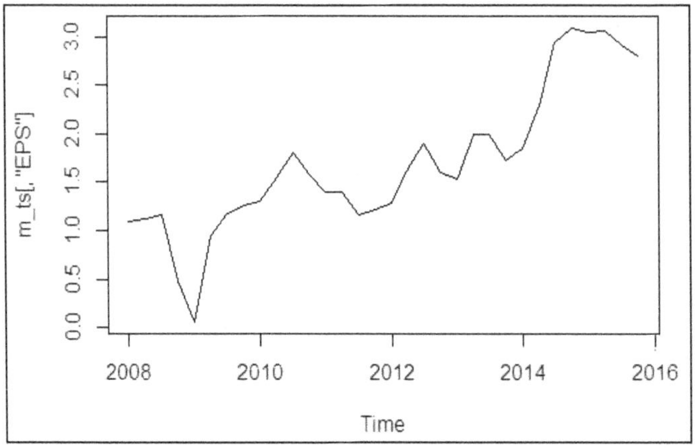

Figure 4: An EPS time series plot

How it works...

To explore time series data trends, you can create a time series plot with the `plot.ts` function. At the start of this recipe, we made a time series plot of Taiwan Semiconductor's total income. From the figure, we can easily observe that after revenue dramatically declined in 2009, Taiwan Semiconductor's overall revenue gradually increased every year. In addition to making a plot from a single time series object, one can plot four different indicators in different sub-figures by setting `plot.type` to `multiple`. *Figure 2* shows that all four financial indicators follow a similar trend. Moreover, we can plot all four indicators in the same figure. Lastly, we demonstrated that we can exclusively plot EPS data from an `mts` object.

There's more...

For `xts` objects, one can use the `plot.xts` function to plot time series data. The following command shows how to make a plot of total income from an `xts` object, `m.xts`:

```
> plot.xts(m.xts)
```

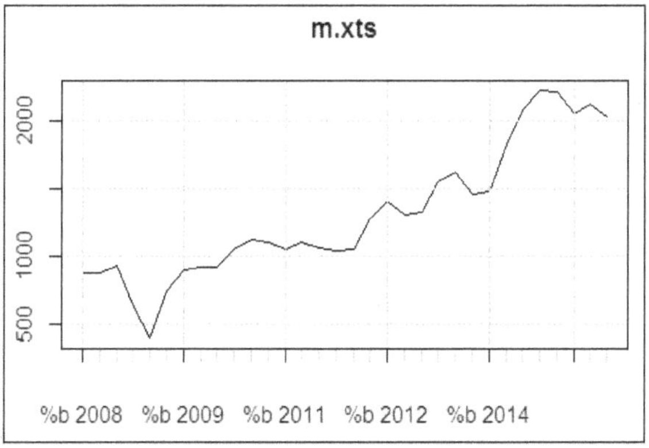

Figure 5: A total income time series plot

Decomposing time series

A seasonal time series is made up of seasonal components, deterministic trend components, and irregular components. In this recipe, we introduce how to use the `decompose` function to destruct a time series into these three parts.

Getting ready

Ensure you have completed the previous recipe by generating a time series object and storing it in two variables: `m` and `m_ts`.

How to do it...

Please perform the following steps to decompose a time series:

1. First, use the `window` function to construct a time series object, `m.sub`, from `m`:

```
> m.sub = window(m, start=c(2012, 1), end=c(2014, 4))
> m.sub
     Qtr1 Qtr2 Qtr3 Qtr4
2012 1055 1281 1414 1313
2013 1328 1559 1626 1458
2014 1482 1830 2090 2225
> plot(m.sub)
```

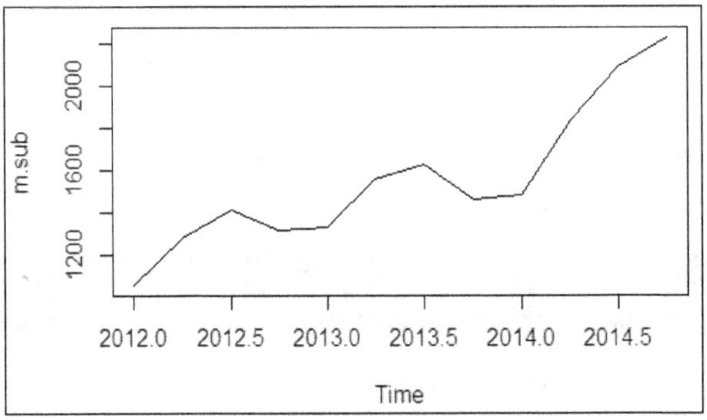

Figure 6: A time series plot in a quarter

2. Use the `decompose` function to destruct the time series object `m.sub`:

```
> components <- decompose(m.sub)
```

3. We can then use the `names` function to list the attributes of `components`:

```
> names(components)
[1] "x"        "seasonal" "trend"    "random"   "figure"
[6] "type"
```

4. Explore the seasonal components from the decomposed time series object:

```
> components$seasonal
          Qtr1       Qtr2       Qtr3       Qtr4
2012 -124.96875   65.28125  127.59375  -67.90625
2013 -124.96875   65.28125  127.59375  -67.90625
2014 -124.96875   65.28125  127.59375  -67.90625
```

5. Use the `plot` function to plot the components of the decomposed time series:

```
> plot(components)
```

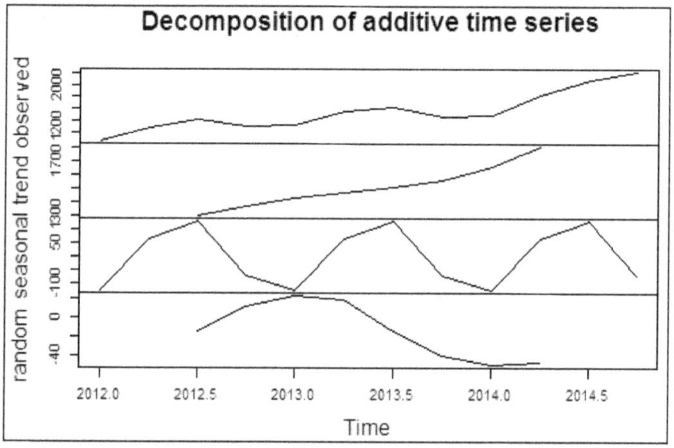

Figure 7: Decomposed time series data

How it works...

From Taiwan Semiconductor's quarterly financial reports, we find that the company generally posts higher profits in the third quarter compared to the first quarter. Based on this observation, we may infer that the data follows a seasonal trend. To support this inference, we can decompose the seasonal components from the time series. To achieve this, we use the `decompose` function to destruct the time series into several distinct components.

For seasonal data, we can destruct a time series into the following three components:

- **T(t)**: It is a deterministic trend component, which shows the linear trend characteristics of the time series
- **S(t)**: It is a seasonal component, which shows the effect on time series periodically
- **e(t)**: It is an irregular component, which captures the white noise of the data

In R, we can use the `decompose` function to destruct a time series into distinct components. We first print Taiwan Semiconductor's time series data in text and a line chart. From the figure, we find the time series data appears to be affected by seasonality. Next, we use the `decompose` function to destruct the time series data. We use additive mode, which takes the following formula: $Y(t) = T(t)+ S(t) + e(t)$. Alternatively, we can set `type = "multiplicative"` to destruct a time series by using a multiplicative model: $Y(t) = T(t)* S(t) + e(t)$. Next, we find the `decompose` function returns a list, in which element `x` is the original dataset, and `"seasonal"`, `"trend"`, and `"random"` correspond to the seasonal component, trend component and irregular component. Last, we can use the `plot` function to plot the decomposed components in separated subfigures.

There's more...

LOWESS (**Locally Weighted Scatterplot Smoothing**) and **LOESS** (**Locally Weighted Smoothing**) are non-parametric methods that fit polynomial regression models to localized data subsets. In the `stl` function, we can use the LOESS method to decompose a periodic time series:

1. In this example, we use the `stl` function to decompose a periodic time series:

```
> comp1 <- stl(m.sub, s.window="periodic")
> names(comp1)
[1] "time.series" "weights"     "call"        "win"
[5] "deg"         "jump"        "inner"       "outer"
```

2. We can also use the `plot` function to examine the decomposed time series components:

```
> plot(ccomp1)
```

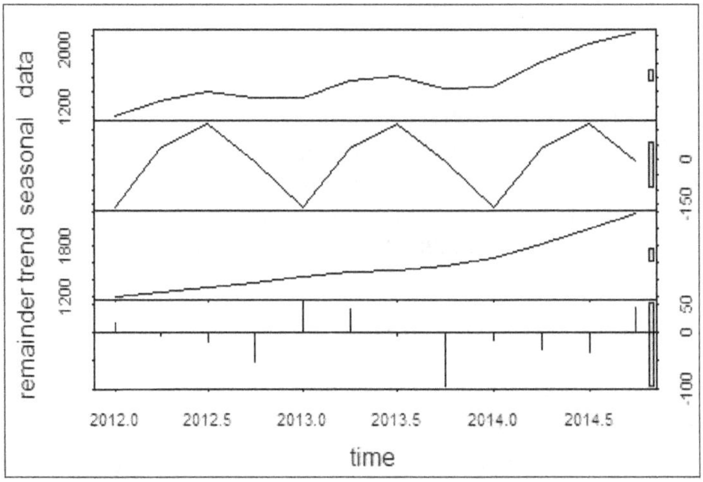

Figure 8: A decomposed time series plot with the LOESS model

Smoothing time series

Time series decomposition allows us to extract distinct components from time series data. The smoothing technique enables us to forecast the future values of time series data. In this recipe, we introduce how to use the `HoltWinters` function to smooth time series data.

Getting ready

Ensure you have completed the previous recipe by generating a time series object and storing it in two variables: `m` and `m_ts`.

How to do it...

Please perform the following steps to smooth time series data:

1. First, use `HoltWinters` to perform Winters exponential smoothing:

```
> m.pre <- HoltWinters(m)
> m.pre
Holt-Winters exponential smoothing with trend and additive
seasonal component.

Call:
HoltWinters(x = m)

Smoothing parameters:
 alpha: 0.8223689
 beta : 0.06468208
 gamma: 1

Coefficients:
         [,1]
a  1964.30088
b    32.33727
s1  -51.47814
s2   17.84420
s3  146.26704
s4   70.69912
```

2. Plot the smoothing result:

```
> plot(m.pre)
```

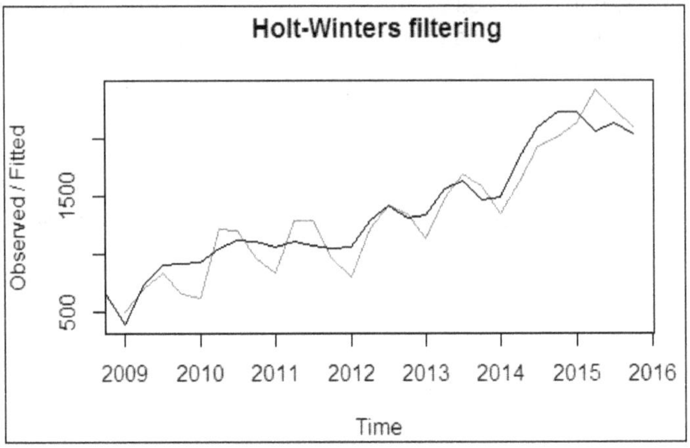

Figure 9: A time series plot with Winters exponential smoothed line

3. We can now find the sum square error of the model:

```
> m.pre$SSE
[1] 733959.4
```

How it works...

At the most basic level, smoothing uses moving averages, which average historical observation values to calculate possible future values. In R, we can use the `HoltWinters` function from the `stat` package to perform time series smoothing. The `HoltWinters` function contains three smoothing methods, which are simple exponential smoothing, Holt exponential smoothing, and Winters exponential smoothing. All three methods use the same function, `HoltWinters`. However, we can invoke them separately based on given `alpha`, `beta`, and `gamma` parameters.

Simple exponential smoothing is performed when data has no trend or seasonal pattern. This method can be represented by the following formula:

$$F_{t+1} = \alpha x_t + (1 - \alpha) F_t$$

In the preceding formula, the prediction value is *F*, the observation is `x`, the weight is α, which is a value between `0` and `1`. From the formula, we find the simple exponential average is a type of weighted prediction; one can make a forecast based on observation samples and previous prediction values. To use simple exponential smoothing, we need to set `beta` and `gamma` to `false` in the `HoltWinters` function.

If the data has a trend pattern but does not have a seasonal component, we can use Holt exponential smoothing. This method estimates the level and the slope of a given time series. We can represent this function with the following formula:

- Forecast equation:

$$F_{t+m} = S_t + b_t m$$

- Level equation:

$$S_t = \alpha x_t + (1-\alpha)(S_{t-1} + b_{t-1})$$

- Trend equation:

$$b_t = \beta(S_t - S_{t-1}) + (1-\beta) b_{t-1}$$

Where **S** denotes the estimate of the level at time *t*, *b* denotes the estimate of slope at time *t*, *α* is the smoothing parameter for the level, and *β* shows the smoothing parameter for the trend. Both *α* and *β* are between `0` and `1`. To use Holt exponential smoothing, we just need to set `gamma` to `false` in the `HoltWinters` function.

However, if time series data contains both trends and seasonal patterns, we can use Winters exponential smoothing to estimate the level, trend, and seasonal component:

- Forecast equation:

$$F_{t+m} = (S_t + b_t m) I_{t-L+m}$$

- Seasonal equation:

$$I_t = \gamma \frac{x_t}{S_t} x_t + (1-\gamma) I_{t-L}$$

- Trend equation:

$$b_t = \beta(S_t - S_{t-1}) + (1-\beta) b_{t-1}$$

- Level equation:

$$S_t = \alpha \frac{x_t}{I_{t-L}} + (1-\alpha)(S_t - b_{t-1})$$

Where S denotes the estimate of the level at time *t*, *b* denotes the estimate of slope at time *t*, **I** denotes the estimate of the seasonal component at time *t*, *α* is the smoothing parameter for the level, *β* shows the smoothing parameter for the trend, and *γ* denotes the smoothing parameter for the trend. All three parameters *α*, *β*, and *γ* are between `0` and `1`. To use Winters exponential smoothing, we need to set `alpha`, `beta`, and `gamma` equal to `true` in the `HoltWinters` function.

As Taiwan Semiconductor's quarterly financial reports exhibit trends and seasonal patterns, we can use the Winters algorithm to smooth the time series data. The Winters algorithm requires seasonal, trend, and level parameters at time *t*. Therefore, we need to apply the `HoltWinters` function on the time series data. Next, when we print the `HoltWinters` object, we see the value of `alpha` is equal to `0.8223689`, the `beta` value is equal to `0.06468208`, and `gamma` is equal to `1`, which means the seasonal component weighs more in short-term predictions.

Moving on, we can use the `plot` function to plot the original dataset and smoothed time series. The line in black denotes the original dataset, and the red line indicates the smoothed time series. The figure shows that the seasonal component has noticeable effects on the time series dataset. Last, we can print the sum of square error of the smoothed time series. This can be used to compare which model performs best in smoothing.

There's more...

Besides using the `HoltWinters` function from the `stat` package, we can use the `fpp` package instead. The `fpp` package contains many prediction functions from time series and the linear model; one can also find the `HoltWinters` function in this package. The following steps show how to fit the time series smoothing model with the `hw` function:

1. First, install and load the `fpp` package:

```
> install.packages("fpp")
> library(fpp)
```

2. We can then smooth the time series data with the Winters algorithm:

```
> fit <- hw(m,seasonal="additive")
> summary(fit)

Forecast method: Holt-Winters' additive method

Model Information:
Holt-Winters' additive method

Call:
```

```
hw(x = m, seasonal = "additive")

  Smoothing parameters:
    alpha = 0.9997
    beta  = 1e-04
    gamma = 1e-04

  Initial states:
    l = 741.2403
    b = 36.2694
    s=-15.2854 87.8547 23.7596 -96.3288

  sigma:  122.9867

     AIC     AICc      BIC
434.8764 441.1373 446.6023

Error measures:
                   ME     RMSE      MAE       MPE     MAPE
Training set 4.643746 122.9867 96.73317 -1.034705 9.364536
                  MASE      ACF1
Training set 0.3867117 0.2207188

Forecasts:
        Point Forecast    Lo 80    Hi 80    Lo 95    Hi 95
2016 Q1       1990.276 1832.663 2147.890 1749.227 2231.326
2016 Q2       2146.609 1923.716 2369.503 1805.724 2487.495
2016 Q3       2246.996 1974.001 2519.991 1829.487 2664.506
2016 Q4       2180.148 1864.897 2495.398 1698.014 2662.281
2017 Q1       2135.420 1782.945 2487.895 1596.356 2674.484
2017 Q2       2291.753 1905.618 2677.888 1701.211 2882.296
2017 Q3       2392.140 1975.046 2809.234 1754.250 3030.030
2017 Q4       2325.291 1879.370 2771.213 1643.313 3007.270
```

3. Lastly, we can plot both the original time series and smoothed time series:

```
> plot.ts(m)
> lines(fitted(fit), col="red")
```

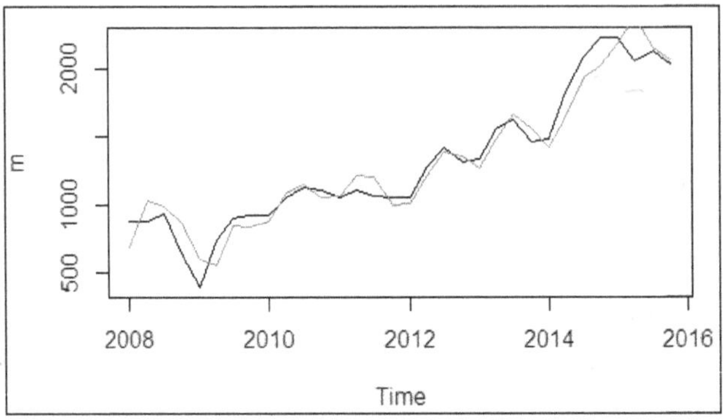

Figure 10: A time series plot with Winters smoothed line

Forecasting time series

After using `HoltWinters` to build a time series smoothing model, we can now forecast future values based on the smoothing model. In this recipe, we introduce how to use the `forecast` function to make a prediction on time series data.

Getting ready

In this recipe, you have to have completed the previous recipe by generating a smoothing model with `HoltWinters` and have it stored in a variable, `m.pre`.

How to do it...

Please perform the following steps to forecast Taiwan Semiconductor's future income:

1. Load the `forecast` package:

```
> library(forecast)
```

2. We can use the `forecast` function to predict the income of the next four quarters:

```
> income.pre <- forecast.HoltWinters(m.pre, h=4)
> summary(income.pre)

Forecast method: HoltWinters
```

```
Model Information:
Holt-Winters exponential smoothing with trend and additive
seasonal component.

Call:
HoltWinters(x = m)

Smoothing parameters:
 alpha: 0.8223689
 beta : 0.06468208
 gamma: 1

Coefficients:
          [,1]
a  1964.30088
b    32.33727
s1  -51.47814
s2   17.84420
s3  146.26704
s4   70.69912

Error measures:
                   ME     RMSE      MAE     MPE     MAPE
Training set 36.86055 168.9165 145.4323 5.76714 14.77564
                  MASE      ACF1
Training set 0.6000302 0.2497206

Forecasts:
        Point Forecast    Lo 80    Hi 80    Lo 95    Hi 95
2016 Q1       1945.160 1730.025 2160.295 1616.140 2274.180
2016 Q2       2046.820 1760.876 2332.763 1609.506 2484.133
2016 Q3       2207.580 1858.743 2556.416 1674.080 2741.079
2016 Q4       2164.349 1756.533 2572.165 1540.649 2788.050
```

3. Plot the prediction results:

```
> plot(income.pre)
```

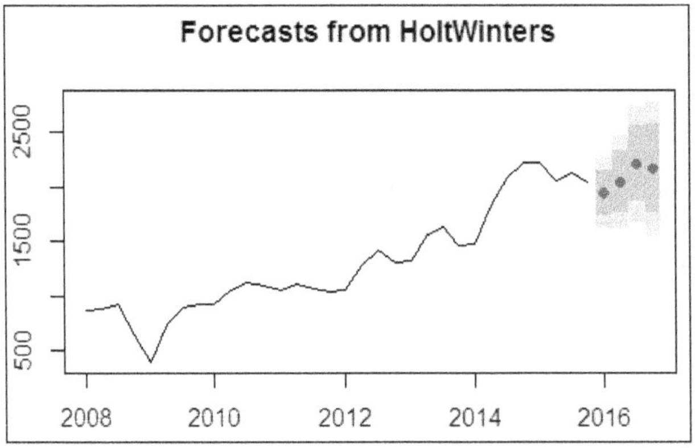

Figure 11: A time series plot with forecast income

4. To measure our model, we can use the `acf` function to calculate the estimates of the autocorrelation function:

```
> acf(income.pre$residuals)
```

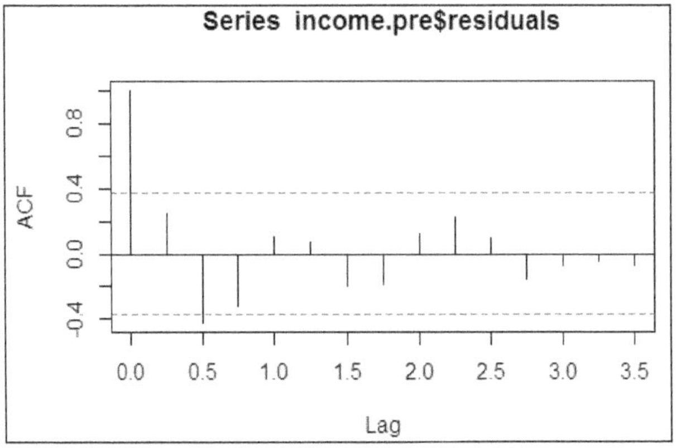

Figure 12: An autocorrelation plot

5. Alternatively, we can perform a Ljung-Box test on the resulting residuals:

```
> Box.test(income.pre$residuals)

  Box-Pierce test

data:  income.pre$residuals
X-squared = 1.7461, df = 1, p-value = 0.1864
```

How it works...

Having produced a Winters exponential smoothing model in the previous recipe, we can now make value predictions based on the smoothing model. We first load the `forecast` package into an R session. Next, we use the `forecast.HoltWinters` function from the `forecast` package to make the prediction. Here, we set `h=4` to predict Taiwan Semiconductor's income for the next four quarters.

We use a `plot` function to plot the prediction results. From the figure, we find a blue section that has been shaded at the end of the time series. The darker shade covers the 80% confidence intervals of our prediction, and the lighter shade shows the 95% confidence intervals.

After building a prediction model, we may need to evaluate the prediction model. Similar to a regression model, we can use residuals, which are the in-sample errors, to assess our model. That is, if we can prove that the residuals follow a white noise distribution, we can conclude that the residuals sequence (or error) is generated from a stochastic process. This means the model is sufficient as it includes most of the message of the time series.

There are many ways that we can evaluate the residual sequence; one of the most intuitive tests is making an autocorrelation plot from the residuals. Here, we use the `acf` function to compute the residuals and chart the autocorrelation plot. The black line indicates the autocorrelation coefficient and the two blue dashed lines denote the confidence interval, which is the upper and lower bounds of the autocorrelation coefficient. If the autocorrelation coefficient quickly decreases and falls between the borders, this means the residuals are white noise. In contrast, if the coefficients are always above or below the boundary, that means the residuals are auto-correlated. In this case, we believe that the residual sequence of the forecast model is not auto-correlated.

Alternatively, one can perform a Ljung-Box test to check whether the residuals are autocorrelated. The Ljung-Box test is made up of the following hypothesis:

$$H_0 : P_1 = P_2 = \ldots = P_k = 0$$
$$H_1 : \exists i \leq m, let\, P_i \neq 0$$

This test uses *Q* statistics to compute the autocorrelation coefficient from a residual sequence, which is formulated as follows:

$$Q = n(n+2)\sum_{i=1}^{k}\frac{\hat{P}_i^2}{n-i} \sim X^2(k)$$

Where *n* is the sample size, $\hat{P}_i$ is the sample autocorrelation at lag *i*, and *k* is the number of lags being tested.

Under the null hypothesis, *Q* statistics follow $X^2(k)$, chi-squared distribution with *k* degrees of freedom. However, if *Q* statistics are above $X^2_{1-a}(k)$, the α quantile of the chi-squared distribution with *k* degrees of freedom, we can reject the null hypothesis.

In this sample, we apply the `Box.test` function on the residual sequence; we find the *p* value is `0.1864`, which means that we cannot reject the null hypothesis. In simple terms, this implies the residuals are white noise, and it proves that our model is sufficient in value forecasting.

There's more...

Following the previous example, we can also make income forecasts by using the prediction function within the `fpp` package:

1. First, load the `fpp` package:

```
> library(fpp)
```

2. We can obtain the smoothing model from the time series with the `hw` function:

```
> fit <- hw(m,seasonal="additive")
```

3. Overlay our forecast on a time series plot:

```
> plot(fit)
```

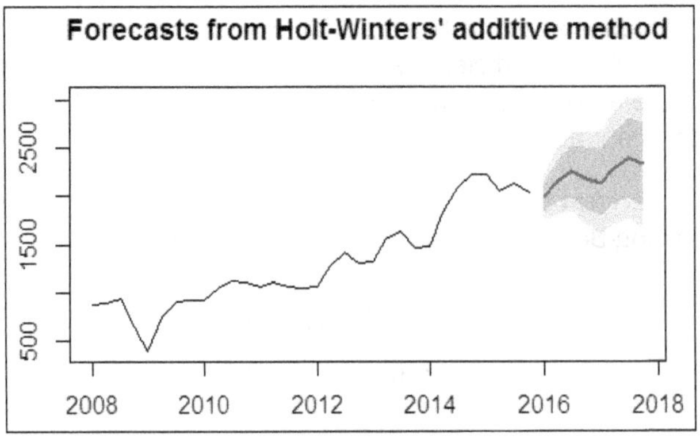

Figure 13: Forecasts from Holt-Winters additive method

Selecting an ARIMA model

Using the exponential smoothing method requires that residuals are non-correlated. However, in real-life cases, it is quite unlikely that none of the continuous values correlate with each other. Instead, one can use ARIMA in R to build a time series model that takes autocorrelation into consideration. In this recipe, we introduce how to use ARIMA to build a smoothing model.

Getting ready

In this recipe, we use time series data simulated from an ARIMA process.

How to do it...

Please perform the following steps to select the ARIMA model's parameters:

1. First, simulate an ARIMA process and generate time series data with the `arima.sim` function:

```
> set.seed(123)
> ts.sim <- arima.sim(list(order = c(1,1,0), ar = 0.7), n = 100)
> plot(ts.sim)
```

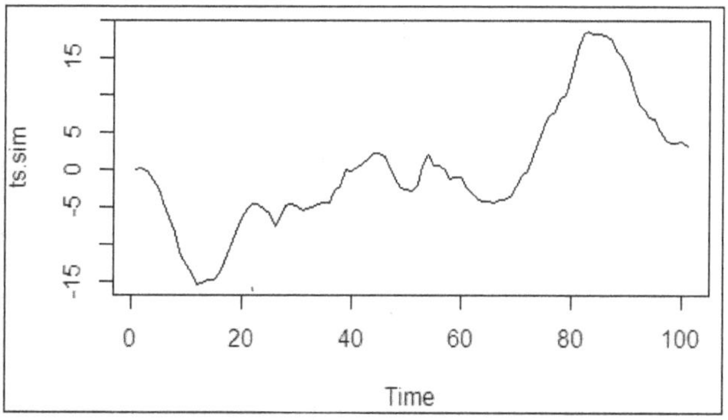

Figure 14: Simulated time series data

2. We can then take the difference of the time series:

```
> ts.sim.diff <- diff(ts.sim)
```

3. Plot the differenced time series:

```
> plot(ts.sim.diff)
```

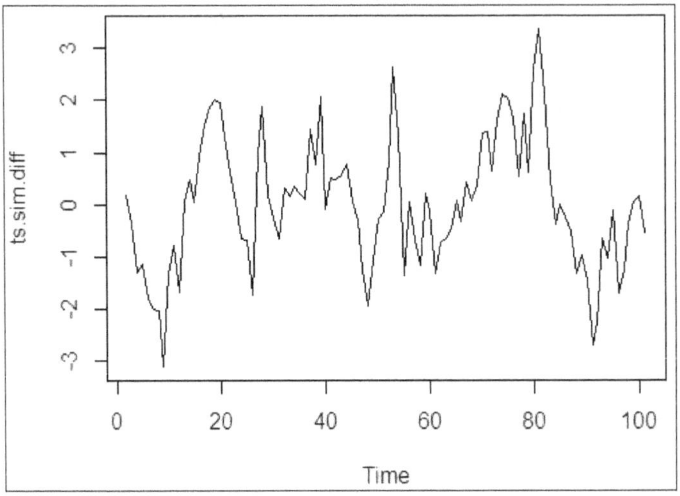

Figure 15: A differenced time series plot

4. Use the `acf` function to make the autocorrelation plot:

```
> acf(ts.sim.diff)
```

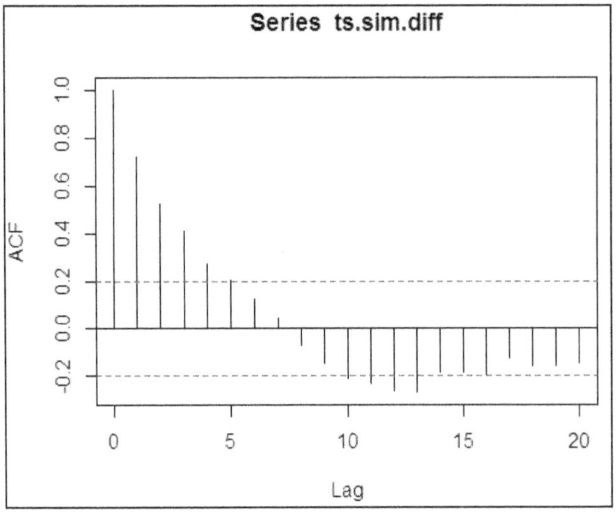

Figure 16: An autocorrelation plot

5. You can make a partial autocorrelation plot with the `pacf` function:

```
> pacf(ts.sim.diff)
```

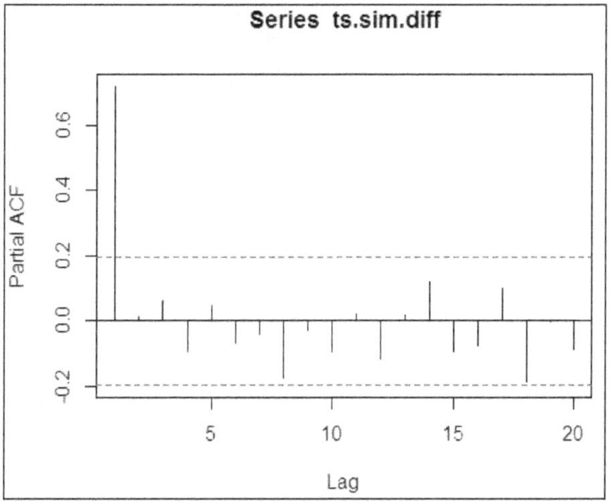

Figure 17: A partial autocorrelation plot

How it works...

The **autoregressive integrated moving average model** (known as **ARIMA**) is also known as the Box-Jenkins model, named after statisticians George Box and Gwilym Jenkins. The purpose of an ARIMA model is to find the best fit of a time series model to past values of a time series.

An ARIMA model can be formulated as *ARIMA(p,d,q)*, where *p* is the order of the autoregressive model, *d* denotes the degree of differencing, and *q* indicates the moving average model. Generally, it takes the following five steps to fit time series to an ARIMA model:

1. Visualize the time series.
2. Difference the non-stationary time series to a stationary time series.
3. Plot ACF and PACF charts to find optimum *p* and *q*.
4. Build the ARIMA model.
5. Make predictions.

In this recipe, we introduce how to follow the first three steps to determine the autoregressive, differencing, and moving average term of the ARIMA model.

We first use the `arima.sim` function with an autoregressive model coefficient equal to `0.7` and of order with *p=1*, *d=1*, and *q=0* to generate 100 data points. We then visualize the time series with the `plot` function. From the figure, we find a non-stationary time series. Therefore, we difference the time series to a stationary time series with the `diff` function. We can now plot the differenced series. From the resulting figure, we find the non-stationary time series has been removed.

In the next step, we use the `acf` function to examine the autocorrelation plot of the differenced series. As autocorrelation coefficients tail off gradually, one can consider the differenced time series as a stationarized series. Therefore, we can determine the differencing parameter *d* is equal to `1`.

Next, we can make autocorrelation and partial autocorrelation plots determine the parameters of *p* and *q*, which correspond to the order of the autoregressive model and the moving average model. Here, we can consider the following ACF and PACF behavior to determine the parameters:

Model	**ACF**	**PACF**
AR(p)	Tails off gradually	Cuts off after *p* lags
MA(q)_	Cuts off after *q* lags	Tails off gradually
ARMA(p,q)	Tails off gradually	Tails off gradually

The autocorrelation plot tails off gradually and *PACF* cuts off after one lag. Thus, we can determine that it is the best fit with the AR model, and it is very likely that *p* equals `1` and *q* equals `0`.

There's more...

To plot a time series along with its ACF and/or its PACF, you can use the `tsdisplay` or `ggtsdisplay` functions:

```
> tsdisplay(ts.sim.diff)
```

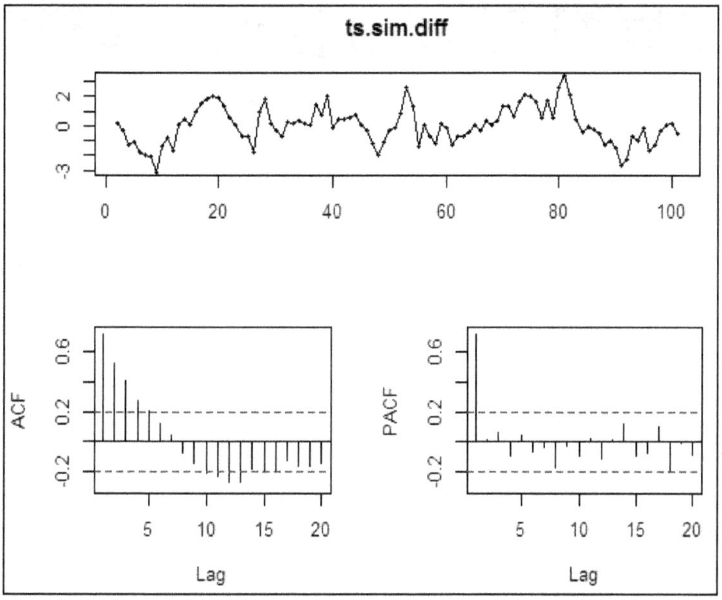

Figure 18: A time series plot with autocorrelation and partial autocorrelation plots

```
> ggtsdisplay(ts.sim.diff)
```

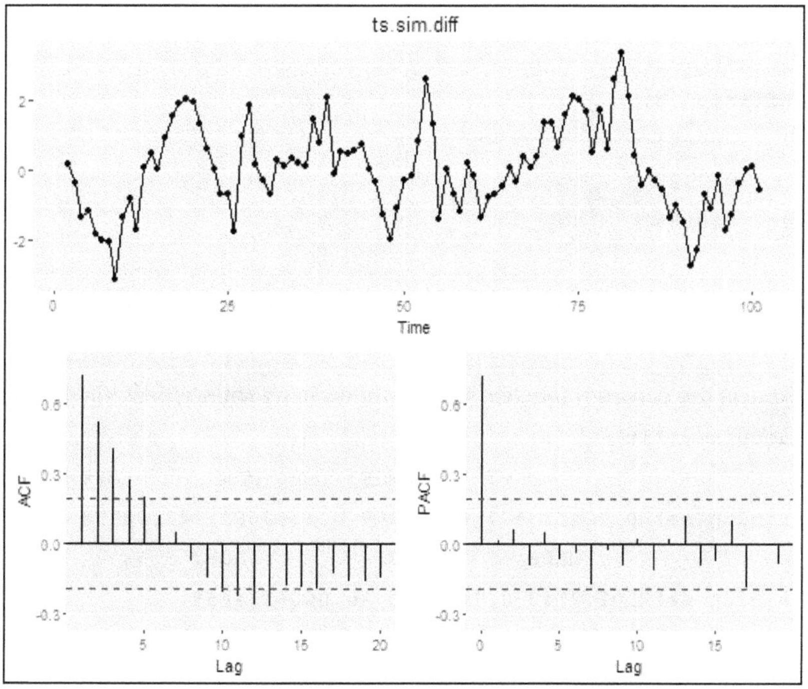

Figure 19: A time series plot made with ggtsdisplay

Creating an ARIMA model

After determining the optimum *p*, *d*, and *q* parameters for an ARIMA model, we can now create an ARIMA model with the `Arima` function.

Getting ready

Ensure you have completed the previous recipe by generating a time series object and storing it in a variable, `ts.sim`.

How to do it...

Please perform the following steps to build an ARIMA model:

1. First, we can create an ARIMA model with time series `ts.sim`, with parameters *p=1, d=1, q=0*:

```
> library(forecast)
> fit <- Arima(ts.sim, order=c(1,1,0))
> fit
Series: ts.sim
ARIMA(1,1,0)

Coefficients:
         ar1
      0.7128
s.e.  0.0685

sigma^2 estimated as 0.7603:  log likelihood=-128.04
AIC=260.09   AICc=260.21   BIC=265.3
```

2. Next, use the accuracy function to print the training set errors of the model:

```
> accuracy(fit)
                      ME      RMSE       MAE       MPE
Training set 0.004938457 0.863265 0.6849681 -41.98798
                 MAPE      MASE          ACF1
Training set 102.2542 0.7038325 -0.0006134658
```

3. In addition to determining the optimum parameter with the ACF and PACF plots, we can use the `auto.arima` function to fit the optimum model:

```
> auto.arima(ts.sim, ic="bic")
Series: ts.sim
ARIMA(1,1,0)

Coefficients:
         ar1
      0.7128
s.e.  0.0685

sigma^2 estimated as 0.7603:  log likelihood=-128.04
AIC=260.09   AICc=260.21   BIC=265.3
```

How it works...

To create an ARIMA model, we can choose to model a non-seasonal series with *ARIMA(p,d,q)*, where *p* is the order of the autoregressive model, *d* denotes the degree of differencing, and *q* indicates the moving average model. Alternatively, if the series has a seasonal pattern, we need to model the series with a seasonal ARIMA. This can be formulated as *ARIMA(p,d,q)(P,D,Q)m*, where *P*, *D*, and *Q* refer to the autoregressive, differencing, and the moving average model of the seasonal part, respectively, and *m* represents the number of periods in each season.

In the previous recipe, we determined the *p*, *d*, and *q* parameters for the ARIMA model. Therefore, we can use the `Arima` function from the `forecast` package to create an ARIMA model based on estimated parameters (*p=1*, *d=1*, and *q=0*). By printing the model, we can find **Akaike Information Criterion** (**AIC**) and **Bayesian Information Criterion** (**BIC**). Typically, the lower the BIC or AIC value, the better the fit. Thus, we can use these two criteria to decide on our model. Moreover, we can use the `accuracy` function to obtain the accuracy measure for our model. Without providing the test set, the function produces the training set (in-sample) errors of the model.

However, if we determine the degree of differencing with autocorrelation and a partial autocorrelation plot, the selection of parameters is rather subjective. Instead, we can use the `auto.arima` function from the `forecast` package, which automatically finds the best-fitted ARIMA model. In this sample, we set the selection criteria as `bic`. It then automatically finds *p=1*, *d=1*, and *q=0* as the model parameters.

There's more...

Besides using the `Arima` function in the `forecast` package, one can use the `arima` function in the `stat` package to create an ARIMA model:

```
> fit2 <- arima(ts.sim)
> summary(fit2)

Call:
arima(x = ts.sim)

Coefficients:
      intercept
        -0.0543
s.e.     0.8053

sigma^2 estimated as 65.5:  log likelihood = -354.51,  aic = 713.02

Training set error measures:
                         ME    RMSE      MAE MPE MAPE     MASE
Training set -1.796052e-12 8.09349 6.121936 Inf  Inf 6.290538
                  ACF1
Training set 0.9873896
```

For further information on the accuracy measures of the ARIMA model, please refer to *Another look at measures of forecast accuracy* by *Hyndman, R.J. and Koehler, A.B. (2006)*. International Journal of Forecasting, 22(4), 679-688.

Forecasting with an ARIMA model

Based on our fitted ARIMA model, we can predict future values. In this recipe, we will introduce how to forecast future values with the `forecast.Arima` function in the `forecast` package.

Getting ready

Ensure you have completed the previous recipe by generating an ARIMA model and storing the model in a variable, `fit`.

How to do it...

Please perform the following steps to forecast future values with `forecast.Arima`:

1. First, use `forecast.Arima` to generate the prediction of future values:

```
> fit.predict <- forecast.Arima(fit)
```

2. We can then use the `summary` function to obtain the summary of our prediction:

```
> summary(fit.predict)

Forecast method: ARIMA(1,1,0)

Model Information:
Series: ts.sim
ARIMA(1,1,0)

Coefficients:
         ar1
      0.7128
s.e.  0.0685

sigma^2 estimated as 0.7603:  log likelihood=-128.04
AIC=260.09   AICc=260.21   BIC=265.3

Error measures:
                      ME     RMSE       MAE       MPE
Training set 0.004938457 0.863265 0.6849681 -41.98798
                 MAPE      MASE          ACF1
Training set 102.2542 0.7038325 -0.0006134658

Forecasts:
    Point Forecast      Lo 80    Hi 80      Lo 95     Hi 95
102       2.879852  1.7624142 3.997289  1.1708784  4.588825
103       2.605201  0.3889232 4.821479 -0.7843035  5.994706
104       2.409429 -0.9178536 5.736711 -2.6792104  7.498068
105       2.269881 -2.1349164 6.674678 -4.4666750  9.006436
106       2.170410 -3.2593426 7.600162 -6.1336799 10.474500
```

```
107        2.099507 -4.2965989  8.495612  -7.6824925 11.881506
108        2.048966 -5.2551824  9.353115  -9.1217650 13.219697
109        2.012941 -6.1441993 10.170080 -10.4623283 14.488209
110        1.987261 -6.9722993 10.946822 -11.7152043 15.689727
111        1.968957 -7.7472568 11.685171 -12.8907097 16.828624
```

3. Then, we can plot our forecast as a line chart:

```
> plot.forecast(fit.predict)
```

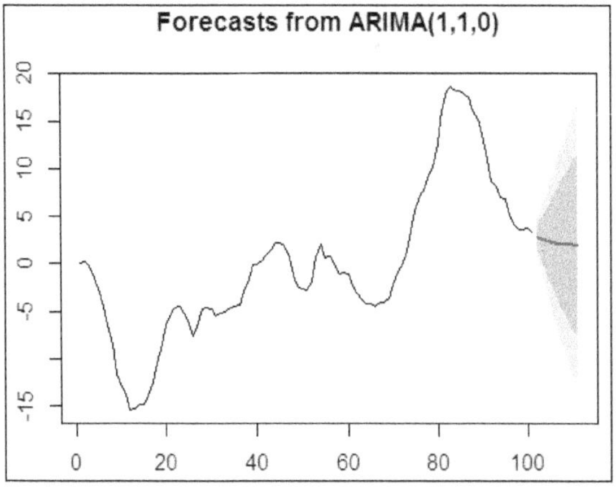

Figure 20: An ARIMA forecast time series plot

4. Moving on, we can evaluate our model with an autocorrelation plot:

```
> acf(fit.predict$residuals)
```

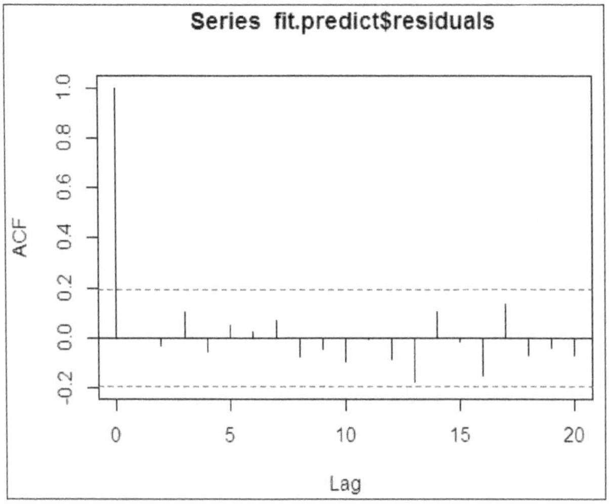

Figure 21: An autocorrelation plot of the ARIMA model

5. Also, we can apply the Ljung-Box test on fitted residuals:

```
> Box.test(fit.predict$residuals)

    Box-Pierce test

data:  fit.predict$residuals
X-squared = 3.801e-05, df = 1, p-value = 0.9951
```

How it works...

In this recipe, we first use the forecast function to make a prediction based on the fitted ARIMA model. We can then print the estimated value under 80% and 95% confidence intervals. Furthermore, we can use the summary function to obtain the model information. The model information reveals the error measures of the fitted model. Last, we can use `plot.forecast` to plot the time series data and its prediction results in a blue shade. The darker blue shade covers the forecast under the 95% confidence interval, and the lighter shade includes the forecast under the 80% confidence interval.

After generating the prediction model, we have to evaluate the prediction model. Similar to an exponential smoothing model, we can use the `acf` function to compute the residuals and make the autocorrelation plot. As the autocorrelation coefficient quickly decreases and falls in between the borders, this means the residuals are white noise.

Moreover, one can perform a Ljung-Box test to check whether the residuals are auto-correlated. In this sample, we apply the `Box.test` function on the residual sequence; we find the *p* value is `0.9951`, which means that we cannot reject the null hypothesis. In simple terms, the residuals are white noise, and it proves that our model is adequate in value forecasting.

There's more...

Besides using the `acf` function and `box.test` function to perform the evaluation process, you can use the `tsdiag` function to diagnose time series fit from different perspectives with a single function:

```
> tsdiag(fit)
```

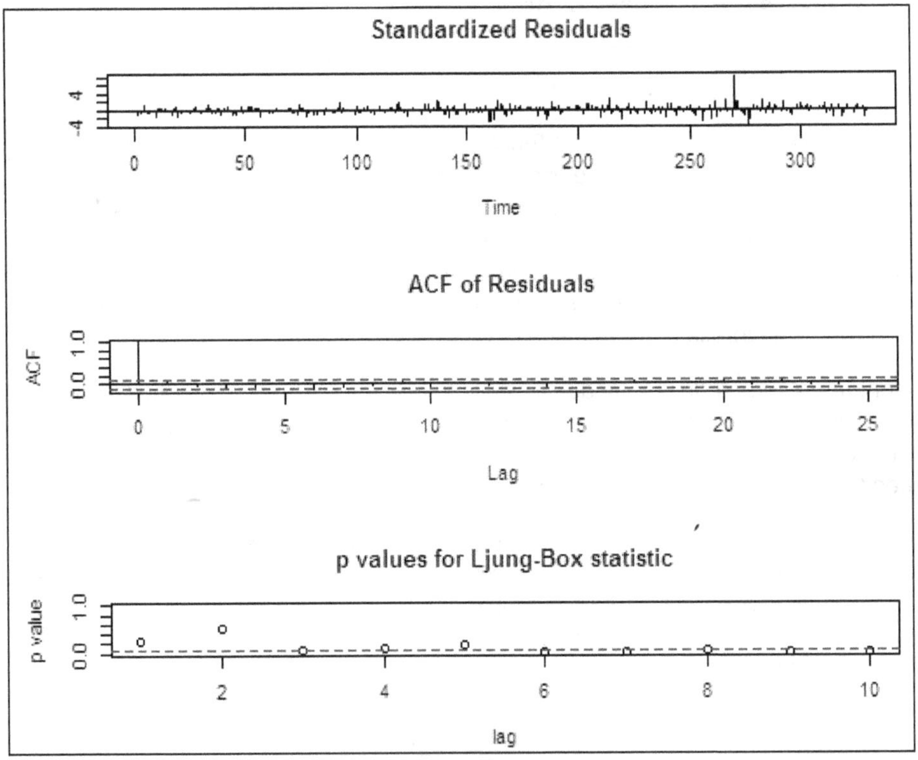

Figure 22: A time series diagnostic plot

Predicting stock prices with an ARIMA model

As the historical prices of a stock are also a time series, we can thus build an ARIMA model to forecast future prices of a given stock. In this recipe, we introduce how to load historical prices with the `quantmod` package, and make predictions on stock prices with ARIMA.

Getting ready

In this recipe, we use the example of stock price prediction to review all the concepts we have covered in previous topics. You will require knowledge of how to create an ARIMA model and make predictions based on a built model to follow this recipe.

How to do it...

Please perform the following steps to predict Facebook's stock price with the ARIMA model:

1. First, install and load the `quantmod` package:

```
> install.packages("quantmod")
> library(quantmod)
```

2. Download the historical prices of Facebook Inc from Yahoo with `quantmod`:

```
> getSymbols("FB",src="yahoo", from="2015-01-01")
```

3. Next, plot the historical stock prices as a line chart:

```
> plot(FB)
```

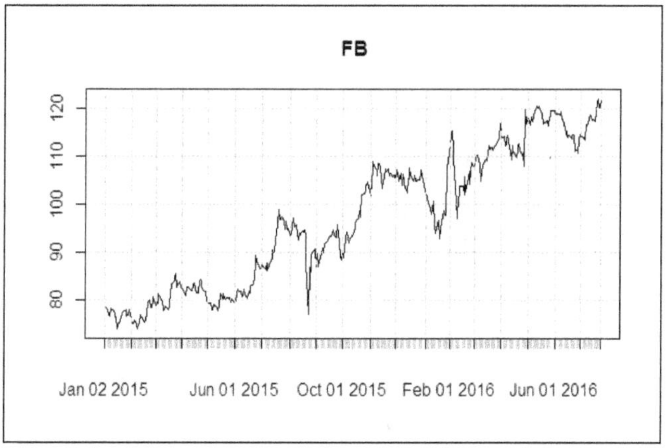

Figure 23: A historical stock price chart of Facebook Inc.

4. Use the `auto.arima` function to find the best-fitted model:

```
> fit <- auto.arima(FB$FB.Close, ic="bic")
> fit
Series: FB$FB.Close
ARIMA(0,1,0) with drift

Coefficients:
```

```
          drift
         0.0962
s.e.     0.0986

sigma^2 estimated as 3.212:  log likelihood=-658.26
AIC=1320.52   AICc=1320.56   BIC=1328.11
```

5. Plot both the historical prices and fitted line chart in the same plot:

```
> plot(as.ts(FB$FB.Close) )
> lines(fitted(fit), col="red")
```

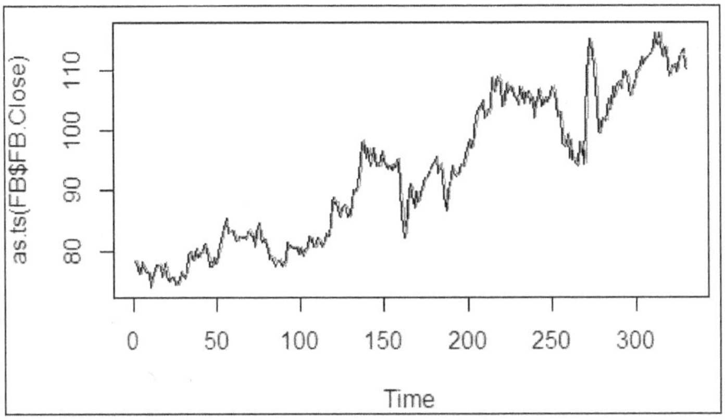

Figure 24: A historical price chart and fitted line chart

6. Moving on, you can make a prediction of future stock prices with the `forecast.Arima` function:

```
> fit.forecast <- forecast.Arima(fit)
> fit.forecast
    Point Forecast    Lo 80    Hi 80    Lo 95    Hi 95
331       110.1962 107.8996 112.4928 106.6838 113.7086
332       110.2924 107.0445 113.5403 105.3251 115.2597
333       110.3886 106.4107 114.3665 104.3049 116.4723
334       110.4848 105.8915 115.0781 103.4600 117.5096
335       110.5810 105.4456 115.7164 102.7270 118.4350
336       110.6772 105.0516 116.3028 102.0736 119.2808
337       110.7734 104.6971 116.8497 101.4805 120.0664
338       110.8696 104.3737 117.3655 100.9350 120.8042
339       110.9658 104.0759 117.8557 100.4286 121.5030
340       111.0620 103.7994 118.3246  99.9548 122.1692
```

7. Lastly, you can make a forecast plot:

```
> plot(fit.forecast)
```

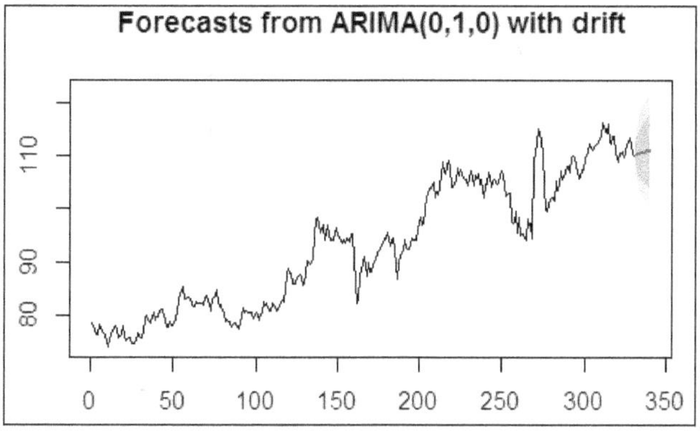

Figure 25: A FB stock price forecast chart

How it works...

This recipe demonstrates how to make a stock price prediction based on an ARIMA model. Also, it summarizes what we have covered about ARIMA in previous recipes. At the start of the recipe, we install and load the `quantmod` package. The `quantmod` package provides various quantitative financial modeling methods and APIs from diverse sources. To retrieve historical stock prices, we use the `getSymbols` function to obtain the historical stock prices of Facebook Inc. since January 1, 2015. Then, `quantmod` loads the stock prices of Facebook and stores them in a variable, `FB`. Next, we use the `plot` function to plot the historical stock price of Facebook as a line chart. From the line chart, we can see sustained growth in Facebook's stock price.

Moving on, we create an ARIMA model to fit the time series of the close price. In this example, the `auto.arima` function finds a non-seasonal ARIMA model with parameters *p=0*, *d=1*, and *q=0*. Next, we plot both the time series as a black line, and the fitted line as a red line on the same plot, to see whether the fitted line matches the original time series data.

After we have built the ARIMA model, we can use this model to forecast future stock prices. Here, we use `forecast.arima` to predict the future price of the stock, and store our prediction in a variable, `fit.forecast`. Last, we can use the `plot` function to plot our prediction in a blue shade within the time series plot.

There's more...

As `quantmod` loads historical stock prices into an `xts` object, one can use the `lag` function to shift the time series. We can calculate the return on the stock and chart it in a histogram. From the histogram, we find the distribution of the performance of the stock:

```
> FB_Return<-diff(FB$FB.Close)/lag(FB$FB.Close,k=-1)*100
> head(FB_Return)
             FB.Close
2015-01-02         NA
2015-01-05 -1.6546224
2015-01-06 -1.3657255
2015-01-07  0.0000000
2015-01-08  2.6112658
2015-01-09 -0.5735167
> hist(FB_Return)
```

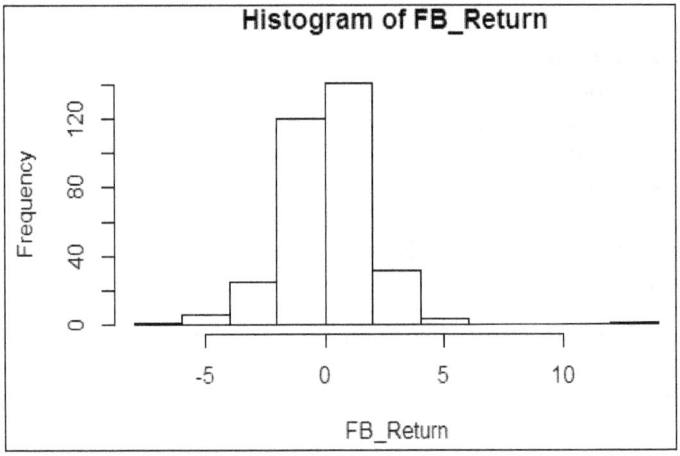

Figure 26: A histogram of daily return on stock price

11
Supervised Machine Learning

This chapter covers the following topics:

- Fitting a linear regression model with `lm`
- Summarizing linear model fits
- Using linear regression to predict unknown values
- Measuring the performance of the regression model
- Performing a multiple regression analysis
- Selecting the best-fitted regression model with stepwise regression
- Applying the Gaussian model for generalized linear regression
- Performing a logistic regression analysis
- Building a classification model with recursive partitioning trees
- Visualizing a recursive partitioning tree
- Measuring model performance with a confusion matrix
- Measuring prediction performance using `ROCR`

Introduction

The aim of a supervised machine learning method is to build a learning model from training data that includes input data containing known labels or results. The model can predict the characteristics or patterns of unseen instances. In general, the input of the training model is made by a pair of input vectors and expected values. If the output variable of an objective function is continuous (but can also be binary), the learning method is regarded as regression analysis. Alternatively, if the desired output is categorical, the learning process is considered as a classification method.

Regression analysis is often employed to model and analyze the relationship between a dependent (response) variable and one or more independent (predictor) variables. One can use regression to build a prediction model that first finds the best-fitted model with minimized squared error of input data. The fitted model can then be further applied to data for continuous value prediction. For example, one may use regression to predict the price range of a rental property based on historical rental prices.

On the other hand, one can use the classification method to identify the category of new observations (testing dataset) based on a classification model built from the training dataset, of which the categories are already known. Similar to regression, classification is categorized as a supervised learning method as it employs known answers (labels) of the training dataset to predict the answer (label) of the testing dataset. For example, one may use the classification method to predict whether a customer is likely to purchase a product.

In this chapter, we will first cover regression analysis. We will explore how to fit a regression model with `lm` and obtain the summary of the fit. Next, we will use the fitted line to predict the value of the new instance. Furthermore, we can measure the prediction performance of the regression fit. We will then introduce generalized linear models (`glm`) and demonstrate how to fit Gaussian and binomial dependent variables into these models. Moving on, we will cover classification models. First, we will build a classification model with a recursive partitioning tree. We will then visualize the built model, and finally assess the performance of the fitted model in confusion and ROC metrics.

Fitting a linear regression model with lm

The simplest model in regression is linear regression, which is best used when there is only one predictor variable, and the relationship between the response variable and independent variable is linear. In this recipe, we demonstrate how to fit the model to data using the `lm` function.

Getting ready

Download the house rental dataset from `https://raw.githubusercontent.com/ywchiu/rcookbook/master/chapter11/house_rental.csv` first, and ensure you have installed R on your operating system.

How to do it...

Perform the following steps to fit data into a simple linear regression model:

1. Read the house rental data into an R session:

```
> house <- read.csv('house_rental.csv', header=TRUE)
```

2. Fit the independent variable `Sqft` and dependent variable `Price` to `glm`:

```
> lmfit <- lm(Price ~ Sqft, data=house)
> lmfit

Call:
lm(formula = Price ~ Sqft, data = house)

Coefficients:
(Intercept)         Sqft
    3425.13        38.33
```

3. Visualize the fitted line with the `plot` and `abline` functions:

```
> plot(Price ~ Sqft, data=house)
> abline(lmfit, col="red")
```

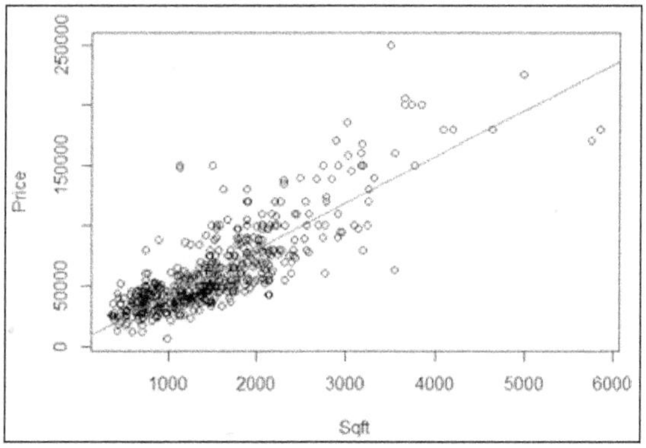

Figure 1: A simple linear regression model

How it works...

The regression model has the form *response ~ terms*, where response is the response vector and terms are a series of terms that specify a predictor. We can illustrate a simple regression model in the formula $y = \alpha + \beta x$, where α is the intercept while the slope, β, describes the change in *y* when *x* changes. By using the least squares method, we can estimate $\beta = \frac{cov[x,y]}{var[x]}$ and $\alpha = \overline{y} - \beta\overline{x}$ (where $\overline{y}$ indicates mean value of *y* and $\overline{x}$ denotes mean value of *x*).

In this recipe, we first input the independent variable `Sqft` and dependent variable `Price` in to the `lm` function and assign the built model to fit. From the output, we find the coefficient of the fitted model shows the intercept equals `3425.13` and coefficient equals `38.33`.

There's more...

The relationship between some predictor variables and response variables can be modeled as an *n*th order polynomial. For instance, we can illustrate the second order polynomial regression model in the formula $y = \alpha + \beta x + cx^2$, where α is the intercept while β illustrates regression coefficients. The following example shows how to fit `anscombe` data into a second order polynomial regression model.

Perform the following steps to fit polynomial regression model with `lm`:

1. Load the `anscombe` dataset:

   ```
   > data(anscombe)
   ```

2. Make a scatterplot of variables `x1` and `y2`:

   ```
   > plot(y2 ~ x1, data=anscombe)
   ```

3. Apply the poly function by specifying `2` in the argument:

   ```
   > lmfit2 <- lm(y2~poly(x1,2), data=anscombe)
   > lines(sort(anscombe$x1), lmfit2$fit[order(anscombe$x1)], col =
   "red")
   ```

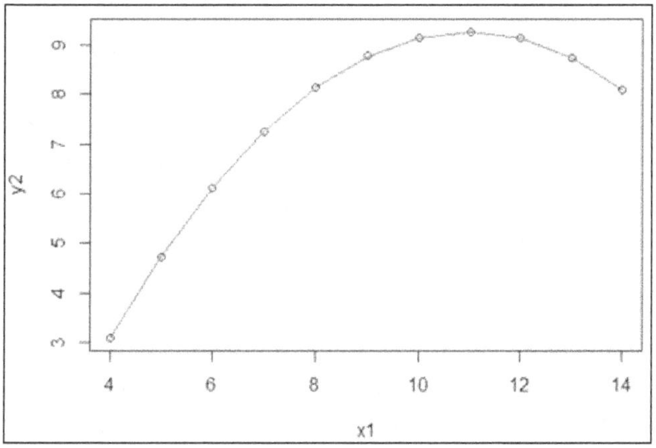

Figure 2: A second order polynomial regression model

Summarizing linear model fits

The `summary` function can be used to obtain the formatted coefficient, standard errors, degree of freedom, and other summarized information of a fitted model. This recipe introduces how to obtain overall model information by using the `summary` function.

Getting ready

You need to have completed the previous recipe by fitting the house rental data into a regression model, and have the fitted model assigned to the variable `lmfit`.

How to do it...

Perform the following steps to summarize the linear regression model:

1. Compute a detailed summary of the fitted model, `lmfit`:

```
> summary(lmfit)
Call:
lm(formula = Price ~ Sqft, data = house)

Residuals:
   Min      1Q Median      3Q     Max
-76819 -12388  -3093   10024 112227

Coefficients:
            Estimate Std. Error t value Pr(>|t|)
```

```
(Intercept) 3425.133   1766.646    1.939     0.053 .
Sqft          38.334      1.034   37.090    <2e-16 ***
Signif. codes:  0 '***' 0.001 '**' 0.01 '*' 0.05 '.' 0.1 ' ' 1

Residual standard error: 20130 on 643 degrees of freedom
Multiple R-squared:  0.6815, Adjusted R-squared:  0.681
F-statistic:  1376 on 1 and 643 DF,  p-value: < 2.2e-16
```

How it works...

The `summary` function is a generic function used to produce summary statistics. In this case, it computes and returns a list of the summary statistics of the fitted linear model. Here, it will output information such as residuals, coefficient standard error R-squared, f-statistic, and degree, of freedom. In the `Call` section, the function called to generate the fitted model is displayed. In the `Residuals` section, it provides a quick summary (`Min`, `1Q median`, `3Q`, and `Max`) of distribution. In the `Coefficients` section, each coefficient is a Gaussian random variable. Within this section, the `Estimate` represents the mean distribution of the variable; `Std.Error` displays the standard error of the variable; `t value` is the estimate divided by `Std.Error`; and `p-value` indicates the probability of getting a value larger than the *t* value. In this sample, as the `p-value` of *Sqft (2e-16)* is much less than 0.05, we can say with a 95% probability of being correct that `Sqft` has a meaning addition to the fitted model.

`Residual standard error` outputs the standard deviation of residuals, while the degree of freedom indicates the differences between the observation in training samples and the number used in the model. `Multiple R-squared` is obtained by dividing the sum of squares. One can use `R-squared` to measure how close the data is to fitting the regression line. Usually, the higher the R-squared value, the better the model fits your data. However, it does not necessarily indicate whether the regression model is adequate. That means you might get a good model with low R-squared, or you can have a bad model with high R-squared. As multiple `R-squared` ignores degrees of freedom, the calculated score is biased. To make the calculation fair, `Adjusted R-squared` (`0.681`) uses an unbiased estimate, and will be slightly less than `Multiple R-squared` (`0.6815`). `F-statistic` is retrieved by performing an F-test on the model. A p-value equal to `2.2e-16` (< 0.05) rejects the null hypothesis (no linear correlation between variables) and indicates that the observed *F* is greater than the critical *F* value. In other words, the result shows that there is significant positive linear correlation between `Sqft` and `Price`.

There's more...

For more information on the parameters used for obtaining a summary of the fitted model, you can use the `help` function or `?` to view the help page:

```
> ?summary.lm
```

Alternatively, you can use the following functions to display the properties of the model:

```
> coefficients(lmfit) # Extract model coefficients
> confint(lmfit, level=0.95) # Computes confidence intervals for model parameters
> fitted(lmfit) # Extract model fitted values
> residuals(lmfit) # Extract model residuals
> anova(lmfit) # Compute analysis of variance tables for fitted model object
> vcov(lmfit) # Calculate variance-covariance matrix for a fitted model object
> influence(lmfit) # Diagnose quality of regression fits
```

Using linear regression to predict unknown values

With a fitted regression model, we can apply the model to predict unknown values. For regression models, we can express the precision of prediction with prediction intervals and confidence intervals. In the following recipe, we introduce how to predict unknown values under these two measurements.

Getting ready

One needs to have completed the previous recipe by fitting the house rental data into a regression model and have the fitted model assigned to variable `lmfit`.

How to do it...

Perform the following steps to predict values with linear regression:

1. Assign values to be predicted to `newdata`:

   ```
   > newdata <- data.frame(Sqft=c(800, 900, 1000))
   ```

2. Compute the prediction result of the given data:

   ```
   > predict(lmfit ,newdata)
          1        2        3
   34092.60 37926.04 41759.47
   ```

3. On the other hand, you can obtain the coefficient and intercept of the fitted model:

```
> lmfit$coefficients[1]
(Intercept)
   3425.133

> lmfit$coefficients[2]
    Sqft
38.33434
```

4. Executing the following function will yield the same prediction value, which we obtained using the predict function in step 2:

```
> c(800, 900, 1000) * lmfit$coefficients[2] +
lmfit$coefficients[1]
```

5. Moving on, we can compute prediction results using confidence intervals with `level` set as `0.95`:

```
> predict(lmfit, newdata, interval="confidence", level=0.95)
       fit       lwr      upr
1 34092.60 31947.18 36238.02
2 37926.04 35914.92 39937.15
3 41759.47 39870.37 43648.57
```

6. We can obtain the prediction intervals of the given data:

```
> predict(lmfit ,newdata, interval='prediction')
       fit       lwr      upr
1 34092.60 -5489.041 73674.25
2 37926.04 -1648.555 77500.63
3 41759.47  2190.892 81328.05
```

7. Lastly, we can plot the prediction results onto a scatterplot:

```
> plot(Price ~ Sqft, data=house)
> pred.res <- predict(lmfit, data = house, interval="prediction")
> pred.conf <- predict(lmfit, data = house, interval =
'confidence')

> lines(pred.conf[, 'fit'] ~ house$Sqft, col="red",  lty = 1,
lwd=3)
> lines(pred.conf[, 'lwr'] ~ house$Sqft, col="blue", lty = 3,
lwd=3)
```

```
> lines(pred.conf[, 'upr'] ~ house$Sqft, col="blue", lty = 3,
lwd=3)
> lines(pred.res[, 'lwr'] ~ house$Sqft, col="green", lty = 2,
lwd=3)
> lines(pred.res[, 'upr'] ~ house$Sqft, col="green", lty = 2,
lwd=3)
```

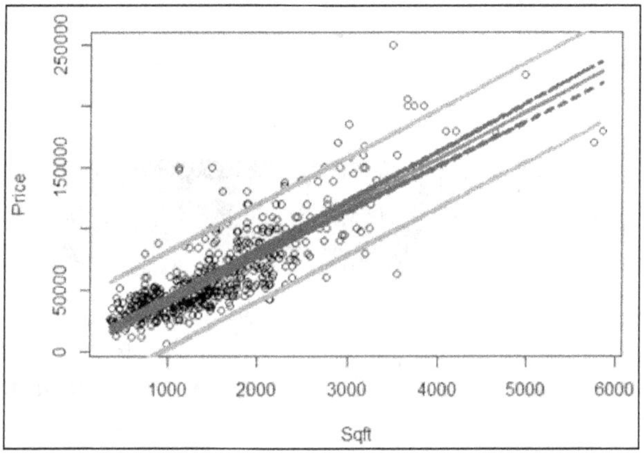

Figure 3: A scatter plot with fitted line, confidence interval, and prediction interval

How it works...

We first build a linear fitted model with variables `Sqft` and `Price`. Next, we assigned values to be predicted to a data frame, `newdata`. It is important to note that the generated model is in the form of `Price ~ Sqft`.

Next, we compute the prediction result by fitting `Sqft` with `800`, `900`, and `1000` into the model. From the output, we obtain the prediction prices as `34092.60`, `37926.04`, and `41759.47`. On the other hand, one can also compute the prediction result with model coefficient and intercept.

Next, we compute the prediction result using a confidence interval by specifying `confidence` in the argument `interval`. From the output of row 1, we get a fitted `Price` of input `Sqft=800` equals to `34092.60`, and a 95% confidence interval (set `0.95` in the argument `level`) of the mean `Price` for `Sqft=800` is between `31947.18` and `36238.02`. In addition, rows 2 and 3 give the prediction result of `Price` with input `Sqft=900` and `Sqft=1000`.

Next, we compute the prediction result using a prediction interval by specifying `prediction` in the argument `interval`. From the output of row 1, we see fitted `y1` of input `Sqft=800` equals `34092.60`, and the 95% prediction interval of `Price` for `Sqft=800` is between `-5489.041` and `73674.25`. Rows 2 and 3 output the prediction result of `Price` with input `Sqft=900` and `Sqft=1000`.

Last, we plot the fitted line, confidence interval and prediction interval on the sample plot. As Figure 3 shows, the dots represent data points, and the red line indicates the regression line. The blue line alongside the red line shows the confidence interval at a 95% confidence interval, while the green line on the border shows the upper and lower bounds of the prediction interval.

There's more...

For those interested in the difference between prediction intervals and confidence intervals, refer to the Wikipedia entry *Contrast with confidence intervals* (`http://en.wikipedia.org/wiki/Prediction_interval#Contrast_with_confidence_intervals`).

Measuring the performance of the regression model

To measure the performance of a regression model, we can calculate the distance from the predicted output and actual output as a quantifier of model performance. In this calculation, we often use **root mean square error** (**RMSE**) and **relative square error** (**RSE**) as common measurements. In the following recipe, we illustrate how to compute these measurements from a built regression model.

Getting ready

You need to have completed the previous recipe by fitting the house rental data into a regression model and have the fitted model assigned to the variable `lmfit`.

How to do it...

Perform the following steps to measure the performance of the regression model:

1. Retrieve predicted values by using the predict function:

```
> predicted <- predict(lmfit, data=house)
```

2. Calculate the root mean square error:

```
> actual <- house$Sqft
> rmse <- (mean((predicted - actual)^2))^0.5
> rmse
[1] 66894.34
```

3. Calculate the relative square error:

```
> mu <- mean(actual)
> rse <- mean((predicted - actual)^2) / mean((mu - actual)^2)
> rse
[1] 7610.695
```

How it works...

The measurement of regression model performance employs the distance between the predicted value and actual value. We often use the following three measurements, which are root mean square error, relative square error, and `R-squared` as the quantifier of regression model performance. In this recipe, we compute the predicted value using the `predict` function, and begin to compute RMSE. RMSE is used to measure the error rate of a regression model, and it is formulated in the following equation:

$$RMSE = \sqrt{\frac{\sum_{i=1}^{n} (p_i - a_i)^2}{n}}$$

Here a is the actual target, p is the predicted value, and n is the number of observations. In this example, we obtain `66894.34` as the root mean square error of the regression model.

As RMSE can only compare models that have errors measured in the same units, we can use the RSE instead to compare models that have errors measured in different units. The relative square error can be formulated in the following equation:

$$RSE = \frac{\sum_{i=1}^{n} (p_i - a_i)^2}{\sum_{i=1}^{n} (\bar{a} - a_i)^2}$$

In this example, we get `7610.695` as the relative squared error of the model.

There's more...

Besides calculating RMSE and RSE independently, we can use the machine learning evaluation metrics package, `MLmetrics`, to measure the performance of a regression model:

1. First, install and load the `MLmetrics` package:

```
> install.packages("MLmetrics")
> library(MLmetrics)
```

2. Retrieve the **root relative squared error loss** (**RRSE**), **relative squared error loss** (**RSE**), and RMSE of the fitted model:

```
> y_rrse <- RRSE(y_pred = predicted, y_true = actual)
> y_rse <- y_rrse^2
> y_rmse <- RMSE(y_pred = predicted, y_true = actual)

> y_rrse
[1] 87.2393
> y_rse
[1] 7610.695
> y_rmse
[1] 66894.34
```

Performing a multiple regression analysis

Multiple regression analysis can be regarded as an extension of simple linear regression. One can use multiple regression analysis to predict the value of a dependent variable based on multiple independent variables. In the following recipe, we demonstrate how to predict house rental prices with multiple variables.

Getting ready

One needs to have completed the previous recipe by downloading the house rental data into a variable `house`.

How to do it...

Perform the following steps to fit the house rental dataset into a multiple regression model:

1. First, you need to fit `Sqft`, `Floor`, `TotalFloor`, `Bedroom`, `Living.Room`, and `Bathroom` into a linear regression model:

```
> fit <- lm(Price ~ Sqft + Floor + TotalFloor  + Bedroom + Living.
Room + Bathroom, data=house)
```

2. Obtain the summary of the fitted model:

```
> summary(fit)

Call:
```

```
lm(formula = Price ~ Sqft + Floor + TotalFloor + Bedroom +
Living.Room + Bathroom, data = house)


Residuals:
   Min     1Q Median     3Q    Max -71429 -11314  -2075   8480
104063


Coefficients:
             Estimate Std. Error t value Pr(>|t|)
(Intercept)  2669.268   3665.690   0.728    0.467
Sqft           37.884      1.615  23.456  < 2e-16 ***
Floor        1025.887    241.543   4.247 2.49e-05 ***
TotalFloor    135.451    197.033   0.687    0.492
Bedroom     -1775.956   1079.401  -1.645    0.100
Living.Room -3368.837   2078.469  -1.621    0.106
Bathroom     2777.701   1835.922   1.513    0.131
Signif. codes:  0 '***' 0.001 '**' 0.01 '*' 0.05 '.' 0.1 ' ' 1

Residual standard error: 19560 on 638 degrees of freedom
Multiple R-squared:  0.7015, Adjusted R-squared:  0.6987
F-statistic: 249.9 on 6 and 638 DF,  p-value: < 2.2e-16
```

3. Now, you can predict the possible house rental price with the fitted model:

```
> predict(fit, house[1:3,])
       1        2        3
45550.75 81335.21 44368.01
```

How it works...

A multiple regression is a regression model with two or more independent variables. Unlike simple linear regression, which is modeled with a straight line, multiple regression analysis is modeled as a function with multiple independent variables. The model has the form: *response ~ terms1 + terms2 + ... termsn*, where *n* is the number of independent variables.

In this example, we first fit the regression model with `Sqft`, `Floor`, `TotalFloor`, `Bedroom`, `Living.Room`, and `Bathroom` as independent variables, and with `Price` as a dependent variable. We can then use the `summary` function to obtain a summary of the fitted model. From the coefficient table, we find `Sqft` has p-value equals to `2e-16`, and `Floor` has p-value equals to `2.49e-05`; we can thus reject the null hypothesis such that `Sqft` and `Floor` has no effect on the fitted model. Or, in other words, we can say with a 95% probability of being correct that `Sqft` and `Floor` have a meaning addition to the fitted model. `F-statistic` shows `249.9`, with p-value equals to `2.2e-16` (<0.05) rejects the null hypothesis ($H_0 : \beta_1 = \beta_2 = \ldots = \beta_n = 0$) that there is no linear correlation between variables.

Last, one can use the predict function to predict the possible price of the first three rental houses with the fitted model, `fit`.

There's more...

Diagnostics are methods to evaluate the assumptions of the regression, which can be used to determine whether a fitted model adequately represents the data. In the following example, we introduce how to diagnose a regression model through the use of a diagnostic plot.

Plot the diagnostic plot of regression model:

```
> par(mfrow=c(2,2))
> plot(fit)
```

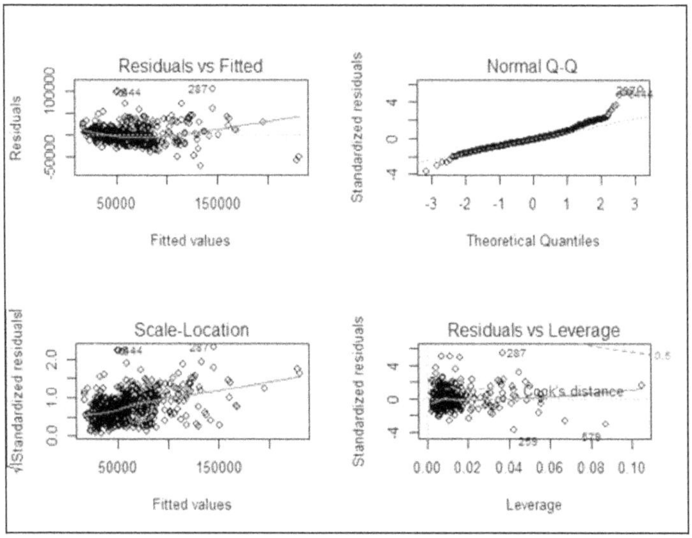

Figure 4: A diagnostic plot of the fitted model

The `plot` function generates four diagnostic plots of a regression model. The upper-left plot shows residuals versus fitted values. Within the plot, residuals represent the vertical distance from a point to the regression line. If all points fall exactly on the regression line, all residuals will fall exactly on the grey dotted line. The red line within the plot is a smoothed curve in regard to residuals, and if all dots fall exactly on the regression line, the position of the red line should exactly match the grey dotted line.

The upper-right plot shows the normal Q-Q plot of residuals. This plot verifies the assumption that residuals are normally distributed. Thus, if the residuals are normally distributed, they should lie exactly on the grey dotted line.

The **Scale-Location** plot in the bottom left measures the square root of the standardized residuals against fitted value. Therefore, if all dots lie on the regression line, the value of **y** should be close to **0**. As it is assumed that the variance of residuals does not change the distribution substantially, if the assumption is correct, the red line should be relatively flat.

The bottom-right plot shows standardized residuals versus leverage. Leverage is a measurement of how each single data point influences the regression. It is a measurement of distance from centroid of regression and level of isolation (measured by whether it has neighbors). Also, you can find the contour of Cook's distance, which is affected by high leverage and large residuals. One can use this to measure how regression would change if a single point is deleted. The red line is smoothed in regard to standardized residuals. For a perfect fit regression, the red line should be close to the dashed line with no points over **0.5** in Cook's distance.

Selecting the best-fitted regression model with stepwise regression

In order to find the best-fitted regression model, one can perform stepwise regression to step-wisely add or remove a term from a fitted model, and finally output a model with the least AIC. In the following recipe, we demonstrate how to perform stepwise regression with the step function.

Getting ready

You need to have completed the previous recipe by fitting house rental data into a multiple regression model, `fit`.

How to do it...

Perform the following steps to search for the best-fitted regression model with the `step` function:

1. First, you can use `step` to select the optimum model with backward elimination:

```
> step(fit, direction="backward")
Start:  AIC=12753.77
Price ~ Sqft + Floor + TotalFloor + Bedroom + Living.Room +
Bathroom

              Df  Sum of Sq        RSS   AIC
- TotalFloor   1 1.8081e+08 2.4428e+11 12752
<none>                      2.4410e+11 12754
- Bathroom     1 8.7580e+08 2.4497e+11 12754
- Living.Room  1 1.0051e+09 2.4510e+11 12754
- Bedroom      1 1.0357e+09 2.4513e+11 12754
- Floor        1 6.9016e+09 2.5100e+11 12770
- Sqft         1 2.1050e+11 4.5459e+11 13153

Step:  AIC=12752.25
Price ~ Sqft + Floor + Bedroom + Living.Room + Bathroom

              Df  Sum of Sq        RSS   AIC
<none>                      2.4428e+11 12752
- Bathroom     1 8.1619e+08 2.4509e+11 12752
- Living.Room  1 1.0233e+09 2.4530e+11 12753
- Bedroom      1 1.1225e+09 2.4540e+11 12753
- Floor        1 1.1700e+10 2.5598e+11 12780
- Sqft         1 2.3530e+11 4.7958e+11 13185

Call:
lm(formula = Price ~ Sqft + Floor + Bedroom + Living.Room +
Bathroom,
    data = house)

Coefficients:
(Intercept)         Sqft         Floor        Bedroom
```

```
     3522.17        38.22      1116.95     -1841.61
 Living.Room     Bathroom
    -3398.44      2672.11
```

2. Use fit data with variables that result in the least AIC:

```
> fit2 <- lm(Price ~ Sqft + Floor + Bedroom + Living.Room +
Bathroom, data=house)
```

How it works...

AIC is formulated as $AIC=2k-2ln(L)$, where k is the number of estimated parameters and L is the maximum number of the likelihood function of the model. The purpose of AIC is to prevent you from selecting an over-fitted model. The measure itself does not have much meaning, and you should use this standard to compare similar models and choose the one with the least AIC. Here, we perform the step regression to find the model with the least AIC.

In this recipe, we perform a backward elimination with the step function by setting the argument `direction="backward"`. In the first step, the model includes all independent variables in the first place. The model then step-wisely eliminates one independent variable at a time. If the model cannot reject the null hypothesis that a variable has a zero coefficient, the variable is removed from the model. Last, we find the fitted model without independent variable `TotalFloor` (AIC: 12752) fits better than the model with `TotalFloor` (AIC: 12754).

There's more...

In the previous steps, we have shown the model without variable `TotalFloor` fits better than the model with the `TotalFloor` variable. We may then be interested in testing whether the coefficient of `TotalFloor` is equal to `0`. Therefore, we can conduct a partial F-test by comparing the **sum of square error** (**SSE**) between `fit1` and `fit2` with an `anova` function:

```
> anova(fit1, fit2)
Analysis of Variance Table


Model 1: Price ~ Sqft + Floor + TotalFloor + Bedroom + Living.Room +
Bathroom
Model 2: Price ~ Sqft + Floor + Bedroom + Living.Room + Bathroom
  Res.Df        RSS Df  Sum of Sq      F Pr(>F)
1    638 2.4410e+11
2    639 2.4428e+11 -1 -180811489 0.4726  0.492
```

The output shows F is equal to 0.4726 (p-value equals to 0.492). Therefore, we cannot reject the null hypothesis (the coefficient of TotalFloor equals to 0) with a 5% level of significance. As a result, we can infer that the TotalFloor variable does not contribute significant information to Price compared with the rest of the independent variables.

Applying the Gaussian model for generalized linear regression

A **generalized linear model** (**GLM**) is a generalization of linear regression, which can include a link function to make a linear prediction. As a default setting, the family object for glm is Gaussian, which makes glm perform exactly the same as lm. In this recipe, we first demonstrate how to fit the model to data using the glm function, and then show that glm with a Gaussian model performs exactly the same as lm.

Getting ready

You need to have completed the previous recipe by downloading the house rental data into a variable, house. Also, you need to fit the house rental data into a multiple regression model, fit.

How to do it...

Perform the following steps to fit the generalized linear regression model with the Gaussian model:

1. Fit independent variables Sqft, Floor, TotalFloor, Bedroom, Living.Room, and Bathroom to glm:

```
> glmfit <- glm(Price ~ Sqft + Floor + TotalFloor  + Bedroom +
Living.Room + Bathroom, data=house, family=gaussian())
> summary(fit)
```

2. Use anova to compare the two fitted models:

```
> anova(fit, glmfit)
Analysis of Variance Table

Model 1: Price ~ Sqft + Floor + TotalFloor + Bedroom + Living.Room
+ Bathroom
Model 2: Price ~ Sqft + Floor + TotalFloor + Bedroom + Living.Room
+ Bathroom
  Res.Df       RSS Df   Sum of Sq F Pr(>F)
1    638 2.441e+11
2    638 2.441e+11  0 -9.1553e-05
```

How it works...

The `glm` function fits a model to data in a similar fashion to the `lm` function. The only difference is that you can specify a different link function in the parameter, `family` (you may use `family` in the console to find different types of the `link` function). In this recipe, we first input the independent variables `Sqft`, `Floor`, `TotalFloor`, `Bedroom`, `Living.Room`, and `Bathroom` and dependent variable `Price` to the `glm` function, and assign the built model to `glmfit`. You can use the built model for further prediction. By applying the summary function to the two different models, it reveals that the residuals and coefficients of the two output summaries are exactly the same.

Finally, we further compare the two fitted models with the `anova` function. The `anova` results reveal that the two models are similar, with the same residual degrees of freedom (`Res.DF`) and residual sum of squares (`RSS Df`).

See also

- For a comparison of generalized linear models with linear models, please refer to *Modern Applied Statistics with S* by *Venables W. N. and Ripley B. D., Springer* (2002).

Performing a logistic regression analysis

In the previous examples, we have discussed how to use fit data into linear model of continuous variables. In addition, we can use the `logit` model in generalized linear models to predict categorical variables. In this recipe, we will demonstrate how to perform binomial logistic regression to create a classification model that can predict binary responses on a given a set of predictors.

Getting ready

Download the house rental dataset from `https://github.com/ywchiu/rcookbook/blob/master/chapter11/customer.csv` first, and ensure you have installed R on your operating system.

How to do it...

Perform the following steps to fit a generalized linear regression model with the `logit` model:

1. Read `customer.csv` into an R session:

```
> customer = read.csv('customer.csv', header=TRUE)
> str(customer)
```

```
'data.frame':100 obs. of  5 variables:
 $ CustomerID : int  1 2 3 4 5 6 7 8 9 10 ...
 $ gender     : Factor w/ 2 levels "F","M": 1 2 1 1 2 2 1 1 2 2
...
 $ age        : int  36 26 21 49 42 49 47 50 26 40 ...
 $ visit.times: int  5 3 2 5 4 1 4 1 2 3 ...
 $ buy        : Factor w/ 2 levels "no","yes": 2 1 2 2 1 1 2 1 1 1
...
```

2. Fit the customer data into an R session:

```
> logitfit = glm(buy ~ visit.times + age + gender, data=customer,
family=binomial(logit))
```

3. Use the `summary` function to obtain the summary of fitted model:

```
> summary(logitfit)

Call:
glm(formula = buy ~ visit.times + age + gender, family =
binomial(logit),
    data = customer)

Deviance Residuals:
   Min       1Q   Median       3Q      Max
-1.909    0.000    0.000    0.000    1.245

Coefficients:
             Estimate Std. Error z value Pr(>|z|)
(Intercept)   26.5278    18.6925   1.419    0.156
visit.times    9.7809     6.1264   1.597    0.110
age           -1.1396     0.7592  -1.501    0.133
genderM      -71.0222  4170.8348  -0.017    0.986

(Dispersion parameter for binomial family taken to be 1)

    Null deviance: 133.7496  on 99  degrees of freedom
Residual deviance:   7.1936  on 96  degrees of freedom
AIC: 15.194

Number of Fisher Scoring iterations: 21
```

4. Use the `predict` function to obtain the prediction result:

```
>pr <- predict(fit, customer, type="response")
```

5. Finally, one can use the table function to retrieve the confusion matrix:

```
> table(customer$buy, ifelse(pr > 0.5, 'yes', 'no'))

      no yes
  no  60   1
  yes  1  38
```

How it works...

Similar to linear regression analysis, logistic regression is used to predict the response (dependent variable) based on a set of predicators (independent variables). Linear regression is used to predict continuous variables, while logistic regression analysis is best suited to predicting variables of binary outcome.

A logistic model is a type of generalized linear regression model, and its response follows binomial distribution. To let the probability of the response always take a value between `0` and `1`, the logistic regression model uses the log odd ratio (`logit`) to transform the dependent variable. `logit` can be formulated as the following equation:

$$ln\left(\frac{p(x)}{1-p(x)}\right) = b_0 + b_2 x$$

Here the ratio of *p(x)* to *1-p(x)* refers to the odd ratio, and b_0 and b_1 represent the coefficients. Following the `logit` equation, we can formulate the probability of the response in the following equation:

$$p(x) = \frac{1}{1+e^{-(b_0+b_1 x)}}$$

Plotting the logistic function into a line chart yields the following graph, in which the response of a `logit` model always takes a value between `0` and `1`:

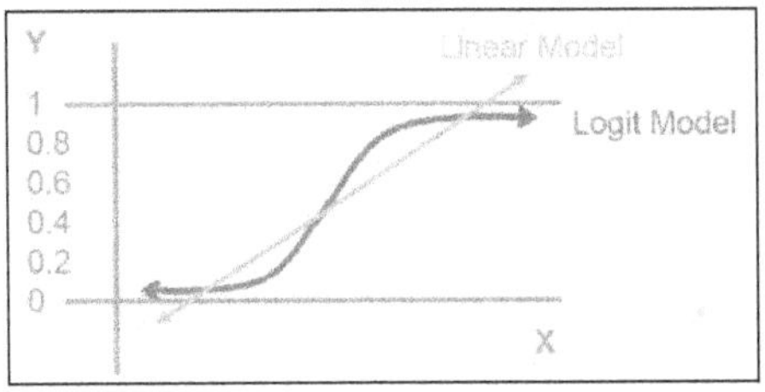

Figure 5: A comparison plot between the logit and linear models

In this recipe, we first load `customer.csv` into an R session. This dataset contains customer features such as ID, gender, age, number of visits, and whether the customer purchased the product. As a sales manager of a company, we would like to build a classification model that can predict whether a customer will purchase the product or not.

Next, we would like to build the classification model based on the formula *buy ~ visit.times + age + gender*. As the outcome of buying behavior is binary, we can perform logistic regression to fit the data into the model. To perform logistic regression, we just need to use `binomial(logit)` as the link function to the `glm` function.

We then use the summary function to obtain the fitted model. From the output, we find the call of function, deviance residuals similar to other regression analysis output. At the coefficient section, the output shows the coefficients, standard error z-statistics, and the associated p-value. Here, we find no variable is statistically significant different from 0 at the 5% level of significance. On the other hand, we find that for every one unit change in `visit.times`, the log odds of buying the product increase by 9.78. We also find that an increase in age, or a male customer, decreases the log odds of buying the product. In the coefficient table, we find the null, deviance residuals, and AIC; we can choose to use a step function to find an optimum model.

We can now predict whether a customer will purchase a product using the predict function and fitted model. Here, we set `type="response"` to obtain the probability that a customer will buy the product. Last, we use the table function to obtain a confusion matrix of our prediction result. From the confusion matrix, we find the prediction model only made two mistakes out of one hundred predictions.

See also

If you would like to know how many alternative links you can use, refer to the `family` document via the `help` function:

```
> ?family
```

Building a classification model with recursive partitioning trees

In the previous recipe, we introduced how to use logistic regression to build a classification model. We now cover how to use a recursive partitioning tree to predict customer behavior. A classification tree uses split condition to predict class labels based on one or multiple input variables. The classification process starts from the root node of the tree; at each node, the process will check whether the input value should recursively continue to the right or left sub-branch according to the split condition, and stops when meeting any leaf (terminal) nodes of the decision tree. In this recipe, we introduce how to apply a recursive partitioning tree on the shopping cart dataset.

Getting ready

Download the house rental dataset from `https://github.com/ywchiu/rcookbook/blob/master/chapter11/customer.csv` first, and ensure you have installed R on your operating system.

How to do it...

Perform the following steps to build a classification tree with `rpart`:

1. Load the `rpart` package:

```
> library(rpart)
```

2. Split the customer data into a training dataset and testing dataset:

```
> set.seed(33)
> idx <- sample(c(1,2),nrow(customer),prob = c(0.8,0.2),
replace=TRUE)
> trainset <- customer[idx == 1, ]
> testset  <- customer[idx == 2, ]
```

3. Use `rpart` to fit the learning model:

```
> fit <- rpart(buy ~ gender + age + visit.times, data = trainset)
```

4. Type `fit` to retrieve the node detail of the classification tree:

```
> fit
n= 80

node), split, n, loss, yval, (yprob)
```

```
        * denotes terminal node

 1) root 80 30 no (0.6250000 0.3750000)
   2) gender=M 42  0 no (1.0000000 0.0000000) *
   3) gender=F 38  8 yes (0.2105263 0.7894737)
     6) visit.times< 2.5 14  6 no (0.5714286 0.4285714) *
     7) visit.times>=2.5 24  0 yes (0.0000000 1.0000000) *
```

5. Use the `printcp` function to examine the complexity parameter:

```
> printcp(fit)

Classification tree:
rpart(formula = buy ~ gender + age + visit.times, data = trainset)

Variables actually used in tree construction:
[1] gender      visit.times

Root node error: 30/80 = 0.375

n= 80

        CP nsplit rel error  xerror     xstd
1 0.733333      0   1.00000 1.00000 0.144338
2 0.066667      1   0.26667 0.26667 0.089443
3 0.010000      2   0.20000 0.20000 0.078528
```

6. Use the `plotcp` function to plot cost complexity parameters:

```
> plotcp(fit)
```

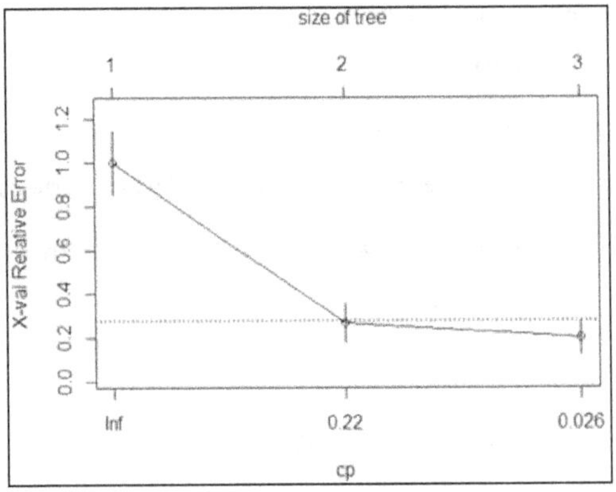

Figure 6: A cost complexity plot

7. Use the `summary` function to examine the built model:

```
> summary(fit)
```

How it works...

In this recipe, we use a recursive partitioning tree from the `rpart` package to build a tree-based classification model. The recursive portioning tree includes two processes: recursive and partitioning. During the process of decision induction, we have to consider a statistic evaluation question (or simply a yes/no question) to partition data into different partitions in accordance with the assessment result. Then, as we have determined the child node, we can repeatedly perform the splitting until stop criteria are satisfied.

For example, the data (shown in Figure 7) in the root node can be partitioned into two groups in regard to the question of whether the gender is equal to M. If yes, the data is divided into the left-hand side; otherwise, it is split into the right-hand side. We continue to partition left-hand side data with the question of whether `visit.times` is lower than `2.5`:

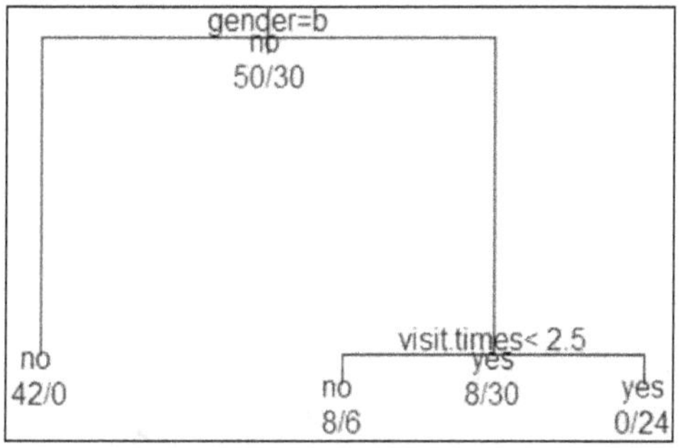

Figure 7: Recursive partitioning tree

We first load the `rpart` package with the `library` function. Next, we build a classification model using the `buy` variable as a classification category (the `class` label) and the remaining variables as input features.

After the model is built, we type the variable name of the built model, `fit`, to display the tree node details. In the printed node detail, `n` indicates the sample size, loss indicates the misclassification cost, `yval` represents the classified membership (`no` or `yes` in this case), and `yprob` stands for the probabilities of two classes (the left value refers to the probability of reaching the `no` label, and the right value refers to probability of reaching the `yes` label).

We then use the `printcp` function to print the complexity parameters of the built tree model. From the output of `printcp`, we find the value of **complexity parameter** (**CP**), which serves as a penalty to control tree size. In short, the greater the CP value, the fewer the number of splits (`nsplit`). The output value (the `rel` error) represents the average deviance of the current tree divided by the average deviance of the null tree. `xerror` represents the relative error estimated by 10 fold classification. `xstd` stands for the standard error of the relative error.

To make the CP (complexity parameter) table more readable, we use `plotcp` to generate an information graphic of the CP table. As per *Figure 6*, the bottom *x* axis illustrates the CP value, the *y* axis illustrates the relative error, and the upper *x* axis displays the size of the tree. The dotted line indicates the upper limit of one standard deviation. From the figure, we can determine that the minimum cross-validation error occurs when the tree is at the `3` size.

We can also use the `summary` function to display the function `call`, the complexity parameter table for the fitted tree model, variable importance (which helps identify the most important variable for tree classification, summing to 100), and detailed information of each node.

The advantage of using a decision tree is that it is very flexible and easy to interpret. It works on both classification and regression problems, and it is also nonparametric. Thus, we do not have to worry whether the data is linearly separable. The main disadvantage of decision trees is that they tend to be biased and over-fitted. However, you can overcome the bias problem by using a conditional inference tree, and solve the overfitting problem by adopting the random forest method or tree pruning.

See also

- For more information about the `rpart`, `printcp`, and `summary` functions, use the `help` function:

  ```
  > ?rpart
  > ?printcp
  > ?summary.rpart
  ```

- `C50` is another package that provides a decision tree and a rule-based model. If you are interested in the package, you may refer to the document at `http://cran.r-project.org/web/packages/C50/C50.pdf`.

Visualizing a recursive partitioning tree

From the last recipe, we learned how to print the classification tree in text format. To make the tree more readable, we can use the `plot` function to obtain the graphical display of a built classification tree.

Getting ready

You need to have the previous recipe completed by generating a classification model, and assign the model into variable fit.

How to do it...

Perform the following steps to visualize the classification tree:

1. Use the `plot` and `text` functions to plot the classification tree:

```
> plot(fit, margin= 0.1)
> text(fit, all=TRUE, use.n = TRUE)
```

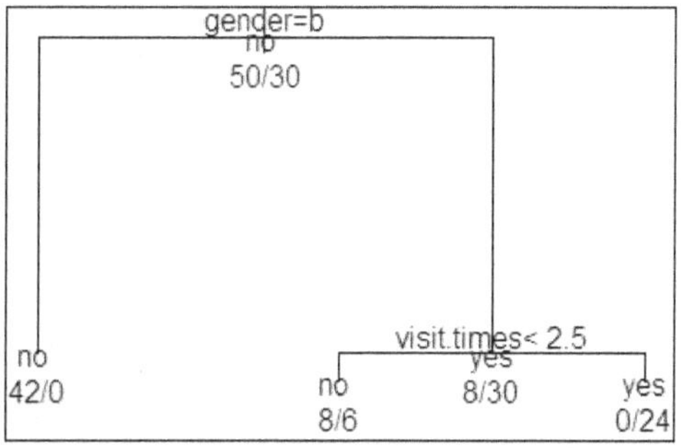

Figure 8: The classification tree of the customer dataset

2. You can also specify the `uniform`, `branch`, and `margin` parameters to adjust the layout:

```
> plot(fit, uniform=TRUE, branch=0.6, margin=0.1)
> text(fit, all=TRUE, use.n = TRUE)
```

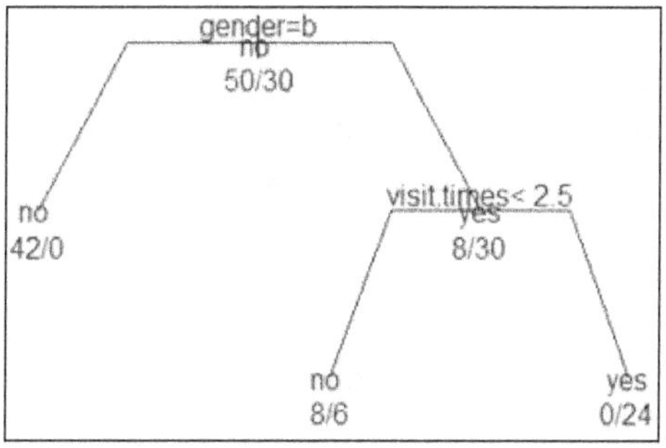

Figure 9: The recursive portioning tree in a different layout

How it works...

Here, we demonstrate how to use the `plot` function to graphically display a classification tree. The `plot` function can simply visualize the classification tree, and we can then use the text function to add text to the plot.

In the figure, we assign `margin=0.1` as a parameter to add extra white space around the border to prevent the displayed text being truncated by the margin. The figure shows that the length of the branches displays the relative magnitude of the drop in deviance. We then use the `text` function to add labels for the nodes and branches. By default, the `text` function will add a split condition on each split, and add a category label in each terminal node. In order to add extra information to the tree plot, we set the `all` parameter equals to `TRUE` to add a label to all nodes. In addition, we add a parameter by specifying `use.n = TRUE` to add extra information, which shows that the actual number of observations falls into two different categories (`no` and `yes`).

In Figure 9, we set the option `branch` to `0.6` to add a shoulder to each plotted branch. In addition, in order to display branches of an equal length rather than the relative magnitude of the drop in deviance, we set the option `uniform` to `TRUE`. As a result, the figure shows a classification tree with short shoulders and branches of equal length.

See also

- One may use `?plot.rpart` to read more about the plotting of the classification tree. This document also includes information on how to specify the `uniform`, `branch`, `compress`, `nspace`, `margin`, and `minbranch` parameters to adjust the layout of the classification tree.

Measuring model performance with a confusion matrix

To measure the performance of a classification model, we can first generate a classification table based on our predicted label and actual label. We then use a confusion matrix to obtain performance measures such as precision, recall, specificity, and accuracy. In this recipe, we will demonstrate how to retrieve a confusion matrix using the `caret` package.

Getting ready

You need to have the previous recipes completed by generating a classification model, and assign the model to the variable `fit`.

How to do it...

Perform the following steps to generate classification measurement:

1. Predict labels using the fitted model, `fit`:

```
> pred = predict(fit, testset[,! names(testset) %in% c("buy")], 
type="class")
```

2. Generate a classification table:

```
> table(pred, testset[,c("buy")])

pred  no yes
  no  11   1
  yes  0   8
```

3. Lastly, generate a confusion matrix using prediction results and actual labels from the testing dataset:

```
> confusionMatrix(pred, testset[,c("buy")])
Confusion Matrix and Statistics

          Reference
Prediction no yes
       no  11   1
       yes  0   8

               Accuracy : 0.95
                 95% CI : (0.7513, 0.9987)
    No Information Rate : 0.55
    P-Value [Acc > NIR] : 0.0001114

                  Kappa : 0.898
 Mcnemar's Test P-Value : 1.0000000

            Sensitivity : 1.0000
            Specificity : 0.8889
         Pos Pred Value : 0.9167
         Neg Pred Value : 1.0000
             Prevalence : 0.5500
         Detection Rate : 0.5500
   Detection Prevalence : 0.6000
      Balanced Accuracy : 0.9444

       'Positive' Class : no
```

How it works...

In this recipe, we demonstrate how to obtain a confusion matrix to measure the performance of a classification model. First, we use the `predict` function to extract predicted labels from the classification model using the testing dataset, `testset`. We then perform the table function to obtain the classification table, based on predicted labels and actual labels.

A confusion table can be described as follows:

	Positive (Reference)	**Negative (Reference)**
Positive (Prediction)	True Positive	False Positive
Negative (Prediction)	False Negative	True Negative

From this confusion matrix, we can use the `confusionMatrix` function from the `caret` package to generate the performance of the classification model in all measurements. We list the formula of each measurement:

- **Accuracy**: $(ACC) = \frac{TP+TN}{P+N}$
- **95% CI**: This denotes the 95% confidence interval
- **No Information Rate (NIR)**: This stands for the largest proportion of observed classes
- **p-value [ACC > NIR]**: The p-value is computed under a one-sided binomial test to find whether the accuracy rate is greater than the no information rate (the rate of the largest class)
- **Kappa**: This statistic estimates the Kappa statistic for model accuracy
- **Mcnemar's test p-value**: The Mcnemar's test p-value is computed by a Mcnemar's test, which is used to assess whether there is significant difference between a (positive, negative) pair and a (negative, positive) pair
- $Sensitivity = \frac{TP}{TP+FN}$
- $Specificity = \frac{TN}{FP+TN}$
- $Prevalence = \frac{TP+FN}{TP+FP+FN+TN}$

- $$PPV = \frac{Sensitivity + Prevalence}{\left(Sensitivity * Prevalence + \left(1 - Specificity\right) * \left(1 - Prevalence\right)\right)}$$
- $$NPV = \frac{Specificity + \left(1 - Prevalence\right)}{\left(\left(1 - Sensitivity\right) * Prevalence + Specificity * \left(1 - Prevalence\right)\right)}$$
- $$Detection\ Rate = \frac{TP}{TP + FP + FN + TN}$$
- $$Detection\ Prevalence = \frac{TP + FP}{TP + FP + FN + TN}$$
- $$Balanced\ Accuracy = \frac{Sensitivity + Specificity}{2}$$

Measuring prediction performance using ROCR

One obstacle to using a confusion matrix to assess the classification model is that you have to arbitrarily select a threshold to determine the value of the matrix. A possible way to determine the threshold is to visualize the confusion matrix under different thresholds in a **Receiver Operating Characteristic** (**ROC**) curve.

An ROC curve is a plot that illustrates the performance of a binary classifier system, and plots the true positive rate against the false positive rate for different cut points. We most commonly use this plot to calculate the **area under curve** (**AUC**), to measure the performance of a classification model. In this recipe, we demonstrate how to illustrate a ROC curve and calculate the AUC to measure the performance of a classification model.

Getting ready

You need to have the previous recipes completed by generating a classification model, and assign the model into variable fit.

How to do it...

Perform the following steps to generate two different classification examples with different costs:

1. First, install and load the `ROCR` package:

```
> install.packages("ROCR")
> library(ROCR)
```

2. Make predictions based on the trained model on the testing dataset with `probability` set to `TRUE`:

```
> pred2 <- predict(fit,testset[, !names(testset) %in% c("buy")],
probability=TRUE)
```

3. Obtain the probability of labels with `"yes"`:

```
> pred.to.roc = pred2[,c("yes")]
```

4. Use the `prediction` function to generate a prediction result:

```
> pred.rocr = prediction(pred.to.roc, testset$churn)
```

5. Use the `performance` function to obtain performance measurement:

```
> perf.rocr = performance(pred.rocr, measure = "auc", x.measure =
"cutoff")
> perf.tpr.rocr = performance(pred.rocr, "tpr","fpr")
```

6. Visualize the ROC curve using the `plot` function:

```
> plot(perf.tpr.rocr, colorize=T,main=paste("AUC:",(perf.rocr@y.
values)))
```

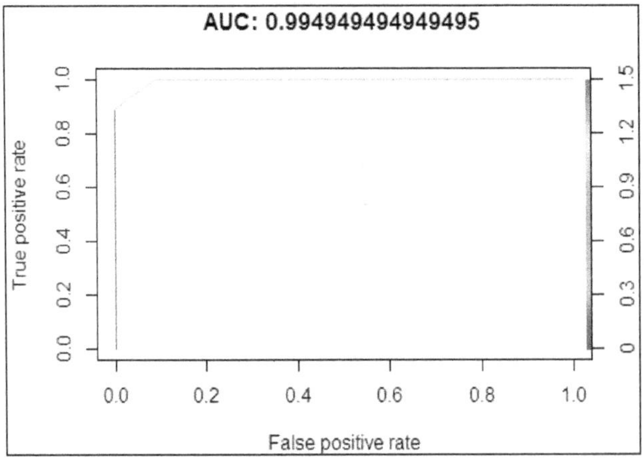

Figure 10: The ROC curve of the model

How it works...

In this recipe, we demonstrate how to generate an ROC curve to illustrate the performance of a binary classifier. First, we install and load the `ROCR` library. We then use `rpart` from the `rpart` package to train a classification model, and use the model to predict labels for the testing dataset. Next, we use the `prediction` function (from the `ROCR` package) to generate prediction results. We adapt the `performance` function to obtain the performance measurement of the true positive rate against the false positive rate. Finally, we use the `plot` function to visualize the ROC plot, and add the value of AUC on the title. In this example, the AUC value is `0.994`, which indicates the `rpart` classifier performs well in classifying the `customer` dataset.

See also...

For those interested in the concept and terminology of ROC, you can refer to `http://en.wikipedia.org/wiki/Receiver_operating_characteristic`.

12
Unsupervised Machine Learning

This chapter covers the following topics:

- Clustering data with hierarchical clustering
- Cutting trees into clusters
- Clustering data with the k-means method
- Clustering data with the density-based method
- Extracting silhouette information from clustering
- Comparing clustering methods
- Recognizing digits using density-based clustering methods
- Grouping similar text documents with k-means clustering methods
- Performing dimension reduction with Principal Component Analysis (PCA)
- Determining the number of principal components using a scree plot
- Determining the number of principal components using the Kaiser method
- Visualizing multivariate data using a biplot

Introduction

The unsupervised machine learning method focuses on revealing the hidden structure of unlabeled data. A key difference between unsupervised learning and supervised learning is that the latter method employs labeled data as learners. Therefore, one can evaluate the model based on *known answers*. In contrast, one cannot evaluate unsupervised learning as it does not have any *known answers*. Mostly, unsupervised learning focuses on two main areas: clustering and dimension reduction.

Clustering is a technique used to group similar objects (close in terms of distance) together in the same group (cluster). Clustering analysis does not use any label information, but simply uses the similarity between data features to group them into clusters.

Dimension reduction is a technique that focuses on removing irrelevant and redundant data to reduce the computational cost and avoid overfitting; you can reduce the features into a smaller subset without a significant loss of information. Dimension reduction can be divided into two parts: feature extraction and feature selection. **Feature extraction** is a technique that uses a lower dimension space to represent data of a higher dimension space. **Feature selection** is used to find the most relevant variables of a prediction model.

In this chapter, we first cover how to use the hierarchical clustering method to cluster hotel location data in a dendrogram. We can use the `cuttree` function to segment data into different groups, and then demonstrate how k-means and the density method work. Next, we can validate the quality of the cluster with silhouette and within the sum of squares measure, and also compare how each method works. Moreover, we can demonstrate how DBSCAN works in recognizing digits. We will also illustrate an example of how to cluster text documents with the k-means method.

Moving on, we discuss how the dimension reduction method works, and first describe how to perform PCA on economy freedom data. We can then use both a scree plot and the Kaiser method to determine the number of principal components. Last, we can visualize the multivariate data with `biplot`.

Clustering data with hierarchical clustering

Hierarchical clustering adopts either an agglomerative or divisive method to build a hierarchy of clusters. Regardless of which approach is adopted, both initially use a distance similarity measure to combine clusters or split clusters. The recursive process continues until there is only one cluster left or one cannot split more clusters. Eventually, we can use a dendrogram to represent the hierarchy of clusters. In this recipe, we will demonstrate how to cluster hotel location data with hierarchical clustering.

Getting ready

In this recipe, we will perform hierarchical clustering on hotel location data to identify whether the hotels are located in the same district. You can download the data from the following GitHub link:

https://github.com/ywchiu/rcookbook/raw/master/chapter12/taipei_hotel.csv

How to do it...

Please perform the following steps to cluster location data into a hierarchy of clusters:

1. First, load the data from `taipei_hotel.csv` and save it into `hotel`:

```
> hotel <- read.csv('taipei_hotel.csv', header=TRUE)
> str(hotel)
```

2. You can then visualize the dataset by its longitude and latitude:

```
> plot(hotel$lon, hotel$lat, col=hotel$district)
```

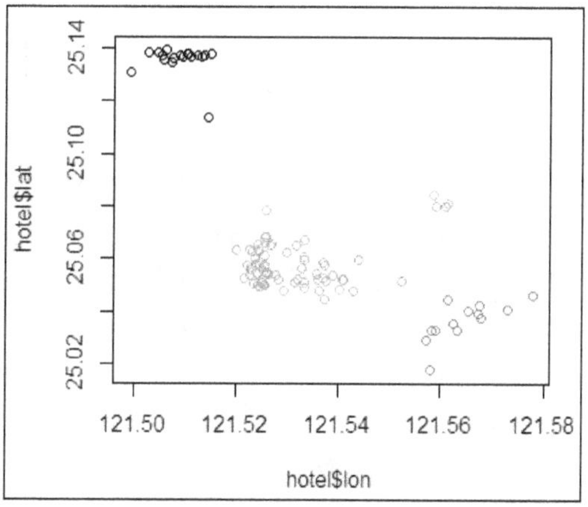

Figure 1: The scatter plot of the hotel location dataset

3. Use agglomerative hierarchical clustering to cluster the geographical data of the hotels:

```
> hotel.dist <- dist(hotel[,c('lat', 'lon')] , method="euclidean")
> hc <- hclust(hotel.dist, method="ward.D2")
> hc

Call:
```

```
hclust(d = hotel.dist, method = "ward.D2")

Cluster method   : ward.D2
Distance         : euclidean
Number of objects: 102
```

4. Last, you can use the `plot` function to plot the dendrogram:

```
> plot(hc, hang = -0.01, cex = 0.7)
```

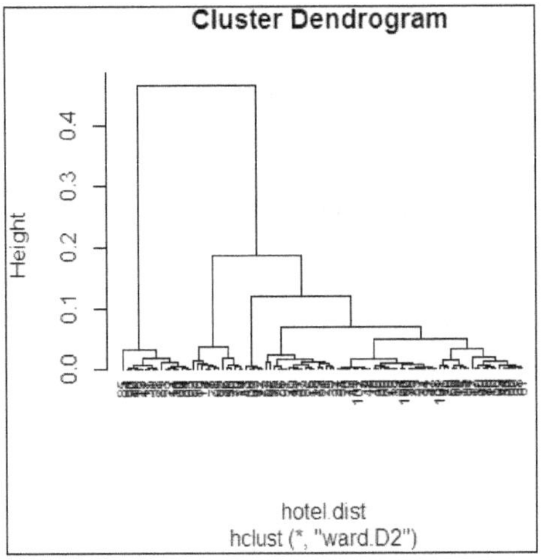

Figure 2: The dendrogram of hierarchical clustering using "ward.D2"

5. In addition, you can use the `single` method to perform hierarchical clustering and see how the generated dendrogram differs from the previous version:

```
> hc2 = hclust(hotel.dist, method="single")
> plot(hc2, hang = -0.01, cex = 0.7)
```

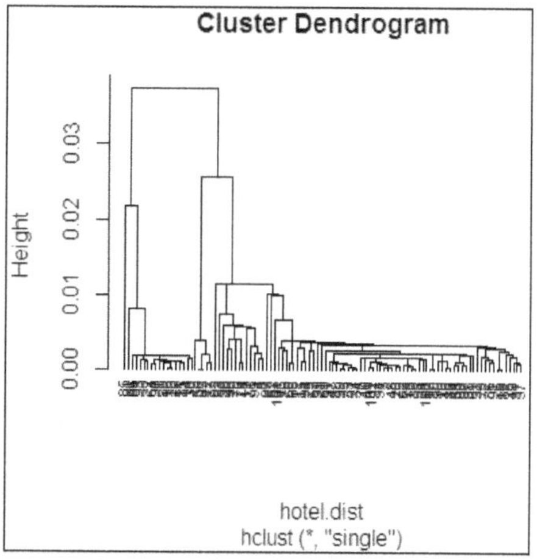

Figure 3: The dendrogram of hierarchical clustering using "single"

How it works...

Hierarchical clustering is a clustering technique that tries to build a hierarchy of clusters iteratively. Generally, there are two approaches to building hierarchical clusters:

- **Agglomerative hierarchical clustering**: This is a *bottom-up* approach. Each observation starts in its own cluster. We can then compute the similarity (or distance) between each cluster and merge the most two similar at each iteration until there is only one cluster left.
- **Divisive hierarchical clustering**: This is a *top-down* approach. All observations start in one cluster, and we then split the cluster into the two least dissimilar clusters recursively until there is one cluster for each observation.

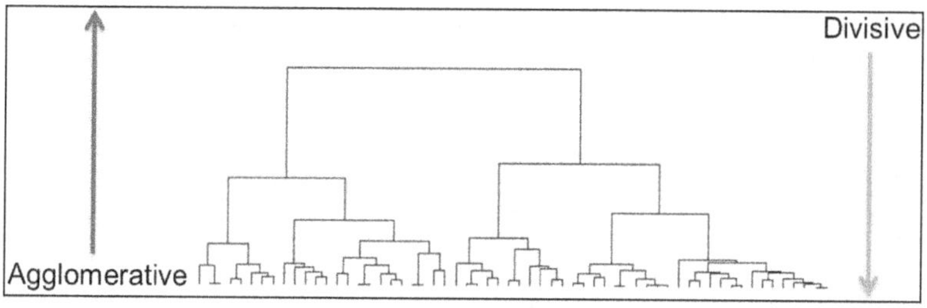

Figure 4: Two approaches of hierarchical clustering

Before performing hierarchical clustering, we need to determine how similar the two clusters are. Here, we list some common distance functions used for similarity measurement:

- **Single-linkage**: The shortest distance between two points in each cluster:

$$dist(C_i, C_j) = \min_{a \in C_i, b \in C_j} dist(a, b)$$

- **Complete-linkage**: The longest distance between two points in each cluster:

$$dist(C_i, C_j) = \max_{a \in C_i, b \in C_j} dist(a, b)$$

- **Average-linkage**: The average distance between two points in each cluster (where $|C_i|$ is the size of cluster C_i and $|C_j|$ is the size of cluster C_j):

$$dist(C_i, C_j) = \frac{1}{|C_i||C_j|} \sum_{a \in C_i, b \in C_j} dist(a, b)$$

- **Ward-method**: The sum of squared distance from each point to the mean of the merged clusters (where μ is the mean vector of $C_i \cup C_j$):

$$dist(C_i, C_j) = \sum_{a \in C_i \cup C_j} \|a - \mu\|$$

In this recipe, we perform hierarchical clustering on customer data. First, we load data from `Taipei_hotel.csv`, and load the data into the `hotel` dataframe. Within the data, we find five variables of hotel location information, which are: address, latitude, longitude, title, and district.

We then perform hierarchical clustering using the `hclust` function. We use Euclidean distance as distance metrics, and use Ward's minimum variance method to perform agglomerative clustering.

Last, we use the `plot` function to plot the dendrogram of hierarchical clusters in *Figure 2*. We specify hang to display labels at the bottom of the dendrogram, and use `cex` to shrink the label to 70% of normal size. In order to compare the differences using the `ward.D2` and `single` methods to generate a hierarchy of clusters, we draw another dendrogram using `single` in *Figure 3*.

There's more...

You can choose a different distance measure and method when performing hierarchical clustering. For more details, refer to the documents for `dist` and `hclust`:

```
> ? dist
> ? hclust
```

In this recipe, we use `hclust` to perform agglomerative hierarchical clustering. If you would like to perform divisive hierarchical clustering, you can use the `diana` function:

1. First, use `diana` to perform divisive hierarchical clustering:

   ```
   > dv = diana(hotel, metric = "euclidean")
   ```

2. Use `summary` to obtain summary information:

   ```
   > summary(dv)
   ```

3. Last, plot a dendrogram and banner with the `plot` function:

   ```
   > plot(dv)
   ```

Cutting tree into clusters

In a dendrogram, we can see the hierarchy of clusters, but we have not grouped data into different clusters yet. However, we can determine how many clusters are within the dendrogram and *cut* the dendrogram at a certain tree height to separate data into different groups. In this recipe, we demonstrate how to use the `cutree` function to separate data into a given number of clusters.

Getting ready

In order to perform the `cutree` function, one needs to have completed the previous recipe by generating an hclust object, `hc`.

How to do it...

Please perform the following steps to cut the hierarchy of clusters into a given number of clusters:

1. First, categorize the data into three groups:

   ```
   > fit <- cutree(hc, k = 3)
   ```

2. You can then examine the cluster labels for the data:

   ```
   > fit
   ```

3. Count the number of data points within each cluster:

```
> table(fit)
fit
 1  2  3
18 66 18
```

4. Make a scatter plot with fitted cluster information:

```
> plot(hotel$lon, hotel$lat, col=fit)
```

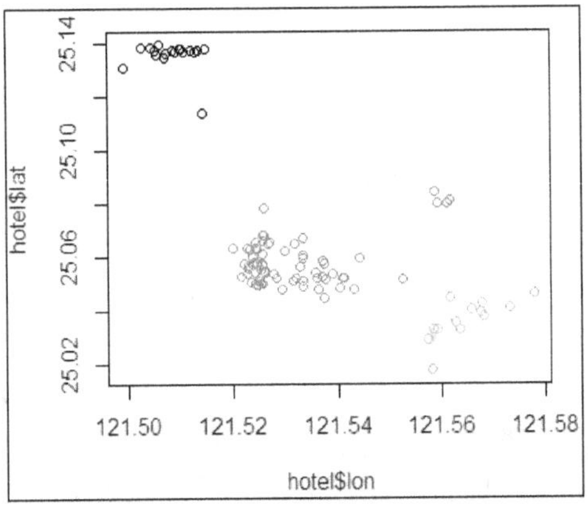

Figure 5: The scatter plot of fitted data

5. Finally, you can visualize how data is clustered with the red rectangle border:

```
> plot(hc)
> rect.hclust(hc, k = 3 , border="red")
```

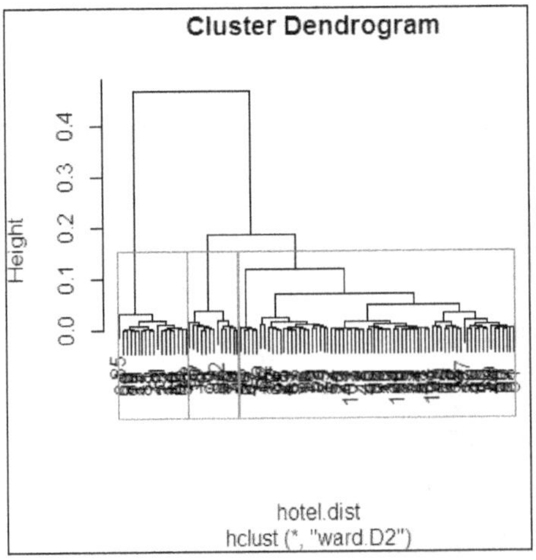

Figure 6: Using red rectangles to distinguish different clusters within the dendrogram

How it works...

In this recipe, we determine there should be three clusters within the tree. Therefore, we specify the number of clusters as `3` in the `cutree` function. Besides using the number of clusters to cut the tree, we can specify the `height` as the cut tree parameter.

Next, we output the cluster labels of the data and use the `table` function to count the number of data points within each cluster. From the table, we find that most of the data is in cluster 2. Last, we can draw red rectangles around the clusters to show how data is categorized into the three clusters with the `rect.hclust` function.

There's more...

Besides drawing rectangles around all hierarchical clusters, you can place a red rectangle around a certain cluster:

```
> plot(hc)
> rect.hclust(hc, k = 3 , which =2, border="red")
```

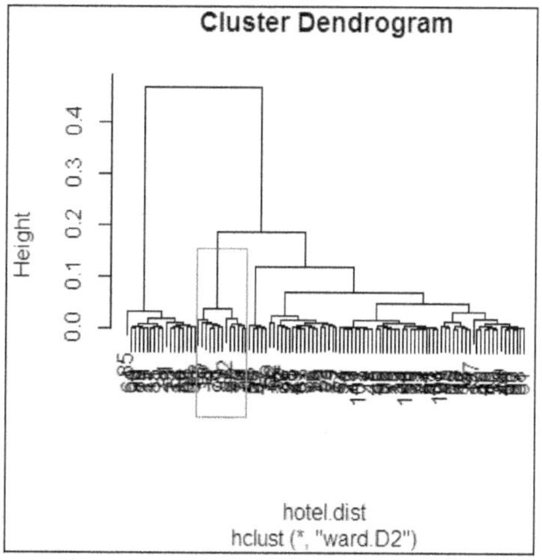

Figure 7: Drawing a red rectangle around a certain cluster

Clustering data with the k-means method

K-means clustering is a method of partitioning clustering. The goal of the algorithm is to partition *n* objects into *k* clusters, in which each object belongs to the cluster with the nearest mean. Unlike hierarchical clustering, which does not require a user to determine the number of clusters at the beginning, the k-means method does require this to be determined first. However, k-means clustering is much faster than hierarchical clustering as the construction of a hierarchical tree is very time-consuming. In this recipe, we will demonstrate how to perform k-means clustering on the hotel location dataset.

Getting ready

In this recipe, we will continue to use the hotel location dataset as the input data source to perform k-means clustering.

How to do it...

Please perform the following steps to cluster the hotel location dataset with the k-means method:

1. First, use `kmeans` to cluster the customer data:

```
> set.seed(22)
> fit <- kmeans(hotel[,c("lon", "lat")], 3)
> fit
K-means clustering with 3 clusters of sizes 19, 18, 65

Cluster means:
       lon      lat
1 121.5627 25.04905
2 121.5087 25.13559
3 121.5290 25.05657

Clustering vector:
  [1] 2 3 2 1 3 3 3 1 3 3 2 3 3 2 3 2 3 3 1 3 3 2 2 3 3 1 3 2 3 2
3 3 3
 [34] 3 2 3 3 3 3 3 3 1 1 3 3 3 3 3 3 3 3 3 3 1 1 3 3 3 3 3 1 3
2 3 2
 [67] 1 3 3 1 1 3 3 1 3 3 1 3 1 3 2 3 1 2 2 3 3 3 2 2 3 1 2 3 3 1
3 3 1
```

```
[100] 3 3 3

Within cluster sum of squares by cluster:
[1] 0.0085979703 0.0008661602 0.0050881876
 (between_SS / total_SS =  89.6 %)

Available components:

[1] "cluster"      "centers"      "totss"        "withinss"
[5] "tot.withinss" "betweenss"    "size"         "iter"
[9] "ifault"
```

2. Draw a scatter plot of the data and color the points according to the clusters:

```
> plot(hotel$lon, hotel$lat, col = fit$cluster)
```

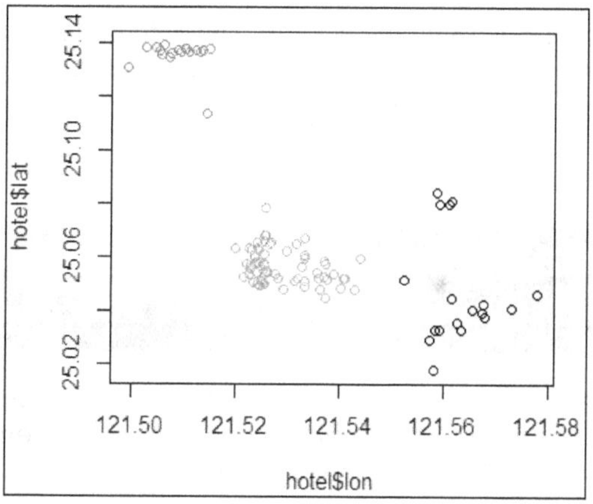

Figure 8: The scatter plot showing data colored according to its cluster label

How it works...

The objective of the k-means algorithm is to minimize the **within-cluster sum of squares** (**WCSS**). Assuming *x* is the given set of observations, $S = \{S_1, S_2, \dots S_k\}$ denotes *k* partitions, and μ_i is the mean of S_i, we can thus formulate the WCSS function as follows:

$$f = \sum_{i=1}^{k} \sum_{x \in S_i} \left\| x - \mu_i \right\|^2$$

The process of k-means clustering can be illustrated by the following steps:

1. Specify the number of clusters *k*.
2. Randomly create *k* partitions.
3. Calculate the center of the partitions.
4. Associate objects to their closest cluster center.
5. Repeat steps 2, 3, and 4 until the WCSS changes very little (or is minimized).

In this recipe, we demonstrate how to use `kmeans` to cluster hotel data. In contrast to hierarchical clustering, k-means clustering requires the user to input the number of *k*. In this example, we use *k=3*. The output of the fitted model shows the size of each cluster, the cluster means of four generated clusters, the cluster vectors in regard to each data point, WCSS by clusters, and other available components. Last, we plot the data points in a scatterplot and use the fitted cluster labels to assign colors with regard to the cluster label.

There's more...

In k-means clustering, you can specify the algorithm used to perform clustering analysis. You can specify either Hartigan-Wong, Lloyd, Forgy, or MacQueen as the clustering algorithm. For more details, please use the `help` function to refer to the document for the `kmeans` function:

```
>help(kmeans)
```

Clustering data with the density-based method

As an alternative to distance measurement, we can use density-based measurement to cluster data. This method finds area with a higher density than the remaining area. One of the most famous methods is DBSCAN. In the following recipe, we demonstrate how to use DBSCAN to perform density-based clustering.

Getting ready

In this recipe, we will continue to use the hotel location dataset as the input data source to perform DBSCAN clustering.

How to do it...

Please perform the following steps to perform density-based clustering:

1. First, install and load the `dbscan` packages:

   ```
   > install.packages("dbscan")
   > library(dbscan)
   ```

2. Cluster data in regard to its density measurement:

```
> fit <- dbscan(hotel.dist, eps = 0.01, minPts = 3)
> fit
DBSCAN clustering for 102 objects.
Parameters: eps = 0.01, minPts = 3
The clustering contains 4 cluster(s) and 3 noise points.

 0  1  2  3  4
 3 17 65 12  5

Available fields: cluster, eps, minPts
```

3. Plot the data in a scatterplot with different cluster labels as color:

```
> plot(hotel$lon,hotel$lat, col=fit$cluster)
```

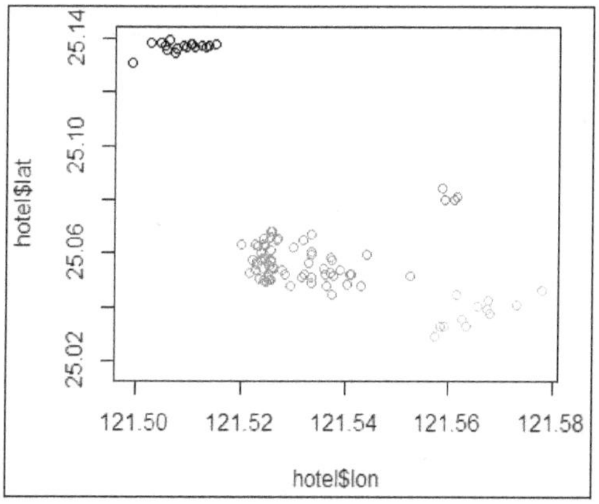

Figure 9: The data scatter plot colored in regard to cluster label

4. You can also use `dbscan` to predict which cluster the data point belongs to. In this example, first make a `data.frame` named `newdata`:

```
> newdata = data.frame(lon= 121.51, lat=25.13)
```

5. You can now predict which cluster the data belongs to:

```
> predict(fit, hotel[,c('lon', 'lat')], newdata)
103
  1
```

How it works...

Density-based clustering uses the idea of density reachability and density connectivity, which makes it very useful in discovering clusters in non-linear shapes. Before discussing the process of density-based clustering, some important background concepts must be explained. Density-based clustering uses two parameters: **Eps** and **MinPts**. **Eps** represents the maximum radius of the neighborhood, while **MinPts** denotes the minimum number of points within the Eps-neighborhood. With these two parameters, we can define a core-point as having more points than **MinPts** within **Eps**. Also, we can define a board-point as having fewer points than **MinPts**, but it is in the neighborhood of a core-point. We can define a core-object if the number of points in the Eps-neighborhood of *p* is more than **MinPts**.

Furthermore, we have to define reachability between two points. We can say that a point *p* is directly density reachable from another point *q* if *q* is within the Eps-neighborhood of *p* and *p* is a core-object. We can then define that a point *p* is generic density reachable from point *q* if there exists a chain of points $p_1, p_2..., p_n$, where $p_1 = q$, $p_n = p$, and p_{i+1} is directly density reachable from p_i with regard to **Eps** and **MinPts** for $1 <= i <= n$.

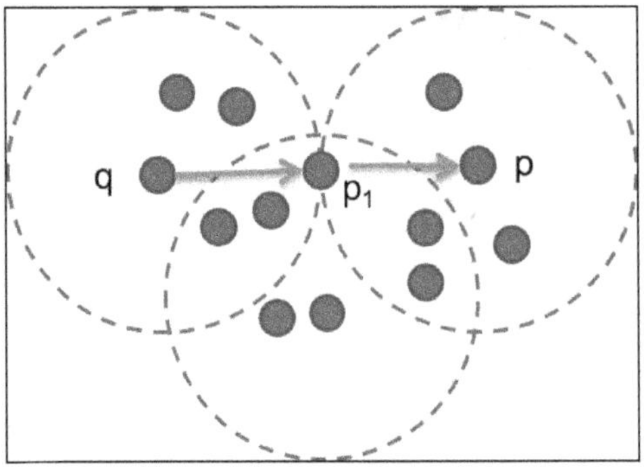

Figure 10: Points p and q are density reachable

From the preliminary concepts of density-based clustering, we can illustrate the process of DBSCAN, the most popular density-based clustering, with the following steps:

1. Randomly select a point *p*.
2. Retrieve all points that are density-reachable from *p* with regard to **Eps** and **MinPts**.
3. If *p* is a core point, then a cluster is formed. Otherwise, if *p* is a board point and no points are density-reachable from *p*, the process will mark the point as noise and continue visiting the next point.
4. Repeat the process until all points have been visited.

In this recipe, we demonstrate how to use the `dbscan` density based method to cluster customer data. First, we install and load the `dbscan` libraries. Next, we perform `dbscan` on the hotel location dataset to cluster data. We specify the reachability distance as `0.01`, minimum reachability number of points as `3`, progress reporting as `null`, and use distance as measurement. The clustering method successfully clusters the data into four clusters with sizes of `17`, `65`, `12`, and `5`. By plotting the points and cluster labels on the plot, we see the four sections of the hotel location are separated in different colors.

The `dbscan` also provides a `predict` function, and one can use this to predict the cluster labels of the input `data.frame`. The point with `lon= 121.51` and `lat=25.13` is classified into cluster `1`.

See also

Besides using the `dbscan` package to perform density-based clustering, one can use the `fpc` package as an alternative. To install and load the `fpc` package, please follow these steps:

```
> install.packages("fpc")
> library(fpc)
```

Extracting silhouette information from clustering

Silhouette information is a measurement to validate a cluster of data. In the previous recipe, we mentioned that the measurement of a cluster involves the calculation of how closely the data is clustered within each cluster, and measuring how far different clusters are apart from each other. The silhouette coefficient combines the measurement of intra-cluster distance and inter-cluster distance. The output value typically ranges from `0` to `1`; the closer to `1`, the better the cluster is. In this recipe, we will introduce how to compute silhouette information.

Getting ready

In order to extract silhouette information from a cluster, one needs to have completed the previous recipe by generating the hotel location dataset.

How to do it...

Please perform the following steps to compute silhouette information:

1. First, install and load the `cluster` package:

   ```
   > install.packages('cluster')
   > library(cluster)
   ```

2. Use `kmeans` to generate a `kmeans` object, `km`:

```
> set.seed(22)
> km <- kmeans(hotel[,c('lon', 'lat')], 3)
```

3. You can then compute silhouette information:

```
> hotel.dist <- dist(hotel[,c('lat', 'lon')] , method="euclidean")
> kms <- silhouette(km$cluster, hotel.dist)
> summary(kms)
Silhouette of 102 units in 3 clusters from silhouette.default(x =
km$cluster, dist = hotel.dist) :
 Cluster sizes and average silhouette widths:
       19        18        65
0.3651140 0.9030704 0.7168364
Individual silhouette widths:
   Min. 1st Qu.  Median    Mean 3rd Qu.    Max.
0.03744 0.56590 0.74860 0.68420 0.80140 0.93890
```

4. Next, plot the silhouette information:

```
> plot(kms)
```

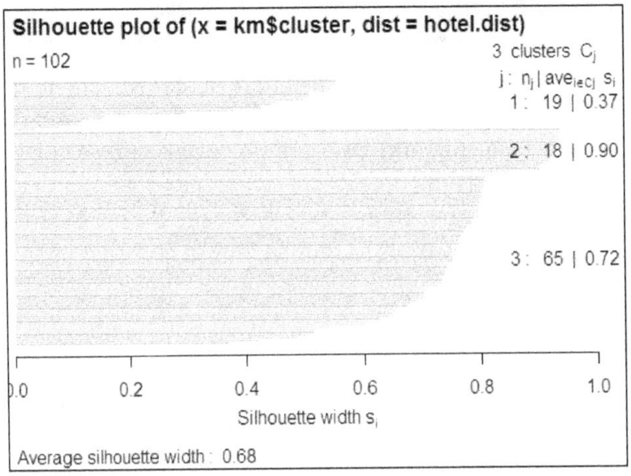

Figure 11: The silhouette plot of the k-means clustering result

How it works...

Silhouette is a measurement that considers how closely related objects are within the cluster and how clusters are separated from each other. Mathematically, we can define the silhouette width for each point x as follows:

$$Silhouette(x) = \frac{b(x) - a(x)}{max([b(x), a(x)])}$$

In the preceding formula *a(x)* is the average distance between x and all other points within the cluster, and *b(x)* is the minimum of the average distances between x and the points in the other clusters.

In this recipe, we demonstrate how to obtain the silhouette value. One can first retrieve silhouette information, which shows cluster sizes, average silhouette widths, and individual silhouette widths. The `silhouette` coefficient is a value ranging from `0` to `1`; the closer to `1`, the better the cluster quality.

Last, we use the `plot` function to draw a silhouette plot. The left-hand side of the plot shows the number of horizontal lines, which represent the number of clusters. The right-hand column shows the mean similarity of the plot of its own cluster minus the mean similarity to the next similar cluster. Average silhouette width is presented at the bottom of the plot.

See also

For those interested in how silhouettes are computed, please refer to the Wikipedia entry for the silhouette value at `http://en.wikipedia.org/wiki/Silhouette_%28clustering%29`.

Comparing clustering methods

After fitting data into clusters using different clustering methods, you may wish to measure the accuracy of the clustering. In most cases, you can use either intra-cluster or inter-cluster metrics as measurements. We now introduce how to compare different clustering methods using `custer.stat` from the `fpc` package.

Getting ready

In order to perform the clustering method comparison, you need to have completed the previous recipe by generating the hotel location dataset.

How to do it...

Perform the following steps to compare clustering methods:

1. First, install and load the `fpc` package:

```
> install.packages("fpc")
> library(fpc)
```

2. You then need to use hierarchical clustering with the single method to cluster customer data and generate the object `hc_single`:

```
> hotel.dist <- dist(hotel[,c('lat', 'lon')] , method="euclidean")
> single_c <- hclust(hotel.dist, method="single")

> hc_single <- cutree(single_c, k = 3)
```

3. Use hierarchical clustering with the complete method to cluster hotel location data and generate the object `hc_complete`:

```
> complete_c <- hclust(hotel.dist,method="complete")
> hc_complete <- cutree(complete_c, k = 3)
```

4. You can then use k-means clustering to cluster customer data and generate the object `km`:

```
> set.seed(22)
> km <- kmeans(hotel[,c('lon', 'lat')], 3)
```

5. Next, retrieve the cluster validation statistics of either clustering method:

```
> cs = cluster.stats(hotel.dist, km$cluster)
```

6. Most often, we focus on using `within.cluster.ss` and `avg.silwidth` to validate the clustering method:

```
> cs[c("within.cluster.ss","avg.silwidth")]
$within.cluster.ss
[1] 0.01455232

$avg.silwidth
[1] 0.6841843
```

7. Finally, we can generate the cluster statistics of each clustering method and list them in a table:

```
> sapply(list(kmeans = km$cluster, hc_single = hc_single, hc_
complete = hc_complete), function(c) cluster.stats(hotel.dist, c)
[c("within.cluster.ss","avg.silwidth")])
                    kmeans     hc_single  hc_complete
within.cluster.ss 0.01455232 0.02550912 0.01484059
avg.silwidth      0.6841843  0.5980621  0.6838705
```

How it works...

In this recipe, we demonstrate how to validate clusters. To validate a clustering method, we often employ two techniques: inter-cluster distance and intra-cluster distance. In these techniques, the higher the inter-cluster distance the better, and the lower the intra-cluster distance the better. In order to calculate related statistics, we can apply `cluster.stat` from the `fpc` package on the fitted clustering object.

From the output, the measurement `within.cluster.ss` stands for within-cluster sum of squares, and `avg.silwidth` represents the average silhouette width. Also, `within.cluster.ss` shows how closely related objects are in clusters; the smaller the value, the more closely related objects are within the cluster. On the other hand, the `silhouette` value usually ranges from `0` to `1`; a value closer to `1` suggests the data is better clustered.

The summary table generated in the last step shows the complete hierarchical clustering method outperforms a single hierarchical clustering method and k-means clustering in `within.cluster.ss` and `avg.silwidth`.

There's more...

The `kmeans` function also outputs statistics (for example, `withinss` and `betweenss`) for users to validate a clustering method:

```
> set.seed(22)
> km <- kmeans(hotel[,c('lon', 'lat')], 3)
> km$withinss
[1] 0.0085979703 0.0008661602 0.0050881876
> km$betweenss
[1] 0.1256586
```

Recognizing digits using the density-based clustering method

We have already covered density-based clustering methods, which are good at handling data without a certain shape. In this recipe, we demonstrate how to use DBSCAN to recognize digits.

Getting ready

In this recipe, we use handwritten digits as clustering input. You can find the figure on the author's GitHub page at `https://github.com/ywchiu/rcookbook/raw/master/chapter12/handwriting.png`.

How to do it...

Perform the following steps to cluster digits with different clustering techniques:

1. First, install and load the `png` package:

```
> install.packages("png")
> library(png)
```

2. Read images from `handwriting.png` and transform the read data into a scatterplot:

```
> img2 = readPNG("handwriting.png", TRUE)
> img3 = img2[,nrow(img2):1]
> b = cbind(as.integer(which(img3 < -1) %% 28), which(img3 < -1) /
28)
> plot(b, xlim=c(1,28), ylim=c(1,28))
```

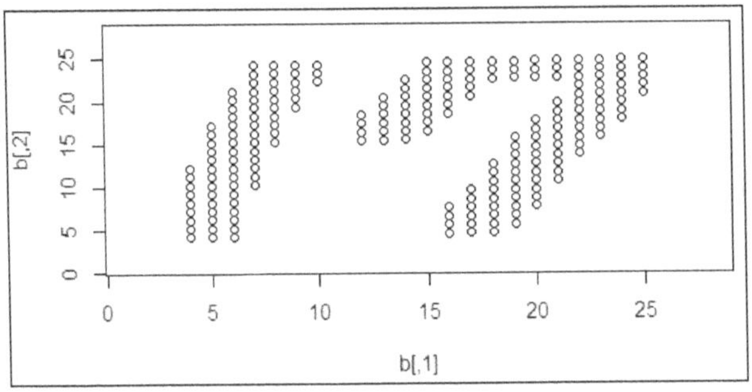

Figure 12: A scatterplot of handwritten digits

3. Perform the k-means clustering method on the handwritten digits:

```
> set.seed(18)
> fit = kmeans(b, 2)
```

```
> plot(b, col=fit$cluster)
> plot(b, col=fit$cluster,  xlim=c(1,28), ylim=c(1,28))
```

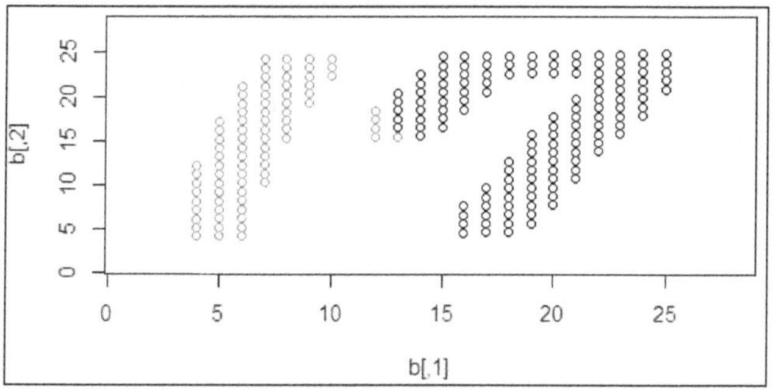

Figure 13: The k-means clustering result on handwriting digits

4. Next, perform the `dbscan` clustering method on the handwritten digits:

```
> ds = dbscan(b, 2)
> ds
DBSCAN clustering for 212 objects.
Parameters: eps = 2, minPts = 5
The clustering contains 2 cluster(s) and 0 noise points.

  1   2
 75 137

Available fields: cluster, eps, minPts
> plot(ds, b,  xlim=c(1,28), ylim=c(1,28))
```

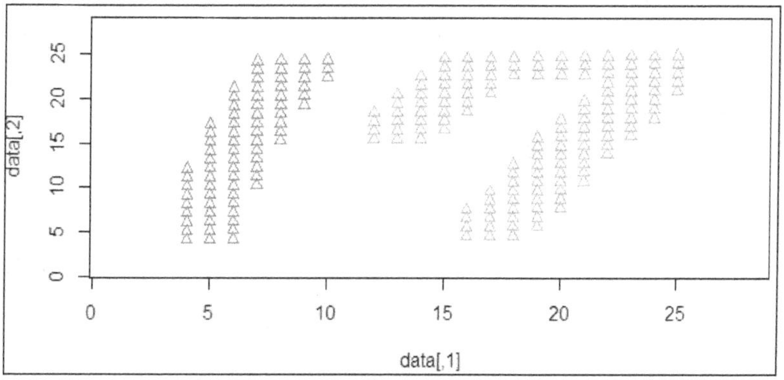

Figure 14: DBSCAN clustering result on handwritten digits

How it works...

In this recipe, we demonstrate how different clustering methods work in regard to a handwriting dataset. The aim of the clustering is to separate the 1 and 7 into different clusters. We perform different techniques to see how data is clustered in regard to the k-means and DBSCAN methods.

To generate the data, we use the Windows application Paint to create a PNG file with dimensions of 28 x 28 pixels. We then read the PNG data using the `readPNG` function and transform the read PNG data points into a scatter plot, which shows the digits 1 and 7 in handwriting.

After the data is read, we perform clustering techniques on the handwritten digits. First, we perform k-means clustering with *k=2* on the dataset. Because k-means clustering employs distance measures, the constructed clusters cover the area of both the 1 and 7. We then perform DBSCAN on the dataset; as this is a density-based clustering technique, it successfully separates the 1 and 7 into different clusters.

See also

If you are interested in how to read various graphic formats in R, you may run the following command:

```
> help(package="png")
```

Grouping similar text documents with k-means clustering methods

Computer programs face limitations in interpreting the meaning of given sentences, and therefore do not know how to group documents based on their similarities. However, if we can convert sentences into a mathematical matrix (document term matrix), a program can compute the distance between each document and group similar ones together.

In this recipe, we demonstrate how to compute the distance between text documents and how we can cluster similar text documents with the k-means method.

Getting ready

In this recipe, we use news titles as clustering input. You can find the data on the author's GitHub page at `https://github.com/ywchiu/rcookbook/raw/master/chapter12/news.RData`.

How to do it...

Perform the following steps to cluster text document with k-means clustering techniques:

1. First, install and load the `tm` and `SnowballC` packages:

```
> install.packages('tm')
> library(tm)
> install.packages('SnowballC')
> library(SnowballC)
```

2. Read the news titles that have been collected from the Internet:

```
> load('news.RData')
```

3. Convert the loaded text documents into the corpus:

```
> doc.vec <- VectorSource(news)
> doc.corpus  <- Corpus(doc.vec)
```

4. Clean the text documents by removing stop words, punctuation, and stemming the document:

```
> doc.corpus  <- tm_map(doc.corpus , removePunctuation)
> doc.corpus  <- tm_map(doc.corpus , removeWords,
stopwords("english"))
> doc.corpus  <- tm_map(doc.corpus , stemDocument)
```

5. Moving on, you can now convert the corpus into a document term matrix:

```
> dtm <- DocumentTermMatrix(doc.corpus)
> inspect(dtm[1:3,1:3] )
<<DocumentTermMatrix (documents: 3, terms: 3)>>
Non-/sparse entries: 1/8
Sparsity           : 89%
Maximal term length: 7
Weighting          : term frequency (tf)

    Terms
Docs adviser agent anoth
   1       0     1     0
   2       0     0     0
   3       0     0     0
```

6. Lastly, perform k-means clustering on the distance matrix of the document term matrix, `dtm`:

```
> set.seed(123)
> fit = kmeans(dist(dtm), 3)
> fit
> news[fit$cluster == 1]
```

How it works...

At the start of the recipe, we install and load both the `tm` and `snowballc` packages into an R session. The `tm` package serves as the text mining package, and `snowballc` is a word stemmer. We can read the data into our R session with the load function. The `tm` package requires a structure corpus to manage a collection of text documents. We thus convert input sentences from text documents into the corpus, `doc.corpus`.

When the corpus is ready, we need to preprocess the text to remove redundant terms. First, we use the `tm_map` function to remove punctuation and stop words from the corpus. Then, as some terms have a morphological form, we need to stem these words to their word stem or root form.

Once we have cleaned up the text documents, we can use the corpus to build a document term matrix. A document term matrix is represented as a matrix with rows as document and columns as terms. The value of the matrix holds the count of a term within each document. By using the `inspect` function, we find the term agent appears once in Document 1.

Last, we can compute the distance matrix over the document term matrix and perform k-means clustering with *k=3* to cluster input sentences into three groups. From the results, we find the news is categorized into three topics: Donald Trump, NBA, and Brexit.

See also

If you are interested in how to use the text mining package `tm` in R, refer to the following document: `https://cran.r-project.org/web/packages/tm/vignettes/tm.pdf`.

Performing dimension reduction with Principal Component Analysis (PCA)

Principal component analysis (**PCA**) is the most widely used linear method in dealing with dimension reduction problems. PCA is useful when data contains many features and there is redundancy (correlation) within these features. To remove redundant features, PCA maps high-dimension data into lower dimensions by reducing features into a smaller number of principal components that account for most of the variance of the original features. In this recipe, we will introduce how to perform dimension reduction with the PCA method.

Getting ready

In this recipe, we will use the economic freedom dataset as our target to perform PCA. The economic freedom (`http://www.heritage.org/index/ranking`) dataset includes global standardized economic freedom measures. You can download the dataset from `https://github.com/ywchiu/rcookbook/raw/master/chapter12/index2015_data.csv`.

How to do it...

Perform the following steps to perform PCA on the economic freedom dataset:

1. First, load the economy freedom dataset:

```
> eco.freedom <- read.csv('index2015_data.csv', header=TRUE)
```

2. Exclude the first column of the economy freedom data:

```
> eco.measure <- eco.freedom[,5:14]
```

3. Perform PCA on the `eco.measure` data:

```
> eco.pca = prcomp(eco.measure, center = TRUE, scale  = TRUE)
> eco.pca
```

4. Obtain a summary from the PCA results:

```
> summary(eco.pca)
```

5. Lastly, use the `predict` function to output the value of the principal components with the first row of data:

```
> predict(eco.pca, newdata=head(eco.measure, 1))
        PC1       PC2        PC3      PC4        PC5        PC6
PC7         PC8
1 0.7042915 1.294181 -1.245911 0.726478 0.3089492 0.5121069
-0.07302236 -0.4793749
        PC9       PC10
1 -0.3758371 -0.1154001
```

How it works...

As the feature selection method may remove some correlated but informative features, we have to consider combining the correlated features into a single feature by using a feature extraction method. PCA is one type of feature extraction; it performs orthogonal transformation to convert possibly correlated variables into principal components. Also, one can use these principal components to identify directions of variance.

The process of PCA follows the steps below:

1. Find the mean vector $\mu = \frac{1}{n}\sum_{i=1}^{n} x_i$, where x_i indicates the data point and *n* denotes the number of points.
2. Compute the covariance matrix with the following equation:

$$c = \frac{1}{n}\sum_{i=1}^{n}(x_i - \mu)(x_i - \mu)^T$$

3. Compute eigenvectors ϕ and corresponding eigenvalues.
4. Rank and choose top-k eigenvectors.
5. Construct a *d x k-dimensional* eigenvector matrix U, where *d* is the number of original dimensions and *k* is the number of eigenvectors.
6. Finally, transform data samples to a new subspace in the equation $y = U^T \cdot x$.

In the following plot, we illustrate that we can use two principal components, ϕ_1 and ϕ_2, to transform the data point from a two-dimensional space to a new 2-dimensional subspace:

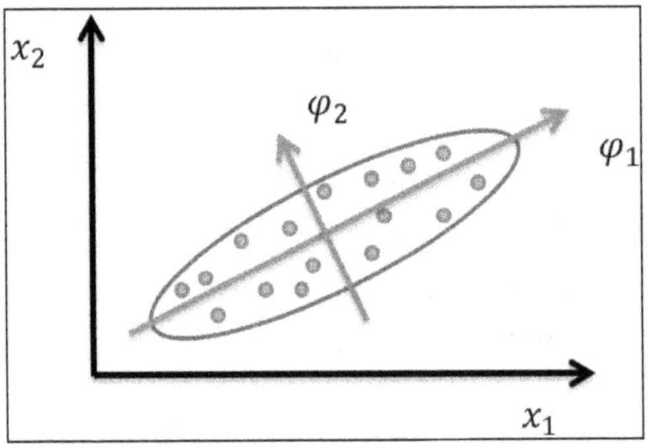

Figure 15: A sample illustration of PCA

In this recipe, we use the `prcomp` function from the stats package to perform PCA on the swiss dataset. First, we remove standardized fertility measures and use the rest of the predictors as input to the function `prcomp`. In addition, we set `eco.measure` as the input dataset. The variable should be shifted to zero center by specifying `center=TRUE`, scale variables into unit variance with the option `scale=TRUE`, and store the output in the variable `eco.pca`.

After printing the value stored in `eco.pca`, we find the standard deviation and rotation of the principal components. Standard deviation indicates the square root of eigenvalues of the covariance/correlation matrix, while the rotation of the principal components shows the coefficient of the linear combination of input features, for example, *PC1 = Property.Rights * 0.39679243 + Freedom.from.Corruption* 0.38401793 + Fiscal.Freedom * -0.02312696 + Gov.t.Spending * -0.11741108 + Business.Freedom * 0.36957438 + Labor.Freedom * 0.21182699 + Monetary.Freedom * 0.31150727 + Trade.Freedom * 0.33282176 + Investment.Freedom * 0.38065185 + Financial.Freedom * 0.38289286*. We find the `Property.Rights` attribute contributes the most for PC1 as it has the highest coefficient.

In addition, we can use the `summary` function to obtain the importance of components. The first row shows the standard deviation of each principal component, the second row shows the proportion of variance explained by each component, and the third row shows the cumulative proportion of explained variance. Finally, we can use the `predict` function to obtain principal components from the input features. Here, we input the first row of the dataset, and retrieve 10 principal components.

There's more...

Another principal component analysis function is `princomp`. In this function, the calculation is performed by using eigen on correlation or covariance matrix instead of single value decomposition used in the `prcomp` function. In general practice, using `prcomp` is preferable; however, we explain how to use `princomp` as follows:

1. First, use `princomp` to perform PCA:

```
> eco.princomp = princomp(eco.measure,
+ center = TRUE,
+ scale  = TRUE)
> swiss.princomp
```

2. You can then obtain summary information:

```
> summary(eco.princomp)
```

3. Last, use the `predict` function to obtain principal components from the input features:

```
> predict(eco.princomp, eco.measure[1,])
```

Determining the number of principal components using a scree plot

As we only need to retain the principal components that account for most of the variance of the original features, we can either use the Kaiser method, a scree plot, or the percentage of variation explained as the selection criteria. The main purpose of a scree plot is to graph the component analysis results as a scree plot and find where the obvious change in slope (elbow) occurs. In this recipe, we will demonstrate how to determine the number of principal components using a scree plot.

Getting ready

Ensure you have completed the previous recipe by generating a principal component object and saving it in variable `eco.pca`.

How to do it...

Perform the following steps to determine the number of principal components with a scree plot:

1. First, generate a bar plot by using `screeplot`:

```
> screeplot(swiss.pca, type="barplot")
```

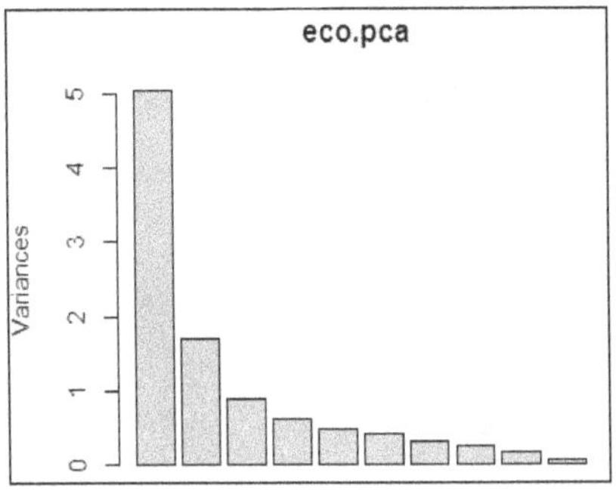

Figure 16: The scree plot in bar plot form

2. You can also generate a line plot by using `screeplot`:

```
> screeplot(eco.pca, type="line")
```

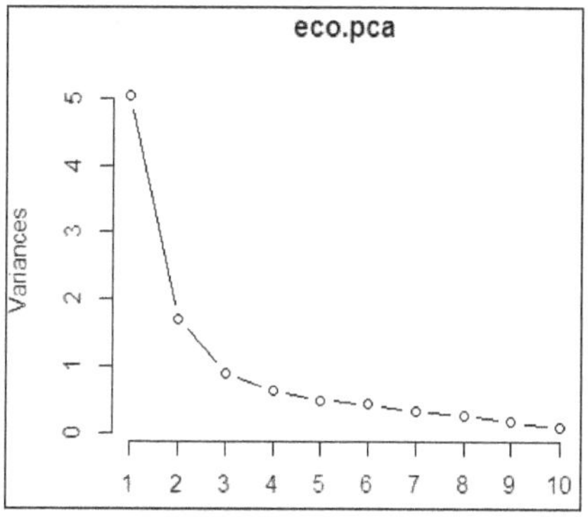

Figure 17: The screeplot in line plot form

How it works...

In this recipe, we demonstrate how to use screeplot to determine the number of principal components. In screeplot, there are two types of plot, namely bar plots and line plots. As both generated scree plots reveal, the obvious change in slope (the so-called elbow or knee) occurs at component 3. As a result, we should retain component 1 and component 2, where the component is in a steep curve before component 3, which is where the flat line trend commences. However, as this method can be ambiguous, you can use other methods (such as the Kaiser method) to determine the number of components.

There's more...

By default, if you use the `plot` function on a generated principal component object, you can also retrieve the scree plot. For more details on `screeplot`, please refer to the following document:

```
> help(screeplot)
```

In addition, you can use `nfactors` to perform parallel analysis and non-graphical solutions to the Cattell screeplot:

```
> install.packages("nFactors")
> library(nFactors)
> ev = eigen(cor(eco.measure))
> ap = parallel(subject=nrow(eco.measure),var=ncol(eco.
measure),rep=100,cent=.05)
```

```
> nS = nScree(x=ev$values, aparallel=ap$eigen$qevpea)
> plotnScree(nS)
```

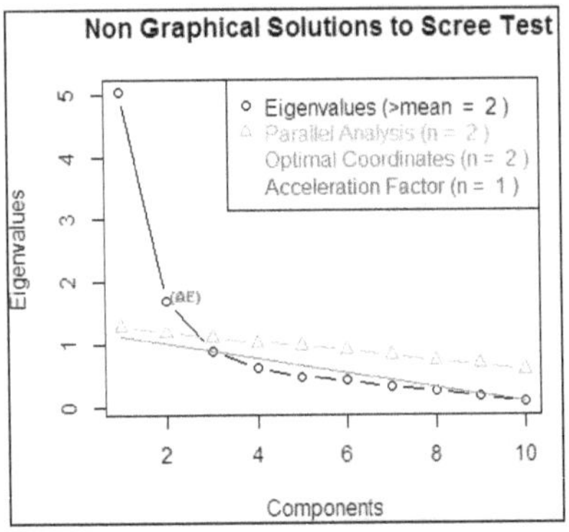

Figure 18: Non-graphical solutions to the screeplot

Determining the number of principal components using the Kaiser method

In addition to a screeplot, we can use the Kaiser method to determine the number of principal components. In this method, the selection criteria retain eigenvalues greater than 1. In this recipe, we demonstrate how to determine the number of principal components using the Kaiser method.

Getting ready

Ensure you have completed the previous recipe by generating a principal component object and saving it in variable `eco.pca`.

How to do it...

Perform the following steps to determine the number of principal components with the Kaiser method:

1. First, obtain the standard deviation from `eco.pca`:

```
> eco.pca$sdev
 [1] 2.2437007 1.3067890 0.9494543 0.7947934 0.6961356 0.6515563
 [7] 0.5674359 0.5098891 0.4015613 0.2694394
```

2. Next, obtain the variance from `swiss.pca`:

```
> eco.pca$sdev ^ 2
 [1] 5.0341927 1.7076975 0.9014634 0.6316965 0.4846048 0.4245256
 [7] 0.3219835 0.2599869 0.1612515 0.0725976
```

3. Select the components with a variance above `1`:

```
> which(eco.pca$sdev ^ 2> 1)
[1] 1 2
```

4. You can also use `screeplot` to select components with a variance above `1`:

```
> screeplot(eco.pca, type="line")
> abline(h=1, col="red", lty= 3)
```

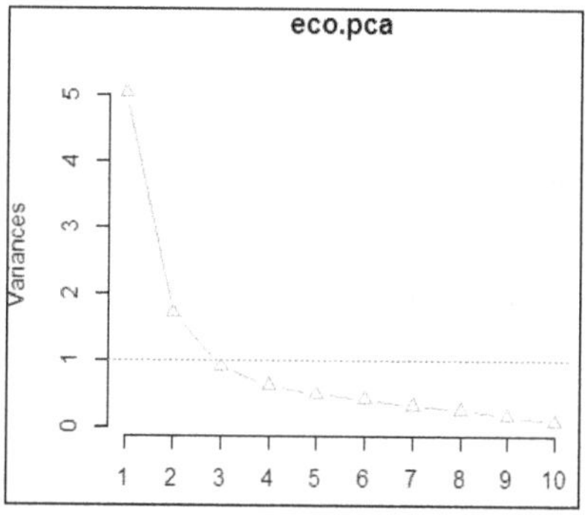

Figure 19: Select components with a variance above 1

How it works...

One can also use the Kaiser method to determine the number of components. As the computed principal component object contains the standard deviation of each component, we can compute variance, as standard deviation is the square root of variance. From the computed variance, we find both components 1 and 2 have a variance above 1. Therefore, we can determine the number of principal components is two (both component 1 and component 2). Also, we can draw a red line on the screeplot (as shown in *Figure 19*) to indicate that we need to retain component 1 and component 2 in this case.

See also

In order to determine which principal components to retain, please refer to the following paper:

Determining the Number of Factors to Retain in EFA: an easy-to-use computer program for carrying out Parallel Analysis. Practical Assessment, Research & Evaluation, 12(2), 1-11 by *Ledesma, R. D., & Valero-Mora, P. (2007).*

Visualizing multivariate data using a biplot

In order to find how data and variables are mapped in regard to the principal component, you can use `biplot`, which plots data and projections of original features on to the first two components. In this recipe, we will demonstrate how to use `biplot` to plot both variables and data on the same figure.

Getting ready

Ensure that you have completed the previous recipe by generating a principal component object and saving it in variable `eco.pca`.

How to do it...

Perform the following steps to create a biplot:

1. Create a scatter plot using component 1 and component 2:

```
> plot(eco.pca$x[,1], eco.pca$x[,2], xlim=c(-6,6), ylim = c(-4,3))
> text(eco.pca$x[,1], eco.pca$x[,2], eco.freedom[,2], cex=0.7,
pos=4, col="red")
```

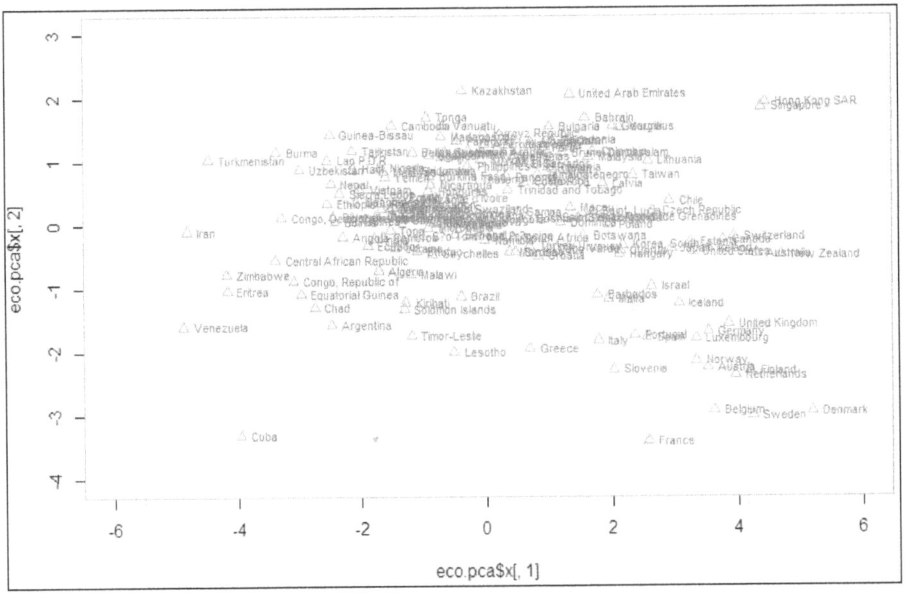

Figure 20: The scatterplot of the first two components from the PCA result

2. In addition, if you would like to add features on the plot, you can create the biplot using a generated principal component object:

```
> rownames(eco.pca$x)  =  as.character(eco.freedom[,2])
> biplot(eco.pca)
```

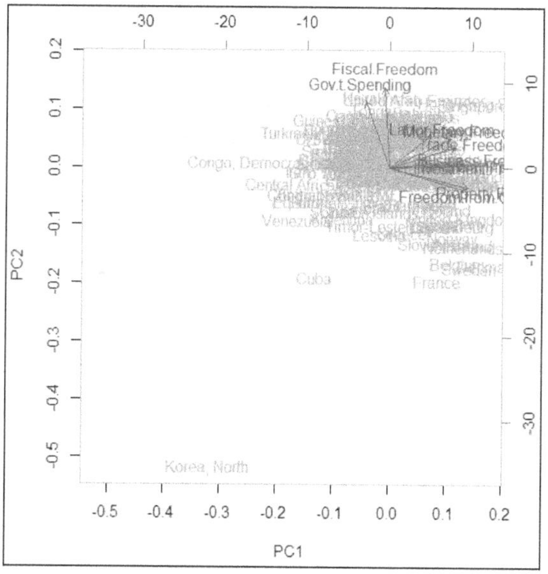

Figure 21: The biplot using the PCA result

How it works...

In this recipe, we demonstrate how to use `biplot` to plot data and projections of original features onto the first two components. In the first step, we demonstrate that we can actually use the first two components to create a scatter plot. Furthermore, if you want to add variables on the same plot, you can use `biplot`. In `biplot`, the provinces with higher fiscal freedom indicators and government spending indicators score higher in PC2. On the other hand, you can see the other indicators score higher in PC1.

See also

Besides `biplot` in the `stats` package, you can also use `ggbiplot`. However, this package is not available from CRAN; you have to first install `devtools` and then install `ggbiplot` from GitHub:

```
> install.packages("devtools")
> library(devtools)
> install_github("ggbiplot", "vqv")
```

```
> library(ggbiplot)
> g <- ggbiplot(eco.pca, obs.scale = 1, var.scale = 1, ellipse = TRUE,
circle = TRUE)
> print(g)
```

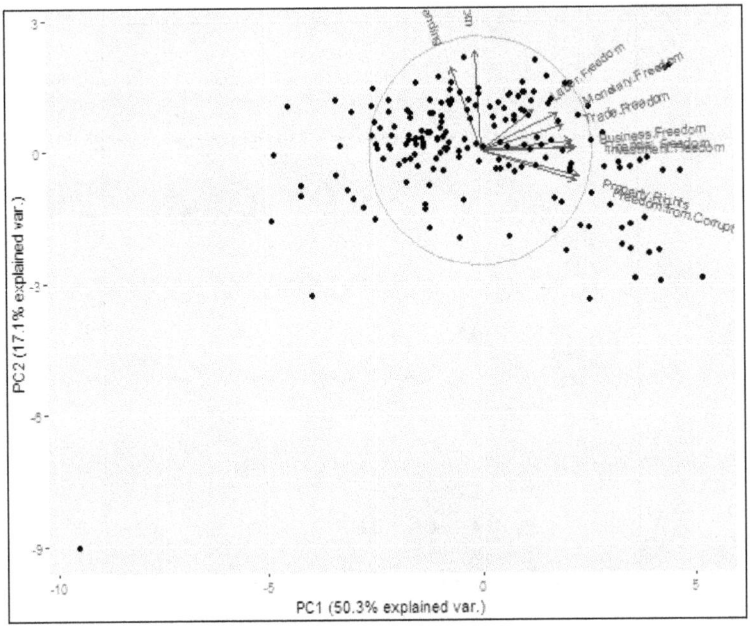

Figure 22: The ggbiplot using the PCA result

Index

A

acf function 352
add_tooltip function 214
aesthetics
  adding, to plot 150-152
aesthetics mapping
  modifying 150-152
agglomerative hierarchical clustering 395
Akaike Information Criterion (AIC) 347
AmeliaView 98
analysis of variance (ANOVA)
  about 282
  reference 286
Apriori rule
  associations, mining with 299-303
area under curve (AUC) 387
argument binding mechanism 6
arguments
  matching 5-7
ARIMA model
  creating 346, 347
  forecasting with 348-351
  selecting 341-344
  stock prices, predicting with 352-355
arrange function 131
arules package 296
association rules
  visualizing 305-307
associations
  displaying 296-299
  mining, with Apriori rule 299-303
axes
  controlling 201-204

B

basic plots
  creating, with ggplot2 146-149
basic syntax 194
Bayesian Information Criterion (BIC) 347
best-fitted regression model
  selecting, with stepwise regression 371-373
binomial random variates
  generating 233-235
biplot
  used, for visualizing multivariate data 422, 423
boot.ci function 264
box.test function 352
Brownian motion 255
BSDA package
  z.test function, finding in 268

C

chi-squared distribution
  sample, generating from 246, 247
classification model
  building, with recursive partitioning trees 379-382
closure
  about 13
  creating, in function 13-15
  passing, to other function 15
clustering
  about 392
  silhouette information, extracting from 405-407
clustering methods
  comparing 407-409

clusters
tree, cutting into 397-399
code chunk
global option, controlling of 190
columns
selecting, with dplyr 125-127
complexity parameter (CP) 381
confidence intervals
obtaining 258-264
confusion matrix
model performance, measuring with 384-386
cont command 32
coord_polar function 157
copy function 110
CRAN (Comprehensive R Archive Network) 2
cSPADE
frequent sequential patterns, mining with 313
CSV file
data, scanning from 44, 45
text data, reading from 43
reading 42, 43
writing 42, 43
cutree function 397

D

data
accessing, from Facebook 62-65
clustering, with density-based method 402-404
clustering, with hierarchical clustering 392-396
clustering, with k-means method 400-402
dropping 87, 88
filtering 83-85
managing, with data.table 106-110
merging 88-90
merging, with dplyr 138-141
missing data, detecting 95-97
missing data, imputing 98-100
reading, from databases 49-52
reading, from Twitter 68-71
reshaping 92-94
sampling, with dplyr 123, 124
scanning, from CSV file 44, 45
slicing, with dplyr 120-122
sorting 90-92
subsetting, with dplyr 120-122
summarizing, with dplyr 134-137
transforming, into transactions 294, 295
databases
data, reading from 49-52
data.frame
enhancing, with data.table 102-105
dataset
sampling from 251-253
data.table
about 102
data.frame, enhancing with 102-105
data, managing with 106-110
fast aggregation, performing with 111-115
large datasets, merging with 115-119
data types
converting 76-78
data variable
renaming 74, 75
date format
working with 78-80
dbscan package 405
debugging function 28
density-based clustering method
used, for recognizing digits 410-412
density-based method
data, clustering with 402-404
desc function 131
descriptive statistics 258
diagnostic plot
used, for diagnosing regression model 370, 371
digits
recognizing, density-based clustering method used 410-412
dimension reduction
about 392
performing, with Principal Component Analysis (PCA) 414-416
dimnames function 75
distance functions, for similarity measurement
average-linkage 396
complete-linkage 396
single-linkage 396
ward-method 396
divisive hierarchical clustering 395

document options
editing 182
download.file function 38
dplyr
about 102
columns, selecting with 125-127
data, merging with 138-141
data, sampling with 123, 124
data, slicing with 120-122
data, subsetting with 120-122
data, summarizing with 134-137
duplicated rows, eliminating with 131, 132
new columns, adding with 133, 134
operations, chaining in 128, 129
rows, arranging with 130, 131
duplicated rows
eliminating, with dplyr 131, 132

E

Eclat
frequent itemsets, mining with 307-310
empirical cumulative distribution function (ECDF) 275
environments
working with 7-10
Eps parameter 404
error message
catching 28
errors
handling, in function 23-27
exact binomial tests
conducting 272, 273
Excel 46
Excel files
working with 46-49

F

Facebook
data, accessing from 62-65
reference 62
facet function 164
faceting 164-166
fast aggregation
performing, with data.table 111-115
feature extraction 392
feature selection 392
f (finish) command 32
force function 17
fpp package 334, 340
fread function 105
frequent itemsets
mining, with Eclat 307-310
frequent sequential patterns
mining, with cSPADE 313
function
closure, creating in 13-15
errors, handling in 23-27

G

Gaussian distribution 243
Gaussian model
applying, for generalized linear regression 374, 375
generalized linear model (GLM) 374
generalized linear regression
Gaussian model, applying for 374, 375
geometric objects
about 153
creating, in ggplot2 154-156
Geospatial Data Abstraction Library (GDAL)
about 176
reference 176
ggplot2
basic plots, creating with 146-149
geometric objects, creating in 154-156
ggtsdisplay function 344
ggvis
interactive graphics, creating with 190-193
interactivity, adding into 208-213
plots, creating with 194-200
global option
controlling, of code chunk 190
grammar 194
graphics package 166
grid.arrange function 171
gridExtra package 171

H

H0 (null hypothesis) 273
H1 (alternative hypothesis) 273

hclust 397
hierarchical clustering
data, clustering with 392-396
Hmisc package, within stat_summary
mean_cl_boot() 161
mean_cl_normal() 161
mean_sdl() 161
median_hilow() 161
HTML (Hypertext Markup Language) 56
Hypertext Transfer Protocol (HTTP) 41

I

inferential statistics 258
infix operators
creating 18, 19
interactive graphics
creating, with ggvis 190-193
interactivity
adding, into ggvis plot 208-213

J

Java Database Connectivity (JDBC) 51
JDBC driver, for MySQL
reference 50

K

Kaiser method
used, for determining number of principal components 420, 421
k-means clustering methods
similar text documents, grouping with 412-414
k-means method
data, clustering with 400-402
knitr package 189
Kolmogorov-Smirnov tests (K-S tests)
performing 274-276

L

lapply function 15
large datasets
merging, with data.table 115-119
LaTeX syntax
using 186
lattice package 166
lazy evaluation
performing 16, 17
legend
controlling 201-204
visual properties, modifying 205
lexical scope
working with 11-13
lexical scoping 10
linear model fits
summarizing 361, 362
linear regression
used, for predicting unknown values 363-366
linear regression model
fitting, with lm 358-360
lm
linear regression model, fitting with 358-360
polynomial regression model, fitting with 360
lm function 100
LOESS (Locally Weighted Smoothing) 330
logistic regression analysis
performing 375-378
LOWESS (Locally Weighted Scatterplot Smoothing) 330
lubridate package 80

M

manova function 291
maps
creating 171-176
mark
tooltip, adding to 214
markdown syntax
writing 183-185
mice package 100
MinPts parameter 404
model performance
measuring, with confusion matrix 384-386
multiple regression analysis
performing 368, 369
multivariate data
visualizing, biplot used 422, 423

N

names<- function 22
new columns
adding, with dplyr 133, 134
n (next) command 32
normal distribution
sample, generating from 239-244
nth order polynomial 360
number of principal components
determining, Kaiser method used 420, 421
determining, scree plot used 418, 419

O

objects command 32
one-way ANOVA
conducting 282-286
open data
downloading 38-41
Open Database Connectivity (ODBC) 51
operations
chaining, in dplyr 128, 129

P

parent.env function 10
parent environment
obtaining 10
Pearson's chi-squared tests
working with 277, 278
performance
measuring, of regression model 366, 367
plots
aesthetics, adding to 150-152
combining 169, 170
creating, with ggvis 194-200
plot.xts function
using 327
Poisson distribution 236
Poisson random variates
generating 236-238
polynomial regression model
fitting, with lm 360
prediction intervals, versus confidence intervals
reference 366
prediction performance
measuring, ROCR used 387, 388
Principal Component Analysis (PCA)
about 414
dimension reduction, performing with 414-416
princomp function 417

Q

qpois 238
Q (quit) command 32

R

R
about 1
advantages 2
reference 2
random samples
generating 228, 229
random seed sequence 230
R code chunk
embedding 186-189
RCurl
used, for downloading file 41, 42
Receiver Operating Characteristic (ROC) curve
about 387
reference 389
records
adding 81, 82
recursive partitioning tree
classification model, building with 379-381
visualizing 382-384
redundant rules
pruning 303, 304
regression model
diagnosing, diagnostic plot used 370, 371
performance, measuring of 366, 367
relative square error (RSE) 366
replacement function
using 20-22
R function
creating 3, 4
debugging 29-35
R language 2

R Markdown reports
creating 178-181
rm function 88
ROCR
used, for measuring
prediction performance 387, 388
root mean square error (RMSE) 366
root relative squared error loss (RRSE) 368
rows
arranging, with dplyr 130, 131
R program 2
R Shiny document
creating 215-219
R Shiny report
publishing 221-226
R user 2

S

sample
generating, from chi-squared
distribution 246, 247
generating, from normal distribution 239-244
generating, from t-distribution 248-250
generating, from uniform
distribution 231-233
sampling package 253
sapply function 15
scale_datetime function 208
scale_logical function 208
scale_ordinal function 208
scales
adjusting 162, 163
using 206, 207
scale_singular function 208
scheme functional programming language 10
scree plot
used, for determining number of principal
components 418, 419
search list
examining 13
SelectorGadget
about 58
reference 58
setkey function 119
Shiny widget
adding, to control rendering 219, 220
silhouette information
about 405
extracting, from clustering 405-407
silhouette value
reference 407
similar text documents
grouping, with k-means
clustering methods 412-414
social network data 62
SQL operations
performing 123
s (step into) command 32
stack function 95
static binding 10
stat package 334
stepwise regression
best-fitted regression model,
selecting with 371-73
stochastic process
about 253
simulating 254, 255
stock prices
predicting, with ARIMA model 352-355
student's t-tests
performing 268-271
subset function 86
sum of square error (SSE) 373
Sweave package 186

T

t-distribution
about 248
sample, generating from 248-250
temporal information
transactions, creating with 310
text data
reading, from CSV file 43
text files
scanning 44, 45
text mining package
reference 414
themes
adjusting 167, 168

time series
decomposing 327-329
forecasting 336-340
smoothing 331-334
time series data
creating 320-322
time series object
plotting 324-326
tooltip
adding, to mark 214
traceback function 33
transactions
creating, with temporal information 310
data, transforming into 294, 295
displaying 296-299
transformations
performing 158-161
tree
cutting, into clusters 397-399
tryCatch function 28
tsdisplay function 344
t-table
reference 251
Twitter
data, reading from 68-71
reference 68
working with 68
two-way ANOVA
conducting 287-291

U

ungroup function 137
uniform distribution
sample, generating from 231-233
uniform distributions 231
unique function 132
unknown values
predicting, linear regression used 363-366
unstack function 94
unsupervised machine learning method 392

V

vega, GitHub page
reference 194

W

web data
scraping 52-56
where command 32
Wilcoxon Rank Sum and Signed Rank test (Mann-Whitney-Wilcoxon)
conducting 279-282
with function 82
within-cluster sum of squares (WCSS) 401
within function 88

X

xts package
using 323

Z

z.test function
finding, in BSDA package 268
Z-tests
performing 265-268

11.621 40690CB00015B/2845 [206132317]

www.ingramcontent.com/pod-product-compliance
Lightning Source LLC
Chambersburg PA
CBHW081458050326
40690CB00015B/2845